Student Resource and Solutions Manual

for

Zill and Cullen's

Differential Equations
with Boundary-Value Problems

Sixth Edition

Warren S. Wright
Loyola Marymount University

Dennis G. Zill
Loyola Marymount University

Carol D. Wright

THOMSON
™
BROOKS/COLE

Australia • Canada • Mexico • Singapore • Spain • United Kingdom • United States

Printed in the United States of America
1 2 3 4 5 6 7 08 07 06 05 04

Printer: Thomson/West

ISBN: 0-534-41888-0

For more information about our products,
contact us at:
Thomson Learning Academic Resource Center
1-800-423-0563

For permission to use material from this text or
product, submit a request online at
http://www.thomsonrights.com.
Any additional questions about permissions can be
submitted by email to **thomsonrights@thomson.com.**

Thomson Higher Education
10 Davis Drive
Belmont, CA 94002-3098
USA

Asia (including India)
Thomson Learning
5 Shenton Way
#01-01 UIC Building
Singapore 068808

Australia/New Zealand
Thomson Learning Australia
102 Dodds Street
Southbank, Victoria 3006
Australia

Canada
Thomson Nelson
1120 Birchmount Road
Toronto, Ontario M1K 5G4
Canada

UK/Europe/Middle East/Africa
Thomson Learning
High Holborn House
50–51 Bedford Road
London WC1R 4LR
United Kingdom

Latin America
Thomson Learning
Seneca, 53
Colonia Polanco
11560 Mexico
D.F. Mexico

Spain (including Portugal)
Thomson Paraninfo
Calle Magallanes, 25
28015 Madrid, Spain

Introduction

This *Student Resource and Solutions Manual* is meant to accompany the text *Differential Equations with Boundary-Value Problems*, Sixth Edition. It is intended to

- provide a list of the terminology, concepts, and skills associated with each section in the text;

- provide solutions for every third problem in the exercise sets, excluding Discussion/Project problems and Computer Lab problems;

- provide hints and suggestions for selected problems throughout the entire set of exercises in each section.

For Chapters 1 – 9 it is also intended to

- provide further elaboration on some of the material discussed in the section;

- review some of the mathematics pertinent to the material being presented (such as integration, synthetic division, and partial fractions);

- furnish extra examples of the material presented in the section;

- illustrate the use of the computer algebra systems *Mathematica* and *Maple* in dealing with the material in the section;

- provide examples illustrating how to solve some of the Computer Lab problems.

At the end of the manual there is a brief introduction to the basic syntax of *Mathematica* along with a summary of some of the commands and functions built into the program.

Table of Contents

1

Introduction to Differential Equations

Definitions and Terminology

The terminology and concepts listed below provide an outline of the main ideas encountered in this section. These can be useful when preparing for a quiz or test.

Terminology and Concepts

- ordinary DE

- partial DE

- order of a DE

- normal form of a DE

- linear DE

- explicit and implicit form of a solution of a DE

- interval of definition of a solution

- trivial solution of a DE

- solution curve

- family of solutions of a DE

- n-parameter family of solutions of an nth-order DE

- particular solution of a DE

- singular solution of a DE

- system of DEs

1

Section 1.1 Definitions and Terminology

The basic skills listed below summarize the more mechanical types of problems encountered in the exercise set for this section.

Basic Skills

- determine the order of a DE
- determine if a DE is linear or nonlinear
- verify that a given function is a solution of a DE
- determine the interval of definition of a solution of a DE
- use an implicit solution to determine an explict solution (this is not always possible)
- identify constant solutions of certain DEs
- verify that a pair of functions is a solution of a system of two DEs

Basic Properties of the Derivative

Definition: The derivative of a function $y = f(x)$ with respect to the independent variable x is

$$f'(x) = \lim_{h \to 0} \frac{f(x+h) - f(x)}{h},$$

whenever this limit exists.

The derivative is also denoted by y' and dy/dx. When the derivative is evaluated at $x = x_0$, denoted by $f'(x_0)$, the resulting value represents the *rate of change* of the function f at x_0. Geometrically, this is the slope of the tangent line to the graph of $y = f(x)$ at the point $(x_0, f(x_0))$. If the function f is differentiable at x_0 then the graph of f at the point $(x_0, f(x_0))$ does not have a sharp corner or vertical tangent line at that point.

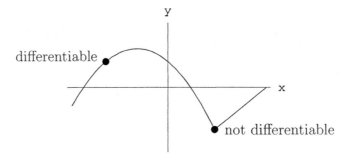

In DEs the independent variable frequently represents time and is denoted by t. In this case, if the position at time t of an object moving on a straight line is given by $s = s(t)$, then the velocity of the object at time t is the first derivative, $v(t) = s'(t)$, and the acceleration is the second derivative of $s(t)$, $a(t) = s''(t)$.

If the first derivative of a function is positive on an interval, then the function is **increasing** on that interval. Correspondingly, when the derivative is negative on an interval, the function is

decreasing on the interval. Relative extrema (maxima or minima) of a function occur where the derivative is either 0 or does not exist. The second derivative of a function can be used to determine concavity. The graph of a function f is **concave up** on an interval if $f''(x) > 0$ on the interval and **concave down** on an interval if $f''(x) < 0$ on the interval.

Hyperbolic Functions   It is unfortunate that many current texts and calculus instructors downplay the hyperbolic functions. These functions are important in applied mathematics and every serious student of differential equations should know their definitions, graphs, and some of their properties. In the text, solutions of DEs will be expressed in terms of hyperbolic functions when appropriate. The six hyperbolic functions are denoted by sinh, cosh, tanh, coth, sech and csch. If x denotes a real variable, the **hyperbolic sine** and **cosine** are defined in terms of exponential functions.

Definitions: $\qquad\qquad \sinh x = \dfrac{e^x - e^{-x}}{2} \qquad$ and $\qquad \cosh x = \dfrac{e^x + e^{-x}}{2} \qquad\qquad$ (1)

The remaining four hyperbolic functions are defined in a manner analogous to trigonometry. For example, the hyperbolic tangent and cotangent are, respectively,

$$\tanh x = \frac{\sinh x}{\cosh x} \qquad \text{and} \qquad \coth x = \frac{\cosh x}{\sinh x}.$$

The graphs of the hyperbolic sine and cosine functions, like the graphs of natural exponential, natural logarithmic, and trigonometric functions, should be memorized.

Graphs:

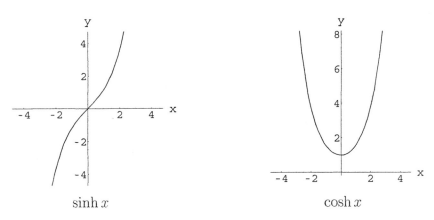

$\sinh x \qquad\qquad\qquad\qquad\qquad\qquad \cosh x$

Some Properties   As can be seen from both the graphs and the definitions in (1),

$$\sinh 0 = 0 \qquad \text{and} \qquad \cosh 0 = 1. \qquad\qquad (2)$$

Also, $\cosh x$ has no real zeros and, unlike the trigonometric functions, no hyperbolic function is periodic. However, a number of other properties similar to properties for trigonometric functions do hold. As you examine the hyperbolic identities (3) through (8) below, determine whether or not the given identity differs from the analogous identity for trigonometric functions.

$$\cosh^2 x - \sinh^2 x = 1 \tag{3}$$

$$\sinh(x_1 \pm x_2) = \sinh x_1 \cosh x_2 \pm \cosh x_1 \sinh x_2 \tag{4}$$

$$\cosh(x_1 \pm x_2) = \cosh x_1 \cosh x_2 \pm \sinh x_1 \sinh x_2 \tag{5}$$

$$\sinh 2x = 2 \sinh x \cosh x \tag{6}$$

$$\cosh 2x = \cosh^2 x + \sinh^2 x \tag{7}$$

The derivatives of $\sinh x$ and $\cosh x$, given below, can be obtained using the definitions in (1).

$$\frac{d}{dx} \sinh x = \cosh x \qquad \text{and} \qquad \frac{d}{dx} \cosh x = \sinh x \tag{8}$$

See Problem 4 in *Chapter 1 in Review* in the text.

Use of Computers   To define a function, such as $f(x) = e^{x^2} + \sin 2x$, use the syntax

> **Clear[f]**   (*Mathematica*)
> **f[x_]:= Exp[x^2]+Sin[2x]**
> **f[x]**

```
f:=x->exp(x^2)+sin(2*x);                    (Maple)
or
f:=unapply(exp(x^2)+sin(2*x), x);
```

If you are not familiar with *Maple* the `unapply` function probably seems like an awkward way to define a function. We will see later that it is a useful command when solving DEs.

Piecewise-defined functions are, as the name suggests, defined in pieces. For example, the function

$$f(x) = \begin{cases} x^2 + 1, & x \le 2 \\ \sin x, & x > 2 \end{cases}$$

is defined in a CAS using:

> **Clear[f]**   (*Mathematica*)
> **f[x_]:= x^2 + 1/; x <= 2**
> **f[x_]:= Sin[x]/; x > 2**

```
f:=x->piecewise(x<=2, x^2+1, x>2,sin(x));      (Maple)
```

The syntax for the first derivative, dy/dx, of $y = f(x)$ is:

$$D[f[x], x] \quad \text{or} \quad f'[x] \qquad \qquad (\textit{Mathematica})$$

```
diff(f(x), x);
```
(*Maple*)

The second derivative, d^2y/dx^2, of $y = f(x)$ is:

$$D[f[x], \{x,2\}] \quad \text{or} \quad f''[x] \qquad \qquad (\textit{Mathematica})$$

```
diff(f(x), x$2);   or   diff(f(x),x,x);
```
(*Maple*)

Exercises 1.1 *Hints, Suggestions, Solutions, and Examples*

3. Fourth order; linear

6. Second order; nonlinear because of R^2

9. Writing the DE in the form $x(dy/dx) + y^2 = 1$, we see that it is nonlinear in y because of y^2. However, writing it in the form $(y^2 - 1)(dx/dy) + x = 0$, we see that it is linear in x.

12. From $y = \frac{6}{5} - \frac{6}{5}e^{-20t}$ we obtain $dy/dt = 24e^{-20t}$, so that

$$\frac{dy}{dt} + 20y = 24e^{-20t} + 20\left(\frac{6}{5} - \frac{6}{5}e^{-20t}\right) = 24.$$

15. The domain of the function, found by solving $x + 2 \geq 0$, is $[-2, \infty)$. From $y' = 1 + 2(x+2)^{-1/2}$ we have

$$(y - x)y' = (y - x)[1 + (2(x+2)^{-1/2}]$$
$$= y - x + 2(y - x)(x+2)^{-1/2}$$
$$= y - x + 2[x + 4(x+2)^{1/2} - x](x+2)^{-1/2}$$
$$= y - x + 8(x+2)^{1/2}(x+2)^{-1/2} = y - x + 8.$$

An interval of definition for the solution of the DE is $(-2, \infty)$ because y' is not defined at $x = -2$.

18. The function is $y = 1/\sqrt{1 - \sin x}$, whose domain is obtained from $1 - \sin x \neq 0$ or $\sin x \neq 1$. Thus, the domain is $\{x \mid x \neq \pi/2 + 2n\pi\}$. From $y' = -\frac{1}{2}(1 - \sin x)^{-3/2}(-\cos x)$ we have

$$2y' = (1 - \sin x)^{-3/2} \cos x = [(1 - \sin x)^{-1/2}]^3 \cos x = y^3 \cos x.$$

An interval of definition for the solution of the DE is $(\pi/2, 5\pi/2)$. Another interval is $(5\pi/2, 9\pi/2)$ and so on.

21. Differentiating $P = c_1 e^t / \left(1 + c_1 e^t\right)$ we obtain

$$\frac{dP}{dt} = \frac{\left(1 + c_1 e^t\right) c_1 e^t - c_1 e^t \cdot c_1 e^t}{(1 + c_1 e^t)^2} = \frac{c_1 e^t}{1 + c_1 e^t} \frac{\left[\left(1 + c_1 e^t\right) - c_1 e^t\right]}{1 + c_1 e^t}$$

$$= \frac{c_1 e^t}{1 + c_1 e^t} \left[1 - \frac{c_1 e^t}{1 + c_1 e^t}\right] = P(1 - P).$$

22. (*Hint*) Assume a function $y = f(x)$ is continuous on $[a, b]$ and that x is any number in the interval. The derivative form of the **Fundamental Theorem of Calculus** is then

$$\frac{d}{dx} \int_a^x f(t)\, dt = f(x).$$

Using the Chain Rule we then have

$$\frac{d}{dx} \int_a^{g(x)} f(t)\, dt = f(g(x))g'(x).$$

24. From $y = c_1 x^{-1} + c_2 x + c_3 x \ln x + 4x^2$ we obtain

$$\frac{dy}{dx} = -c_1 x^{-2} + c_2 + c_3 + c_3 \ln x + 8x,$$

$$\frac{d^2 y}{dx^2} = 2c_1 x^{-3} + c_3 x^{-1} + 8,$$

and

$$\frac{d^3 y}{dx^3} = -6c_1 x^{-4} - c_3 x^{-2},$$

so that

$$x^3 \frac{d^3 y}{dx^3} + 2x^2 \frac{d^2 y}{dx^2} - x \frac{dy}{dx} + y = (-6c_1 + 4c_1 + c_1 + c_1)x^{-1} + (-c_3 + 2c_3 - c_2 - c_3 + c_2)x$$

$$+ (-c_3 + c_3)x \ln x + (16 - 8 + 4)x^2$$

$$= 12x^2.$$

25. (*Hint*) For a piecewise-defined function to be smooth at the point where the pieces are joined the function must be continuous there and the left-hand and right-hand derivatives of the two pieces at the point must be equal.

27. **(a)** From $y = e^{mx}$ we obtain $y' = me^{mx}$. Then $y' + 2y = 0$ implies

$$me^{mx} + 2e^{mx} = (m + 2)e^{mx} = 0.$$

Since $e^{mx} > 0$ for all x, $m = -2$. Thus $y = e^{-2x}$ is a solution.

(b) From $y = e^{mx}$ we obtain $y' = me^{mx}$ and $y'' = m^2 e^{mx}$. Then $y'' - 5y' + 6y = 0$ implies

$$m^2 e^{mx} - 5me^{mx} + 6e^{mx} = (m-2)(m-3)e^{mx} = 0.$$

Since $e^{mx} > 0$ for all x, $m = 2$ and $m = 3$. Thus $y = e^{2x}$ and $y = e^{3x}$ are solutions.

30. Solving $y^2 + 2y - 3 = (y+3)(y-1) = 0$ we see that $y = -3$ and $y = 1$ are constant solutions.

33. From $x = e^{-2t} + 3e^{6t}$ and $y = -e^{-2t} + 5e^{6t}$ we obtain

$$\frac{dx}{dt} = -2e^{-2t} + 18e^{6t} \quad \text{and} \quad \frac{dy}{dt} = 2e^{-2t} + 30e^{6t}.$$

Then

$$x + 3y = (e^{-2t} + 3e^{6t}) + 3(-e^{-2t} + 5e^{6t}) = -2e^{-2t} + 18e^{6t} = \frac{dx}{dt}$$

and

$$5x + 3y = 5(e^{-2t} + 3e^{6t}) + 3(-e^{-2t} + 5e^{6t}) = 2e^{-2t} + 30e^{6t} = \frac{dy}{dt}.$$

41-42. (*Hint*) A function is not differentiable at a point where a tangent line to its graph is vertical.

50. (*Suggestion*) You have y, y', and y''. Eliminate c_1 and c_2 between these expressions.

51-54. (*Suggestion*) In this manual, see the comments **Signs of Derivatives** at the beginning of Section 2.1.

55-56. (*Example*) Use a CAS to verify that $y = 4x^2 \sin(5 \ln x) + 13x \ln x + x$ is a solution of the DE $x^2 y'' - 3xy' + 29y = 338x \ln x$.

```
Clear[y]                                    (Mathematica)
y[x_]:= 4x^2 Sin[5Log[x]] + 13x Log[x] + x
y[x]

x^2 y''[x] − 3x y'[x] + 29y[x] // Simplify
```

```
y := x−>4*x^2*sin(5*ln(x))+13*x*ln(x)+x;     (Maple)
x^2*diff(y(x),x$2)-3*x*diff(y(x),x)+29*y(x);
simplify(%);
```

In both cases the output of the simplified expression is $338x \ln x$, which verifies that the given function is a solution of the DE for $x > 0$.

Section 1.2
Initial-Value Problems

The terminology and concepts listed below provide an outline of the main ideas encountered in this section. These can be useful when preparing for a quiz or test.

Terminology and Concepts

- IVPs

- initial conditions

- how the interval of definition of a solution of an IVP may differ from the domain of the function in the solution

- existence and uniqueness of solutions of IVPs

The basic skills listed below summarize the more mechanical types of problems encountered in the exercise set for this section.

Basic Skills

- given a family of solutions of a DE, find a particular solution satisfying given initial conditions

- determine a region in the plane for which a given DE will have a unique solution whose graph passes through a point in the region

- match graphs of solution curves with given initial conditions

Existence and Uniqueness
Theorem 1.1 is local in nature, so the interval I_0 may not be very large. The words "existence" and "uniqueness" are often not clearly understood. Use of the notation such as $y = y(x)$ and saying that a solution "exists" on the interval I_0 does *not* necessarily mean that we can actually find (meaning display) an explicit or implicit solution of the DE defined on I_0. We know in theory that a solution $y = y(x)$ exists but we may have to approximate it on I_0. The word "uniqueness" basically means we cannot have *two* (or more) solutions $y = y_1(x)$ and $y = y_2(x)$ defined on I_0 such that their solution curves both pass through the initial point (x_0, y_0). In other words, on I_0 we cannot have either of the following situations:

8

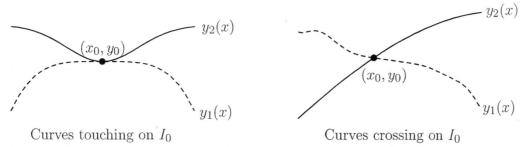

Curves touching on I_0 Curves crossing on I_0

Since any point on the graph of a solution curve may be considered as an initial condition of the DE resulting in that curve, no two solution curves can intersect.

Exercises 1.2   *Hints, Suggestions, Solutions, and Examples*

3. Letting $x = 2$ and solving $1/3 = 1/(4 + c)$ we get $c = -1$. The solution is $y = 1/(x^2 - 1)$. This solution is defined on the interval $(1, \infty)$.

6. Letting $x = 1/2$ and solving $-4 = 1/(1/4 + c)$ we get $c = -1/2$. The solution is $y = 1/(x^2 - 1/2) = 2/(2x^2 - 1)$. This solution is defined on the interval $(-1/\sqrt{2}, 1/\sqrt{2})$.

9. Substituting the initial conditions into $y = c_1 \cos t + c_2 \sin t$ and $y' = c_1 \sin t + c_2 \cos t$ we obtain

$$\frac{\sqrt{3}}{2} c_1 + \frac{1}{2} c_2 = \frac{1}{2}$$

$$-\frac{1}{2} c_1 + \frac{\sqrt{3}}{2} c_2 = 0.$$

Solving, we find $c_1 = \sqrt{3}/4$ and $c_2 = 1/4$. The solution of the IVP is $x = (\sqrt{3}/4) \cos t + (1/4) \sin t$.

11-14. (*Suggestion*) Another two-parameter family of solutions of the DE in these problems is given by $y = c_3 \cosh x + c_4 \sinh x$. You are encouraged to find solutions of these IVPs and then show that these solutions are equivalent to the ones obtained from the family $y = c_1 e^x + c_2 e^{-x}$.

12. Substituting the initial conditions into $y = c_1 e^x + c_2 e^{-x}$ and $y' = c_1 e^x - c_2 e^{-x}$ we obtain

$$e c_1 + e^{-1} c_2 = 0$$

$$e c_1 - e^{-1} c_2 = e.$$

Solving, we find $c_1 = \frac{1}{2}$ and $c_2 = -\frac{1}{2} e^2$. The solution of the initial-value problem is $y = \frac{1}{2} e^x - \frac{1}{2} e^2 e^{-x} = \frac{1}{2} e^x - \frac{1}{2} e^{2-x}$.

15. Two solutions are $y = 0$ and $y = x^3$.

18. For $f(x, y) = \sqrt{xy}$ we have $\partial f / \partial y = \frac{1}{2} \sqrt{x/y}$. Thus, the DE will have a unique solution in any region where $x > 0$ and $y > 0$ or where $x < 0$ and $y < 0$.

21. For $f(x, y) = x^2/(4 - y^2)$ we have $\partial f / \partial y = 2x^2 y/(4 - y^2)^2$. Thus the DE will have a unique solution in any region where $y < -2$, $-2 < y < 2$, or $y > 2$.

9

24. For $f(x,y) = (y+x)/(y-x)$ we have $\partial f/\partial y = -2x/(y-x)^2$. Thus the DE will have a unique solution in any region where $y < x$ or where $y > x$.

27. Both $f(x,y) = \sqrt{y^2-9}$ and $\partial f/\partial y = y/\sqrt{y^2-9}$ are continuous in the regions of the plane determined by $y < -3$ and $y > 3$, with no restrictions on x. Since the point $(2,-3)$ is not in either of these regions, there is no guarantee of a unique solution through $(2,-3)$.

30. (a) Since $\dfrac{d}{dx}\tan(x+c) = \sec^2(x+c) = 1 + \tan^2(x+c)$, we see that $y = \tan(x+c)$ satisfies the DE.

(b) Solving $y(0) = \tan c = 0$ we obtain $c = 0$ and $y = \tan x$. Since $\tan x$ is discontinuous at $x = \pm\pi/2$, the solution is not defined on $(-2,2)$ because this interval contains $\pm\pi/2$.

(c) The largest interval on which the solution can exist is $(-\pi/2, \pi/2)$.

33. (a) Differentiating $3x^2 - y^2 = c$ we get $6x - 2yy' = 0$ or $yy' = 3x$.

(b) Solving $3x^2 - y^2 = 3$ for y we get

$$y = \phi_1(x) = \sqrt{3(x^2-1)}, \qquad 1 < x < \infty,$$
$$y = \phi_2(x) = -\sqrt{3(x^2-1)}, \qquad 1 < x < \infty,$$
$$y = \phi_3(x) = \sqrt{3(x^2-1)}, \qquad -\infty < x < -1,$$
$$y = \phi_4(x) = -\sqrt{3(x^2-1)}, \qquad -\infty < x < -1.$$

(c) Only $y = \phi_3(x)$ satisfies $y(-2) = 3$.

36. We consider the points on the graphs with x-coordinates $x_0 = -1$, $x_0 = 0$, and $x_0 = 1$. We then note that only (d) and (e) satisfy $y(0) = -1$. Since the graph has a horizontal tangent line at $(0,-1)$, $y'(0) = 0$ and the graph satisfies the conditions only in (e).

41. (*Suggestion*) In this manual, see the comments **Signs of Derivatives** at the beginning of Section 2.1.

Section 1.3   Differential Equations as Mathematical Models

The terminology and concepts listed below provide an outline of the main ideas encountered in this section. These can be useful when preparing for a quiz or test.

Terminology and Concepts

- construction of mathematical models

- level of resolution of a model

- use of assumptions and empirical laws

- steps of the modeling process

- model for population growth

- model for radioactive decay

- Newton's law of cooling/warming

- model for the spread of a disease

- model for the draining of a tank

- models for first- and second-order chemical reactions

- models to describe mixtures of differing concentrations of elements in liquid solutions

- models for electrical series circuits

- models for the motion of falling bodies involving Newton's laws of motion

- model for a slipping chain from a frictionless peg

- model for the shape of a flexible cable suspended between two vertical supports

- approaches to DEs: analytical, qualitative, numerical

The basic skills listed below summarize the more mechanical types of problems encountered in the exercise set for this section.

Basic Skills

- set up DEs, sometimes with initial conditions, that can be used to model the the situations listed above

Mixtures   In setting up DEs such as

$$\frac{dA}{dt} + \frac{1}{100} A = 6 \qquad [(8) \text{ in the text}]$$

it is helpful to keep in mind the units. The derivative, dA/dt, is the rate at which the amount of salt (pounds) is changing in time (minutes) and so is measured in pounds per minute. In the derivation of (8) in the text observe that the unit "gal" cancels out in both R_{in} and R_{out} to give "lb/min." Also, note carefully that even though the tank contains 300 gallons of brine, the initial condition associated with (8) is *not* $A(0) = 300$ since $A(t)$, which represents the amount of salt at time t, is measured in pounds, whereas 300 is a measure of the number of gallons of fluid in the tank. Indeed, an initial condition for $A(t)$ is not specified. See the text: Problem 9 in Exercises 1.3.

Exercises 1.3   *Hints, Suggestions, Solutions, and Examples*

3. Let b be the rate of births and d the rate of deaths. Then $b = k_1 P$ and $d = k_2 P^2$. Since $dP/dt = b - d$, the DE is $dP/dt = k_1 P - k_2 P^2$.

6. By inspecting the graph in the text we take T_m to be $T_m(t) = 80 - 30 \cos \pi t/12$. Then the temperature of the body at time t is determined by the DE

$$\frac{dT}{dt} = k \left[T - \left(80 - 30 \cos \frac{\pi}{12} t \right) \right], \quad t > 0.$$

8. (*Hint*) A quantity P is jointly proportional to two quantities x and y if there exists a constant k such that $P = kxy$.

9. The rate at which salt is leaving the tank is

$$(3 \text{ gal/min}) \cdot \left(\frac{A}{300} \text{ lb/gal} \right) = \frac{A}{100} \text{ lb/min}.$$

Thus $dA/dt = A/100$. The initial amount of salt is $A(0) = 50$ lbs.

12. The rate at which salt is entering the tank is

$$R_{in}(c_{in} \text{ lb/gal}) \cdot (r_{in} \text{ gal/min}) = c_{in} r_{in} \text{ lb/min}.$$

Now let $A(t)$ denote the number of pounds of salt and $N(t)$ the number of gallons of brine in the tank at time t. The concentration of salt in the tank as well as in the outflow is $c(t) = x(t)/N(t)$. But the number of gallons of brine in the tank remains steady, is increased, or is decreased depending on whether $r_{in} = r_{out}$, $r_{in} > r_{out}$, or $r_{in} < r_{out}$. In any case, the number of gallons of brine in the tank at time t is $N(t) = N_0 + (r_{in} - r_{out})t$. The output rate of salt is then

$$R_{out} = \left(\frac{A}{N_0 + (r_{in} - r_{out})t} \text{ lb/gal} \right) \cdot (r_{out} \text{ gal/min}) = r_{out} \frac{A}{N_0 + (r_{in} - r_{out})t} \text{ lb/min}.$$

The DE for the amount of salt, $dA/dt = R_{in} - R_{out}$, is

$$\frac{dA}{dt} = c_{in} r_{in} - r_{out} \frac{A}{N_0 + (r_{in} - r_{out})t} \quad \text{or} \quad \frac{dA}{dt} + \frac{r_{out}}{N_0 + (r_{in} - r_{out})t} A = c_{in} r_{in}.$$

15. Since $i = dq/dt$ and $L \, d^2 q/dt^2 + R \, dq/dt = E(t)$, we obtain $L \, di/dt + Ri = E(t)$.

18. Since the barrel in Figure 1.35(b) in the text is submerged an additional y feet below its equilibrium position the number of cubic feet in the additional submerged portion is the volume of the circular cylinder: $\pi \times (\text{radius})^2 \times \text{height}$ or $\pi(s/2)^2 y$. Then we have from Archimedes' principle

$$\text{upward force of water on barrel} = \text{weight of water displaced}$$

$$= (62.4) \times (\text{volume of water displaced})$$

$$= (62.4)\pi(s/2)^2 y = 15.6\pi s^2 y.$$

It then follows from Newton's second law that

$$\frac{w}{g}\frac{d^2y}{dt^2} = -15.6\pi s^2 y \qquad \text{or} \qquad \frac{d^2y}{dt^2} + \frac{15.6\pi s^2 g}{w} y = 0,$$

where $g = 32$ and w is the weight of the barrel in pounds.

20. (*Hint*) If the damping force is proportional to the instantaneous velocity, then there exists a constant β such that this force is equal to $\beta\, dx/dt$.

21. Let $x(t)$ denote the height of the top of the chain at time t with the positive direction upward. The weight of the portion of chain off the ground is $W = (x\text{ ft}) \cdot (1\text{ lb/ft}) = x$. The mass of the chain is $m = W/g = x/32$. The net force is $F = 5 - W = 5 - x$. By Newton's second law,

$$\frac{d}{dt}\left(\frac{x}{32}v\right) = 5 - x \qquad \text{or} \qquad x\frac{dv}{dt} + v\frac{dx}{dt} = 160 - 32x.$$

Thus, the DE is

$$x\frac{d^2x}{dt^2} + \left(\frac{dx}{dt}\right)^2 + 32x = 160.$$

24. The gravitational force on m is $F = -kM_r m/r^2$. Since $M_r = 4\pi\delta r^3/3$ and $M = 4\pi\delta R^3/3$ we have $M_r = r^3 M/R^3$ and

$$F = -k\frac{M_r m}{r^2} = -k\frac{r^3 Mm/R^3}{r^2} = -k\frac{mM}{R^3}r.$$

Now from $F = ma = md^2r/dt^2$ we have

$$m\frac{d^2r}{dt^2} = -k\frac{mM}{R^3}r \qquad \text{or} \qquad \frac{d^2r}{dt^2} = -\frac{kM}{R^3}r.$$

27. The DE is $x'(t) = r - kx(t)$ where $k > 0$.

30. The DE is $dP/dt = kP$, so from Problem 37 in Exercises 1.1, $P = e^{kt}$, and a one-parameter family of solutions is $P = ce^{kt}$.

33. This DE could describe a population that undergoes periodic fluctuations.

36. (a) If ρ is the mass density of the raindrop, then $m = \rho V$ and

$$\frac{dm}{dt} = \rho\frac{dV}{dt} = \rho\frac{d}{dt}\left[\frac{4}{3}\pi r^3\right] = \rho\left(4\pi r^2\frac{dr}{dt}\right) = \rho S\frac{dr}{dt}.$$

If dr/dt is a constant, then $dm/dt = kS$ where $\rho\, dr/dt = k$ or $dr/dt = k/\rho$. Since the radius is decreasing, $k < 0$. Solving $dr/dt = k/\rho$ we get $r = (k/\rho)t + c_0$. Since $r(0) = r_0$, $c_0 = r_0$ and $r = kt/\rho + r_0$.

(b) From Newton's second law, $\dfrac{d}{dt}[mv] = mg$, where v is the velocity of the raindrop. Then

$$m\frac{dv}{dt} + v\frac{dm}{dt} = mg \qquad \text{or} \qquad \rho\left(\frac{4}{3}\pi r^3\right)\frac{dv}{dt} + v(k4\pi r^2) = \rho\left(\frac{4}{3}\pi r^3\right)g.$$

13

Dividing by $4\rho\pi r^3/3$ we get

$$\frac{dv}{dt} + \frac{3k}{\rho r}v = g \quad \text{or} \quad \frac{dv}{dt} + \frac{3k/\rho}{kt/\rho + r_0}v = g, \quad k < 0.$$

39. At time t, when the population is 2 million cells, the DE $P'(t) = 0.15P(t)$ gives the rate of increase at time t. Thus, when $P(t) = 2$ (million cells), the rate of increase is $P'(t) = 0.15(2) = 0.3$ million cells per hour or 300,000 cells per hour.

Chapter 1 in Review

Hints, Suggestions, Solutions, and Examples

3. $\dfrac{d}{dx}(c_1 \cos kx + c_2 \sin kx) = -kc_1 \sin kx + kc_2 \cos kx;$

$\dfrac{d^2}{dx^2}(c_1 \cos kx + c_2 \sin kx) = -k^2 c_1 \cos kx - k^2 c_2 \sin kx = -k^2(c_1 \cos kx + c_2 \sin kx);$

$\dfrac{d^2 y}{dx^2} = -k^2 y \quad \text{or} \quad \dfrac{d^2 y}{dx^2} + k^2 y = 0$

6. $y' = -c_1 e^x \sin x + c_1 e^x \cos x + c_2 e^x \cos x + c_2 e^x \sin x;$

$y'' = -c_1 e^x \cos x - c_1 e^x \sin x - c_1 e^x \sin x + c_1 e^x \cos x - c_2 e^x \sin x + c_2 e^x \cos x + c_2 e^x \cos x + c_2 e^x \sin x$

$\quad = -2c_1 e^x \sin x + 2c_2 e^x \cos x;$

$y'' - 2y' = -2c_1 e^x \cos x - 2c_2 e^x \sin x = -2y; \qquad y'' - 2y' + 2y = 0$

9. b   **12.** a,b,d

15. The slope of the tangent line at (x, y) is y', so the DE is $y' = x^2 + y^2$.

18. **(a)** Differentiating $y^2 - 2y = x^2 - x + c$ we obtain $2yy' - 2y' = 2x - 1$ or $(2y - 2)y' = 2x - 1$.

(b) Setting $x = 0$ and $y = 1$ in the solution we have $1 - 2 = 0 - 0 + c$ or $c = -1$. Thus, an implicit solution of the IVP is $y^2 - 2y = x^2 - x - 1$.

(c) Solving $y^2 - 2y - (x^2 - x - 1) = 0$ by the quadratic formula we get $y = (2 \pm \sqrt{4 + 4(x^2 - x - 1)})/2$
$= 1 \pm \sqrt{x^2 - x} = 1 \pm \sqrt{x(x - 1)}$. Since $x(x - 1) \geq 0$ for $x \leq 0$ or $x \geq 1$, we see that neither $y = 1 + \sqrt{x(x - 1)}$ nor $y = 1 - \sqrt{x(x - 1)}$ is differentiable at $x = 0$. Thus, both functions are explicit solutions of the DE, but neither is a solution of the IVP.

21. (a)

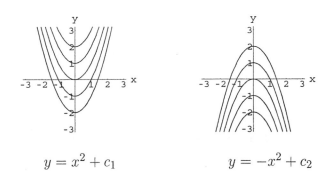

$$y = x^2 + c_1 \qquad\qquad y = -x^2 + c_2$$

(b) When $y = x^2 + c_1$, $y' = 2x$ and $(y')^2 = 4x^2$. When $y = -x^2 + c_2$, $y' = -2x$ and $(y')^2 = 4x^2$.

(c) Pasting together x^2, $x \geq 0$, and $-x^2$, $x \leq 0$, we get $y = \begin{cases} -x^2, & x \leq 0 \\ x^2, & x > 0. \end{cases}$

24. Differentiating $y = x\sin x + (\cos x)\ln(\cos x)$ we get

$$y' = x\cos x + \sin x + \cos x\left(\frac{-\sin x}{\cos x}\right) - (\sin x)\ln(\cos x)$$

$$= x\cos x + \sin x - \sin x - (\sin x)\ln(\cos x)$$

$$= x\cos x - (\sin x)\ln(\cos x)$$

and

$$y'' = -x\sin x + \cos x - \sin x\left(\frac{-\sin x}{\cos x}\right) - (\cos x)\ln(\cos x)$$

$$= -x\sin x + \cos x + \frac{\sin^2 x}{\cos x} - (\cos x)\ln(\cos x)$$

$$= -x\sin x + \frac{\cos^2 x + \sin^2 x}{\cos x} - (\cos x)\ln(\cos x)$$

$$= -x\sin x + \sec x - (\cos x)\ln(\cos x).$$

Thus

$$y'' + y = -x\sin x + \sec x - (\cos x)\ln(\cos x) + x\sin x + (\cos x)\ln(\cos x) = \sec x.$$

To obtain an interval of definition we note that the domain of $\ln x$ is $(0, \infty)$, so we must have $\cos x > 0$. Thus, an interval of definition is $(-\pi/2, \pi/2)$.

27. From the graph we see that estimates for y_0 and y_1 are $y_0 = -3$ and $y_1 = 0$.

30. We are given $d\theta/dt = \omega$. Differentiating $x = r\cos\theta$ we get

$$\frac{dx}{dt} = -r(\sin\theta)\frac{d\theta}{dt} = -r(\sin\theta)\omega = -r\omega\sin\theta$$

and

$$\frac{d^2x}{dt^2} = -r\omega(\cos\theta)\frac{d\theta}{dt} = -r\omega^2\cos\theta = -\omega^2(r\cos\theta) = -\omega^2 x.$$

Thus, $x(t)$ satisfies $d^2x/dt^2 + \omega^2 x = 0$.

Similarly, differentiating $y = r\sin\theta$ we get

$$\frac{dy}{dt} = r(\cos\theta)\frac{d\theta}{dt} = r\omega\cos\theta$$

and

$$\frac{d^2y}{dt^2} = -r\omega(\sin\theta)\frac{d\theta}{dt} = -r\omega^2\sin x = \omega^2 y.$$

Thus, $y(t)$ satisfies $d^2y/dt^2 + \omega^2 y = 0$.

2 First-Order Differential Equations

Section 2.1   Solution Curves Without a Solution

The terminology and concepts listed below provide an outline of the main ideas encountered in this section. These can be useful when preparing for a quiz or test.

Terminology and Concepts

- lineal element
- direction field (slope field)
- autonomous first-order DE
- critical (stationary, equilibrium) point
- equilibrium solution
- phase portrait (phase line)
- asymptotically stable critical point (attractor)
- unstable critical point (repeller)
- semi-stable critical point

The basic skills listed below summarize the more mechanical types of problems encountered in the exercise set for this section.

Basic Skills

- sketch a direction field of a first-order DE by hand
- sketch a solution curve through a given point in a direction field
- find the critical points of an autonomous first-order DE
- draw the phase portrait of an autonomous first-order DE
- classify critical points as asymptotically stable (attractor), unstable (repeller) or semi-stable
- use a phase portrait to sketch solution curves of autonomous first-order DEs

• determine the behavior of solutions of autonomous first-order DEs as $t \to \pm\infty$

Signs of Derivatives   The first derivative and its algebraic sign over an interval plays an important role in the qualitative analysis in Section 2.1. If $y = f(x)$ is a differentiable function and if $dy/dx > 0$ on an interval then f is increasing on the interval, whereas if $dy/dx < 0$ on an interval then f is decreasing on the interval. Similarly, if $y = f(x)$ possesses a second derivative, then the sign of d^2y/dx^2 determines the concavity of the graph of f. If $d^2y/dx^2 > 0$ on an interval then the graph of f is concave upward on the interval, whereas if $d^2y/dx^2 < 0$ on an interval, then the graph of f is concave downward.

Use of Computers   Both *Mathematica* and *Maple* can be used to obtain direction fields of first-order DEs. In *Mathematica*, a graphics package containing the command **PlotVectorField** must first be loaded. (It is important that this be done only once during any session of *Mathematica* and that it be done before the command is first used. If this protocol is not followed, you should save your current file, quit *Mathematica*, and restart the program.) **PlotVectorField** is actually used to graph the vector field corresponding to a vector function so it must be adapted to the case of a first-order DE having the form $dy/dx = f(x,y)$. The routine below plots the direction field for the DE $dy/dx = 0.2xy$ where $-5 < x < 5$ and $-5 < y < 5$.

```
<<Graphics`                                        (Mathematica)
PlotVectorField[{1, 2x y}, {x, -5, 5}, {y, -5, 5},
    Frame->True, Axes->True, AxesLabel->{"x", "y"},
    Plotpoints->15]
```

This example will plot a direction field in which the lineal elements (arrows) have their base at a grid point. The number of grid points in both horizontal and vertical directions is specified by the **PlotPoints** option. The output is shown below in Figure 2.1. To obtain a direction field with linear elements all the same length, the graphics options **ScaleFunction** and **ScaleFactor** can be added as shown below.

```
PlotVectorField[{1, 2x y},                         (Mathematica)
    {x, -5, 5}, {y, -5, 5}, Frame->True, Axes->True,
    AxesLabel->{"x", "y"}, ScaleFunction->(0.5&),
    ScaleFactor->0.5, PlotPoints->9]
```

The output for this command is shown below in Figure 2.2.

Figure 2.1

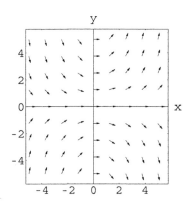

Figure 2.2

In *Maple* the command `fieldplot` is used to obtain a direction field. The routine below plots the direction field for the DE $dy/dx = 0.2xy$.

```
with(plots);                                      (Maple)
fieldplot([1, 0.2*x*y], x=-5..5, y=-5..5);
```

To obtain a direction field with lineal elements all having the same length it is necessary to scale the input by the length of each vector as shown below.

```
with(plots);                                      (Maple)
fieldplot([1/sqrt(1+(0.2*x*y)^2), 0.2*x*y/sqrt(1+(0.2*x*y)^2)],
   x=-5..5, y=-5..5);
```

Exercises 2.1   *Hints, Suggestions, Solutions, and Examples*

3.

6.

9.

12.

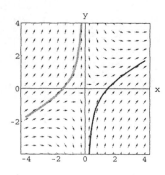

13. (*Hint*) The slope of a lineal element at the point (x_1, y_1) in a grid is $f(y_1)$ and can be approximated by measuring up from the y-axis in Figure 2.14 in the text.

15. (a) The isoclines have the form $y = -x + c$, which are straight lines with slope -1.

(b) The isoclines have the form $x^2 + y^2 = c$, which are circles centered at the origin.

21. Solving $y^2 - 3y = y(y - 3) = 0$ we obtain the critical points 0 and 3. From the phase portrait we see that 0 is asymptotically stable (attractor) and 3 is unstable (repeller).

24. Solving $10 + 3y - y^2 = (5 - y)(2 + y) = 0$ we obtain the critical points -2 and 5. From the phase portrait we see that 5 is asymptotically stable (attractor) and -2 is unstable (repeller).

27. Solving $y \ln(y + 2) = 0$ we obtain the critical points -1 and 0. From the phase portrait we see that -1 is asymptotically stable (attractor) and 0 is unstable (repeller).

30. The critical points are approximately at -2.2, 0.5, and 1.7. Since $f(y) > 0$ for $y < -2.2$ and $0.5 < y < 1.7$, the graph of the solution is increasing on $(-\infty, -2.2)$ and $(0.5, 1.7)$. Since $f(y) < 0$ for $-2.2 < y < 0.5$ and $y > 1.7$, the graph is decreasing on $(-2.2, 0.5)$ and $(1.7, \infty)$.

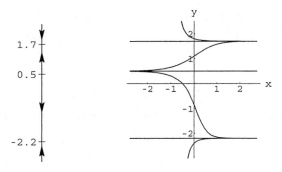

34. (*Suggestion*) Start by assuming that $\displaystyle\lim_{x \to \infty} \frac{dy}{dx} = f(\lim_{x \to \infty} y)$.

Section 2.2 — Separable Variables

The terminology and concepts listed below provide an outline of the main ideas encountered in this section. These can be useful when preparing for a quiz or test.

Terminology and Concepts

- separable DE

The basic skills listed below summarize the more mechanical types of problems encountered in the exercise set for this section.

Basic Skills

- determine whether or not a DE is separable

- use integration to solve a separable DE

- when possible, solve an equation for the dependent variable to find an explicit solution of a DE

Integration   The method of separation of variables is the first of several techniques for solving DEs that require integration. Recall from calculus that an indefinite integral (or antiderivative) of a function $f(x)$ is another function $F(x)$ such that $F'(x) = f(x)$. The most general indefinite integral of f is written:

$$\int f(x)dx = F(x) + c, \tag{1}$$

where $F'(x) = f(x)$ and c is an arbitrary constant. Of the many formulas given in typical tables of integrals there are only a few that appear as a matter of course in typical texts in DEs. If $u = g(x)$ is a differentiable function, then its differential is defined as $du = g'(x)dx$. Each of the following results can be verified by differentiation.

$$\int u^n du = \frac{u^{n+1}}{n+1} + c, \quad n \neq 1 \tag{2}$$

$$\int u^{-1} du = \int \frac{1}{u} du = \ln|u| + c \tag{3}$$

$$\int e^u du = e^u + c \tag{4}$$

$$\int \sin u\, du = -\cos u + c \tag{5}$$

$$\int \cos u\, du = \sin u + c \tag{6}$$

22

$$\int \sec^2 u \, du = \tan u + c \tag{7}$$

$$\int \csc^2 u \, du = -\cot u + c \tag{8}$$

$$\int \frac{1}{a^2 + u^2} \, du = \frac{1}{a} \tan^{-1}\left(\frac{u}{a}\right) + c \tag{9}$$

For example, to evaluate $\int \sqrt{5x-2}\,dx$ we identify $u = 5x - 2$, $du = 5\,dx$, and write

$$\int (5x-2)^{1/2}\,dx = \frac{1}{5}\int \overbrace{(5x-2)^{1/2}}^{u^{1/2}} \underbrace{5\,dx}_{du} = \frac{1}{5}\frac{2}{3}(5x-2)^{3/2} + c = \frac{2}{15}(5x-2)^{3/2} + c.$$

You are encouraged to always check your answer. In the last example, in view of the Chain Rule we have

$$\frac{d}{dx}\left[\frac{2}{15}(5x-2)^{3/2} + c\right] = \frac{2}{15}\frac{3}{2}(5x-2)^{1/2}\cdot 5 = (5x-2)^{1/2}.$$

A formula very similar to (9) uses inverse hyperbolic functions

$$\int \frac{1}{a^2 - u^2}\,du = \frac{1}{a}\tanh^{-1}\left(\frac{u}{a}\right) + c, \quad |u| < a. \tag{10}$$

Every inverse hyperbolic function is a natural logarithm. Hence, if the integrand on the left-hand side of (10) is expanded into partial fractions, then an alternative form of (10) is found to be

$$\int \frac{1}{a^2 - u^2}\,du = \frac{1}{2a}\ln\left|\frac{a+u}{a-u}\right| + c, \quad |u| \neq a. \tag{11}$$

Formulas (10) and (11) are useful in some of the problems in Exercises 3.2.

Integration by Parts   Integration by parts is a method for integrating certain products of functions as well as some inverse functions such as $\ln x$ (the inverse of e^x) and $\sin^{-1} x$. If $u = f(x)$ and $v = g(x)$, then the differentials are $du = f'(x)\,dx$ and $dv = g'(x)\,dx$, respectively. Integration by parts is a two-step procedure that begins with an integration: Integrate the function that you choose to call dv, and then differentiate the other function (by default after dv is chosen, the function remaining is u):

$$\int u \boxed{dv} = \boxed{u}\,\boxed{v} - \int v \boxed{du}. \tag{12}$$

When choosing which portion of the integrand to designate dv you should take account of the fact that this function will need to be integrated, first to obtain v, and then integrated again to obtain $\int v\,du$. Thus, you will normally want to choose dv both so that its integral is known to you and

23

Section 2.2 Separable Variables

so that the integral of dv is no more complicated than v itself. For example, in $\int x^3 \sin x\, dx$ you want to let $dv = \sin x\, dx$ because $\sin x$ is easily integrated and $\int \sin x\, dx = -\cos x + c$ is no more complicated than $\sin x$. [If you chose $dv = x^3\, dx$, $\int x^3\, dx = \frac{1}{4}x^4 + c$ is more complicated than x^3.] There are exceptions however. For example, when finding $\int x^3 \ln x\, dx$, you should let $u = \ln x$ and $dv = x^3\, dx$.

Some Important Integrals   The following integrals require integration by parts.

(a) If $n = 1, 2, 3, \ldots$, the indefinite integral of the form $\int x^n e^{ax}\, dx$, a a constant, demands integration by parts n times. For example, to evaluate $\int x^2 e^{5x}\, dx$ we must use (12) twice. It is also important to remember that for integrals of this type, integration by parts always starts by integrating the exponential function. For example,

differentiate

$$\int x\, \boxed{e^{-2x}}\, dx = \boxed{x}\,\boxed{\left(-\tfrac{1}{2}e^{-2x}\right)} - \int \boxed{1}\cdot\left(-\frac{1}{2}e^{-2x}\right) dx = -\frac{1}{2}xe^{-2x} - \frac{1}{4}e^{-2x} + c.$$

integrate

(b) An integral $\int x^\alpha \ln x\, dx$, $\alpha \neq -1$, begins by integrating x^α. For example,

$$\int x^3 \ln x\, dx = \frac{1}{4}x^4 \ln x - \frac{1}{4}\int x^4 \frac{1}{x}\, dx = \frac{1}{4}x^4 \ln x - \frac{1}{4}\int x^3\, dx = \frac{1}{4}x^4 \ln x - \frac{1}{16}x^4 + c.$$

The special case when $\alpha = -1$, does not require integration by parts. By writing

$$\int x^{-1} \ln x\, dx = \int \ln x\, \frac{dx}{x}$$

and identifying $u = \ln x$, $du = dx/x$, the integral is recognized as the form given in (2) with $n = 1$:

$$\int x^{-1} \ln x\, dx = \int \overset{u^1}{\overbrace{\ln x}}\, \overset{du}{\overbrace{\frac{dx}{x}}} = \frac{1}{2}(\ln x)^2 + c.$$

Verify this last result by differentiating the right-hand side.

(c) You might recall that integrals of the type $\int e^{\alpha x} \sin \beta x\, dx$ or $\int e^{\alpha x} \cos \beta x\, dx$ require integration by parts twice. The unusual feature of these integrals is that in the second application of integration by parts the original integral is recovered on the right-hand side. We then solve for the original integral. It is hard to make a mistake here since we can begin the process by integrating either the exponential function or the trigonometric function. But it is important to remember that if the first integraton by parts is begun by integrating, say, the exponential function, then the second integration by parts must also begin by integrating the exponential function. For example,

24

differentiate

$$\int \boxed{e^{2x}} \sin 3x \, dx = \boxed{\frac{1}{2}e^{2x}} \; \boxed{\sin 3x} - \int \frac{1}{2}e^{2x} \boxed{(3\cos 3x)} \, dx$$

integrate

differentiate

$$= \frac{1}{2}e^{2x}\sin 3x - \frac{3}{2}\int \boxed{e^{2x}}\cos 3x\, dx = \frac{1}{2}e^{2x}\sin 3x - \frac{3}{2}\left[\boxed{\frac{1}{2}e^{2x}} \; \boxed{\cos 3x} - \int \frac{1}{2}e^{2x}\boxed{(-3\sin 3x)}\, dx\right]$$

integrate

$$= \frac{1}{2}e^{2x}\sin 3x - \frac{3}{4}e^{2x}\cos 3x - \frac{9}{4}\int e^{2x}\sin 3x\, dx.$$

If the original integral is denoted by the symbol I then the last result is equivalent to

$$I = \frac{1}{2}e^{2x}\sin 3x - \frac{3}{4}e^{2x}\cos 3x - \frac{9}{4}I.$$

Solving for I then gives $\frac{13}{4}I = \frac{1}{2}e^{2x}\sin 3x - \frac{3}{4}e^{2x}\cos 3x$ or

$$I = \int e^{2x}\sin 3x\, dx = \frac{2}{13}e^{2x}\sin 3x - \frac{3}{13}e^{2x}\cos 3x + c.$$

See **Use of Computers** later in this section for *Mathematica* and *Maple* implementations of both indefinite and definite integration.

Partial Fractions   The following discussion illustrates a procedure for integrating certain rational functions $P(x)/Q(x)$, where we shall assume for this discussion that the degree of $P(x)$ is less than the degree of $Q(x)$. (If the degree $P(x) \geq$ degree $Q(x)$, then start the process with long division.) This method, known as **partial fractions**, consists of expanding such a rational function into simpler component fractions, and then evaluating the integral term-by-term. We begin with an example. When the terms in the sum

$$\frac{2}{x+5} + \frac{1}{x+1} \tag{13}$$

are combined by means of a common denominator, we obtain the single rational expression

$$\frac{3x+7}{(x+5)(x+1)}. \tag{14}$$

Now suppose that we are faced with the problem of evaluating the integral $\int \dfrac{3x+7}{(x+5)(x+1)}\, dx$. Of course, the solution is obvious. We use the equality of (13) and (14) to write:

$$\int \frac{3x+7}{(x+5)(x+1)}\, dx = \int \frac{2}{x+5}\, dx + \int \frac{1}{x+1}\, dx = 2\ln|x+5| + \ln|x+1| + c.$$

25

Section 2.2 Separable Variables

In the discussion that follows, we review the assumptions and the algebra of four cases of partial fraction decomposition.

CASE 1 Nonrepeated Linear Factors

We state the following fact from algebra without proof. If

$$\frac{P(x)}{Q(x)} = \frac{P(x)}{(a_1 x + b_1)(a_2 x + b_2) \cdots (a_n x + b_n)}$$

where all the factors $a_i x + b_i$, $i = 2, \ldots, n$, in the denominator are distinct and the degree of $P(x)$ is less than n, then unique real constants $C_1, C_2, \ldots, C_n$ exist such that

$$\frac{P(x)}{Q(x)} = \frac{C_1}{a_1 x + b_1} + \frac{C_2}{a_2 x + b_2} + \cdots + \frac{C_n}{a_n x + b_n}.$$

For example, to expand $(2x + 1)/(x - 1)(x + 3)$ into individual partial fractions we make the assumption that the integrand can be written as

$$\frac{2x + 1}{(x - 1)(x + 3)} = \frac{A}{x - 1} + \frac{B}{x + 3}.$$

Combining the terms of the right-hand member of the equation over a common denominator gives

$$\frac{2x + 1}{(x - 1)(x + 3)} = \frac{A(x + 3) + B(x - 1)}{(x - 1)(x + 3)}.$$

Since the denominators are identical, the numerators are also identical

$$2x + 1 = A(x + 3) + B(x - 1) = (A + B)x + (3A - B), \tag{15}$$

and the coefficients of the powers of x are the same:

$$2 = A + B \quad \text{and} \quad 1 = 3A - B.$$

These simultaneous equations can now be solved for A and B. The results are $A = \frac{3}{4}$ and $B = \frac{5}{4}$. Therefore,

$$\frac{2x + 1}{(x - 1)(x + 3)} = \frac{\frac{3}{4}}{x - 1} + \frac{\frac{5}{4}}{x + 3}.$$

Note: The numbers A and B in the preceding example can be determined in an alternative manner. Since (15) is an identity, that is, the equality is true for every value of x, it holds for $x = 1$ and $x = -3$, *the zeros of the denominator*. Setting $x = 1$ in (15) gives $3 = 4A$, from which it follows immediately that $A = \frac{3}{4}$. Similarly, by setting $x = -3$ in (15), we obtain $-5 = (-4)B$ or $B = \frac{5}{4}$. In the *Remarks* at the end of Section 7.2 in the text an equivalent and simpler method for evaluating the coefficients in a partial fraction decomposition is given. It is important to be aware that this technique, called the **cover-up method**, works only when the denomintor is a product of linear factors, none of which are repeated.

CASE II Repeated Linear Factors

If

$$\frac{P(x)}{Q(x)} = \frac{P(x)}{(ax+b)^n}$$

where $n > 1$ and the degree of $P(x)$ is less than n, then unique real constants $C_1, C_2, \ldots, C_n$ can be found such that

$$\frac{P(x)}{(ax+b)^n} = \frac{C_1}{ax+b} + \frac{C_2}{(ax+b)^2} + \cdots + \frac{C_n}{(ax+b)^n}.$$

Combining the Cases: When the denominator $Q(x)$ contains distinct as well as repeated inear factors, we combine the two cases. For example, to expand $(6x-1)/x^3(2x-1)$ into partial fractions we write

$$\frac{6x-1}{x^3(2x-1)} = \frac{A}{x} + \frac{B}{x^2} + \frac{C}{x^3} + \frac{D}{2x-1}$$

since x^3 means that the linear term x is used as a factor x three times. Putting the right-hand side over a common denominator and equating numerators gives

$$6x - 1 = Ax^2(2x-1) + Bx(2x-1) + C(2x-1) + Dx^3 \tag{16}$$

$$= (2A+D)x^3 + (-A+2B)x^2 + (-B+2C)x - C. \tag{17}$$

If we set $x = 0$ and $x = \frac{1}{2}$ in (16), we find $C = 1$ and $D = 16$, respectively. Now since the denominator of the original expression has only two distinct zeros, we can find A and B by equating the coefficients of x^3 and x^2 in

$$6x - 1 = 0x^3 + 0x^2 + 6x - 1 = (2A+16)x^3 + (-A+2B)x^2 + (-B+2)x - 1.$$

Thus, $A = -8$ and $B = -4$. Therefore,

$$\frac{6x-1}{x^3(2x-1)} = -\frac{8}{x} - \frac{4}{x^2} + \frac{1}{x^3} + \frac{16}{2x-1}.$$

CASE III Nonrepeated Quadratic Factors

Suppose the denominator of the rational function $P(x)/Q(x)$ can be expressed as a product of distinct quadratic factors $a_i x^2 + b_i x + c_i$, $i = 1, 2, \ldots, n$, that are irreducible (not factorable) over the real numbers. If the degree of $P(x)$ is less than $2n$, we can find unique real constants $A_1, A_2, \ldots, A_n, B_1, B_2, \ldots, B_n$ such that

$$\frac{P(x)}{(a_1 x^2 + b_1 x + c_1)(a_2 x^2 + b_2 x + c_2) \cdots (a_n x^2 + b_n x + c_n)}$$

$$= \frac{A_1 x + B_1}{a_1 x^2 + b_1 x + c} + \frac{A_2 x + B_2}{a_2 x^2 + b_2 x + c_2} + \cdots + \frac{A_n x + B_n}{a_n x^2 + b_n x + c_n}.$$

27

For example, to expand $4x/(x^2+1)(x^2+2x+3)$ into partial fractions we first write

$$\frac{4x}{(x^2+1)(x^2+2x+3)} = \frac{Ax+B}{x^2+1} + \frac{Cx+D}{x^2+2x+3}$$

and, after putting the right-hand side over a common denominator, equate numerators:

$$4x = (Ax+B)(x^2+2x+3) + (Cx+D)(x^2+1)$$

$$= (A+B)x^3 + (2A+B+D)x^2 + (3A+2B+C)x + (3B+D).$$

Since the denominator of the original fraction has no real zeros, we compare coefficients of powers of x:

$$0 = A + C$$

$$0 = 2A + B + D$$

$$4 = 3A + 2B + C$$

$$0 = 3B + D.$$

Solving the equations yields $A = 1$, $B = 1$, $C = -1$, and $D = -3$. Therefore,

$$\frac{4x}{(x^2+1)(x^2+2x+3)} = \frac{x+1}{x^2+1} - \frac{x+3}{x^2+2x+3}.$$

CASE IV   Repeated Quadratic Factors

In the case when the denominator contains a repeated irreducible quadratic polynomial ax^2+bx+c,

$$\frac{P(x)}{Q(x)} = \frac{P(x)}{(ax^2+bx+c)^n},$$

then if $P(x)$ is of degree less than $2n$, we can find real constants $A_1, A_2, \ldots, A_n, B_1, B_2, \ldots, B_n$ such that

$$\frac{P(x)}{(ax^2+bx+c)^n} = \frac{A_1x+B_1}{ax^2+bx+c} + \frac{A_2x+B_2}{(ax^2+bx+c)^2} + \cdots + \frac{A_nx+B_n}{(ax^2+bx+c)^n}.$$

Combining the Cases: The denominator $Q(x)$ in the fraction $(x+3)/x^2(x-5)(x^2+1)^2$ contains a distinct linear factor $x-5$, repeated linear factors x^2, and repeated irreducible quadratic factors $(x^2+1)^2$. To expand into partial fractions, we write

$$\frac{x+3}{x^2(x-5)(x^2+1)^2} = \underbrace{\frac{A}{x} + \frac{B}{x^2}}_{repeated\ linear} + \underbrace{\frac{C}{x-5}}_{linear} + \underbrace{\frac{Dx+E}{x^2+1} + \frac{Fx+G}{(x^2+1)^2}}_{repeated\ quadratic}.$$

Important Note: Although we have used it as a technique of integration, the ability to expand a rational expression $P(x)/Q(x)$ into partial fractions is also used in another context in the study of DEs; *it is absolutely essential* in the study of the Laplace transform in Chapter 7. See Section

7.2 for the syntax used in *Mathematica* and *Maple* to obtain a partial fraction decomposition of a rational function.

Use of Computers   We show below how *Mathematica* and *Maple* can be used to graph a function, graph an equation, find (and name) the solution of an equation, and find the indefinite integral (or antiderivative) and definite integral of a function.

Graphing a Function: To graph a function, such as $f(x) = \sqrt{4 - x^2}$, use the syntax

> **Clear[f]**   (*Mathematica*)
> **f[x_]:= Sqrt[4 - x^2]**
> **f[x]**
> **Plot[f[x], {x, -3, 3}]**
>
> ```
> f:=x->sqrt(4-x^2);                                   (Maple)
> plot(f(x), x = -3..3);
> ```

Because $\sqrt{4 - x^2}$ is not a real number for $x < -2$ or $x > 2$, *Mathematica* returns some error messages when it tries to evaluate the function at points in these intervals; it does however then plot the graph. *Maple*, on the other hand, appears to ignore such problems and simply plots the graph. Since the graph of $f(x)$ is the upper half of the circle $x^2 + y^2 = 4$, you would expect the graph to look like a semicircle. Due, however, to the fact that both programs choose the widths and heights of their plots without regard to the scales on the horizontal and vertical axes, the semicircle may look more like a semi-ellipse. To insure that one unit on each axis has the same length on the computer screen, a graphics option is used as shown below.

> **Plot[f[x], {x, -2, 2}, AspectRatio->2/4]**   (*Mathematica*)
>
> ```
> plot(f(x), x=-3..3, scaling=CONSTRAINED);      (Maple)
> ```

(The **AspectRatio** option in the **Plot** command is the height of the plot divided by its width.)

Both programs provide a large number of graphics options that can be used to control the appearance of plots. Probably the most commonly used of these is an option that specifies the extent of the vertical axis to be shown. This may be necessary to see details of the graph that are obscured if the CAS (as opposed to the user) is allowed to choose the range of the graph shown. Examples are shown below with pictures displayed for the *Mathematica* output. The pictures generated by *Maple* are similar.

> **Plot[(x^2 - 1)Exp[2x^2], {x, -2, 2},**   (*Mathematica*)
> **PlotRange->{-2, 2}]**

29

```
plot((x^2-1)*exp(2*x^2), x = -2..2,        (Maple)
    y = -2..2));
```

without **PlotRange**

with **PlotRange**

Finally, you may want to superimpose the graphs of two or more functions on the same set of coordinate axes. The commands below show the graphs of $f(x) = x^3 - 3x^2 + x + 1$ and $g(x) = 1 - 5x/4$, the tangent line to the graph of $f(x)$ at $x = 3/2$.

```
Clear[f, g]                                 (Mathematica)
f[x_]:= x^3 - 3x^2 + x + 1
g[x_]:= 1 - 5x/4
{f[x], g[x]}
Plot[{f(x), g(x)}, {x, -2, 3}]
```

```
f:=x->x^3-3*x^2+x+1;                        (Maple)
g:=x->1-5*x/4;
plot({f(x),g(x)}, x = -2..3);
```

Graphing an Equation: In addition to graphing functions, a CAS can be used to graph equations that implicitly define functions. Equations such as this frequently arise when solving a DE using separation of variables. For example, the solution of $(1 + e^y)y' = x \cos x + \sin x$, $y(0) = 0$, obtained using separation of variables, is $y + e^y = x \sin x + 1$. We cannot solve this equation for y in terms of elementary functions, but we can graph the equation using **ContourPlot** in *Mathematica* and `contourplot` in *Maple* as shown below.

```
f[x_, y_]:= y + Exp[y] - x Sin[x] - 1       (Mathematica)
f[x, y]
ContourPlot[f[x, y], {x, -15, 15}, {y, -10, 5}, PlotPoints->50,
    Contours->{0}, ContourShading->False, Frame->False,
    Axes->True, AxesOrigin->{0, 0}, AxesLabel->{"x", "y"}]
```

30

Many of the graphics options used above, such as **Frame** and **Axes** are self-explanatory. The default setting for **PlotPoints** is 15 and may result in a graph consisting of straight line segments and sharp corners. The **Contours** options is set to 0 in this case because the equation being graphed is $f(x, y) = y + e^y - x \sin x - 1 = 0$. Alternatively, the function could have been specified as $f(x, y) = y + e^x - x \sin x$ and the **Contours** option set to 1.

In *Maple* we effect the same result using the commands

```
f:=(x,y)->y+exp(y)-x*sin(x)-1;                  (Maple)
with(plots);
contourplot(f(x,y), x=-15..15, y=-10..5,
    contours=[0], grid=[80,80]);
```

The `grid` option is set to numbers that are large enough to insure a fairly smooth-looking graph.

Solving an Equation: Depending on the nature of an equation, there are a number of commands in both *Mathematica* and *Maple* that can be used to find solutions of the equation. If the equation involves only polynomial or rational functions, use **Solve** or **NSolve** in *Mathematica* and `solve` in *Maple*.

$$\text{Solve[2x^2 - 3x - 4==0, x]} \qquad (Mathematica)$$

or

$$\text{NSolve[2x^2 - 3x - 4==0, x]}$$

and

```
solve(2*x^2-3*x-4=0, x);                         (Maple)
```

Both **Solve** and `solve` will give exact solutions when possible. To obtain decimal versions of the solutions use **NSolve** in *Mathematica* and insert a decimal point in one of the numbers in the equation (`2*x^2-3*x-4.=0`) in *Maple*. (This will also work with the **Solve** command in *Mathematica* as an alternative to **NSolve**.)

To solve equations involving more complicated expressions such as $e^x = 3\cos x + 1$ use **FindRoot** in *Mathematica* and `fsolve` in *Maple*. In both programs, either Newton's method or a variation of it is used, so some form of an initial guess should be provided. This can be obtained from graphs and will enable you to select the solution you want in the event there is more than one.

$$\text{Plot[\{Exp[x], 3Cos[x]+1\}, \{x, -5, 5\}]} \qquad (Mathematica)$$
$$\text{FindRoot[Exp[x]==3Cos[x]+1, \{x, 1\}]}$$

31

```
plot([exp(x), 3*cos(x)+1], x=-5..5, y=-3..10);  (Maple)
fsolve(exp(x), 3*cos(x)+1, x=0..2);
```

As shown above, in *Maple*, a range of values can be specified. Other options are available, but this is generally the most effective way to find the solution you are looking for. If there is no solution within this range, no solution will be given. If there are multiple solutions within the range, only one will be given, unless you are solving a polynomial equation, in which case all solutions within the range will be given.

Naming the Solution of an Equation: If you intend to use the solution of an equation in a subsequent computation in your CAS, then it is convenient to have a name for this value so that you do not have to read it from the screen and then type it in. The routines below show how to do this, assigning the name a to the solution of $e^{3x-x^2} \cos x = 0$.

Clear[f]   (*Mathematica*)
f[x_]:= Exp[x - x^2] + Cos[x]
f[x]
Plot[f[x], {x, -5, 5}]
sol = FindRoot[f[x]== 0, {x, 2}]
a = x/.sol

```
f:=x->exp(x-x^2)+cos(x);                         (Maple)
plot(f(x), x=-5..5);
a:=fsolve(f(x)=0, x=1..3);
```

Integrals: *Mathematica* and *Maple* can find both antiderivatives (indefinite integrals) of many functions and definite integrals of most functions. To find an antiderivative of $f(x)$ use

Integrate[f[x], x]   (*Mathematica*)

```
int(f(x), x);                                    (Maple)
```

If you want to further work with the antiderivative in the CAS it can be helpful to define it as a function that can itself be evaluated at a point, graphed, or differentiated. The syntax to accomplish this is

g[x_]:= Evaluate[Integrate[f(x), x]]   (*Mathematica*)
g[x]

```
g:=unapply(int(f(x),x),x);                       (Maple)
```

32

To find the definite integral, $\int_a^b f(x)\,dx$, of f we use

$$\textbf{Integrate}[\text{f}(\text{x}), \{\text{x, a, b}\}] \qquad\qquad (Mathematica)$$

```
int(f(x),x=a..b);
```
$\qquad\qquad (Maple)$

The comands **Integrate** and `int` will try to find exact values of a definite integral by evaluating the antiderivative at the endpoints of the interval. If the CAS is unable to find an antiderivative of $f(x)$ (for example, when $f(x) = \sin(\sin x)$, or if you simply want a numerical approximation of the definite integral, use

$$\textbf{NIntegrate}[\text{f}(\text{x}), \{\text{x, a, b}\}] \qquad\qquad (Mathematica)$$

```
evalf(int(f(x),x=a..b));
```
$\qquad\qquad (Maple)$

Exercises 2.2   *Hints, Suggestions, Solutions, and Examples*

In many of the following problems we will encounter an expression of the form $\ln|g(y)| = f(x) + c$. To solve for $g(y)$ we exponentiate both sides of the equation. This yields $|g(y)| = e^{f(x)+c} = e^c e^{f(x)}$ which implies $g(y) = \pm e^c e^{f(x)}$. Letting $c_1 = \pm e^c$ we obtain $g(y) = c_1 e^{f(x)}$.

3. From $dy = -e^{-3x}\,dx$ we obtain $y = \frac{1}{3}e^{-3x} + c$.

6. From $\dfrac{1}{y^2}\,dy = -2x\,dx$ we obtain $-\dfrac{1}{y} = -x^2 + c$ or $y = \dfrac{1}{x^2 + c_1}$.

9. From $\left(y + 2 + \dfrac{1}{y}\right)dy = x^2 \ln x\,dx$ we obtain $\dfrac{y^2}{2} + 2y + \ln|y| = \dfrac{x^3}{3}\ln|x| - \dfrac{1}{9}x^3 + c$.

12. From $2y\,dy = -\dfrac{\sin 3x}{\cos^3 3x}\,dx$ or $2y\,dy = -\tan 3x \sec^2 3x\,dx$ we obtain $y^2 = -\frac{1}{6}\sec^2 3x + c$.

15. From $\dfrac{1}{S}\,dS = k\,dr$ we obtain $S = ce^{kr}$.

18. From $\dfrac{1}{N}\,dN = \left(te^{t+2} - 1\right)dt$ we obtain $\ln|N| = te^{t+2} - e^{t+2} - t + c$ or $N = c_1 e^{te^{t+2} - e^{t+2} - t}$.

21. From $x\,dx = \dfrac{1}{\sqrt{1 - y^2}}\,dy$ we obtain $\frac{1}{2}x^2 = \sin^{-1} y + c$ or $y = \sin\left(\dfrac{x^2}{2} + c_1\right)$.

24. From $\dfrac{1}{y^2 - 1}\,dy = \dfrac{1}{x^2 - 1}\,dx$ or $\dfrac{1}{2}\left(\dfrac{1}{y - 1} - \dfrac{1}{y + 1}\right)dy = \dfrac{1}{2}\left(\dfrac{1}{x - 1} - \dfrac{1}{x + 1}\right)dx$ we obtain

$\ln|y - 1| - \ln|y + 1| = \ln|x - 1| - \ln|x + 1| + \ln c$ or $\dfrac{y - 1}{y + 1} = \dfrac{c(x - 1)}{x + 1}$. Using $y(2) = 2$ we find

$c = 1$. A solution of the IVP is $\dfrac{y - 1}{y + 1} = \dfrac{x - 1}{x + 1}$ or $y = x$.

27. Separating variables and integrating we obtain

$$\frac{dx}{\sqrt{1 - x^2}} - \frac{dy}{\sqrt{1 - y^2}} = 0 \quad \text{and} \quad \sin^{-1} x - \sin^{-1} y = c.$$

Setting $x = 0$ and $y = \sqrt{3}/2$ we obtain $c = -\pi/3$. Thus, an implicit solution of the IVP is $\sin^{-1} x - \sin^{-1} y = \pi/3$. Solving for y and using an addition formula from trigonometry, we get

$$y = \sin\left(\sin^{-1} x + \frac{\pi}{3}\right) = x\cos\frac{\pi}{3} + \sqrt{1 - x^2}\sin\frac{\pi}{3} = \frac{x}{2} + \frac{\sqrt{3}\sqrt{1 - x^2}}{2}.$$

30. Separating variables, we have

$$\frac{dy}{y^2 - y} = \frac{dx}{x} \quad \text{or} \quad \int \frac{dy}{y(y - 1)} = \ln|x| + c.$$

Using partial fractions, we obtain

$$\int \left(\frac{1}{y - 1} - \frac{1}{y}\right) dy = \ln|x| + c$$

$$\ln|y - 1| - \ln|y| = \ln|x| + c$$

$$\ln\left|\frac{y - 1}{xy}\right| = c$$

$$\frac{y - 1}{xy} = e^c = c_1.$$

Solving for y we get $y = 1/(1 - c_1 x)$. We note by inspection that $y = 0$ is also a solution of the DE.

(a) Setting $x = 0$ and $y = 1$ we have $1 = 1/(1 - 0)$, which is true for all values of c_1. Thus, solutions passing through $(0, 1)$ are $y = 1/(1 - c_1 x)$.

(b) Setting $x = 0$ and $y = 0$ in $y = 1/(1 - c_1 x)$ we get $0 = 1$. Thus, the only solution passing through $(0, 0)$ is $y = 0$.

33. The singular solution $y = 1$ satisfies the initial-value problem.

34

36. Separating variables we obtain $\dfrac{dy}{(y-1)^2 - 0.01} = dx$. Then, from

(11) in this section of the manual with $u = y - 1$ and $a = \frac{1}{10}$, we get

$$5\ln\left|\frac{10y-11}{10y-9}\right| = x + c.$$

Setting $x = 0$ and $y = 1$ we obtain $c = 5\ln 1 = 0$. The solution is

$$5\ln\left|\frac{10y-11}{10y-9}\right| = x.$$

Solving for y we obtain

$$y = \frac{11 + 9e^{x/5}}{10 + 10e^{x/5}}.$$

Alternatively, we can use the fact that

$$\int \frac{dy}{(y-1)^2 - 0.01} = -\frac{1}{0.1}\tanh^{-1}\frac{y-1}{0.1} = -10\tanh^{-1}10(y-1).$$

(We use the inverse hyperbolic tangent because $|y - 1| < 0.1$ or $0.9 < y < 1.1$. This follows from the initial condition $y(0) = 1$.) Solving the above equation for y we get $y = 1 + 0.1\tanh(x/10)$.

39. (a) Separating variables we have $2y\,dy = (2x+1)dx$. Integrating gives $y^2 = x^2 + x + c$. When $y(-2) = -1$ we find $c = -1$, so $y^2 = x^2 + x - 1$ and $y = -\sqrt{x^2 + x - 1}$. The negative square root is chosen because of the initial condition.

(b) From the figure, the largest interval of definition appears to be approximately $(-\infty, -1.65)$.

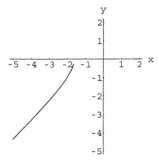

(c) Solving $x^2 + x - 1 = 0$ we get $x = -\frac{1}{2} \pm \frac{1}{2}\sqrt{5}$, so the largest interval of definition is $(-\infty, -\frac{1}{2} - \frac{1}{2}\sqrt{5})$. The right-hand endpoint of the interval is excluded because $y = -\sqrt{x^2 + x - 1}$ is not differentiable at this point.

48. (*Suggestion*) First, explicitly solve the DE and then use **ContourPlot** in *Mathematica* or `contourplot` in *Maple* as described earlier in this section of this manual to obtain level curves. In either CAS the contours option can be set to a list of values for the constant.

50. (*Suggestion*) See the suggestion above for Problem 48.

Section 2.3   **Linear Equations**

The terminology and concepts listed below provide an outline of the main ideas encountered in this section. These can be useful when preparing for a quiz or test.

Terminology and Concepts

- linear first-order DE

- homogeneous linear DE

- variation of parameters as applied to a linear first-order DE

- integrating factor

- general solution of a linear first-order DE

- singular points of a linear first-order DE

- transient term

- special functions

The basic skills listed below summarize the more mechanical types of problems encountered in the exercise set for this section.

Basic Skills

- find the integrating factor for a linear first-order DE

- use the integrating factor to find the general solution of a linear first-order DE

- find transient terms in the general solution of a linear first-order DE

- solve a linear first-order DE satisfying an initial condition

- solve a linear first-order IVP with a piecewise-defined coefficient

- determine an interval of definition of the general solution of a linear first-order DE

An Identity   In precalculus mathematics you learned that the exponential statement $e^N = M$ is equivalent to the logarithmic expression $N = \ln M$. If we substitute this last expression into the first, we obtain the important identity

$$e^{\ln M} = M,$$

which is very useful in this section when computing the integrating factor $e^{\int P(x)\,dx}$. For example, to determine the integrating factor for $(5 + x^2)\dfrac{dy}{dx} - 2xy = x$ we first put the DE into standard form

$$\frac{dy}{dx} - \frac{2x}{5+x^2}\, y = \frac{x}{5+x^2} \quad \text{and identify } P(x) = -2x/(5+x^2):$$

$$e^{-\int \frac{2x}{5+x^2}\, dx} = e^{-\ln|5+x^2|} = e^{\ln|5+x^2|^{-1}} = (5+x^2)^{-1} = \frac{1}{5+x^2}\,.$$

We were able to discard the absolute value sign in $|5+x^2|$ because $5+x^2 > 0$ for all x.

It may take some time to get used to the fact that once the DE has been put into standard form and the integrating factor determined, multiplying both sides of the standard form by the integrating factor recasts the left-hand side of the equation into the derivative of the integrating factor and the dependent variable. Above, we saw that the integrating factor for $\dfrac{dy}{dx} - \dfrac{2x}{5+x^2}\, y = \dfrac{x}{5+x^2}$ is $\dfrac{1}{5+x^2}$. Multiplying the DE by this factor gives

$$\frac{1}{5+x^2}\frac{dy}{dx} - \frac{2x}{(5+x^2)^2}\, y = \frac{x}{(5+x^2)^2}\,. \tag{1}$$

The left-hand side of the last DE can then be written compactly as the derivative of a product of the integrating factor and $y(x)$:

$$\frac{d}{dx}\left[\frac{1}{5+x^2}\, y\right] = \frac{x}{(5+x^2)^2}\,. \tag{2}$$

Until you become comfortable with this notion we suggest that you actually multiply the differential equation $dy/dx + P(x)y = f(x)$ by the integrating factor and then verify using the product rule that

$$e^{\int P(x)\, dx}\frac{dy}{dx} + P(x)e^{\int P(x)\, dx}y = e^{\int P(x)\, dx}f(x) \qquad \text{(as in (1))}$$

and

$$\frac{d}{dx}\left[e^{\int P(x)\, dx}y\right] = e^{\int P(x)\, dx}f(x) \qquad \text{(as in (2))}$$

are formally equivalent. After that, once you have computed the integrating factor, you can jump to the last form and proceed with integration. In the case of (2) the integral of the left-hand side is

$$\int \frac{d}{dx}\left[\frac{1}{5+x^2}\, y\right]dx = \frac{1}{5+x^2}\, y.$$

The integral of the right-hand side is

$$\int \frac{x}{(5+x^2)^2}\, dx = \frac{1}{2}\int (5+x^2)^{-2}(2x\, dx) = -\frac{1}{2}(5+x^2)^{-1} + c.$$

Hence the solution of the DE on $(-\infty, \infty)$ is

$$\frac{1}{5+x^2}\, y = -\frac{1}{2}(5+x^2)^{-1} + c \qquad \text{or} \qquad y = -\frac{1}{2} + c(5+x^2).$$

37

Use of Computers   As we have seen in Section 2.2, a CAS can help you find the solution of a DE by performing some of the more complicated integrations required. In addition to this, a CAS can also frequently find directly the solution of a DE. For example, to solve the differential equation $xy' - xy = e^x$ use

$$\text{Dsolve}[\text{x y}'[\text{x}] - \text{x y}[\text{x}] == \text{Exp}[\text{x}], \text{y}[\text{x}], \text{x}] \qquad (Mathematica)$$

```
dsolve(x*diff(y(x),x)-x*y(x)=exp(x),y(x));     (Maple)
```

To solve an IVP such as $xy' - xy = e^x$, $y(1) = 2$, use

$$\text{Dsolve}[\{\text{x y}'[\text{x}] - \text{x y}[\text{x}] == \text{Exp}[\text{x}], \text{y}[1] == 2\}, \text{y}[\text{x}], \text{x}] \qquad (Mathematica)$$

```
dsolve({x*diff(y(x),x)-x*y(x)=exp(x),y(1)=2},     (Maple)
    y(x));
```

The output is

$$\{\{y[x] \to e^{x-1}(2 + e\ln x)\}\} \qquad (Mathematica)$$

$$y[x] = \left(\ln x + \frac{2}{e}\right)e^x \qquad (Maple)$$

In neither case can the output be used subsequently in the CAS to easily obtain numerical values or graphs of the function which is the solution, $y(x)$, of the IVP. The following routines can be used to assign a specific function name to the solution. This function can then be evaluated at a number, differentiated, graphed, etc.

$$\text{Clear}[\text{y}] \qquad (Mathematica)$$
$$\text{sol} = \text{Flatten}[\text{Dsolve}[\{\text{x}^2 \text{ y}'[\text{x}] - \text{x y}[\text{x}] == \text{y}[\text{x}]^2, \text{y}[1] == 2\}, \text{y}[\text{x}], \text{x}]];$$
$$\text{y}[\text{x\_}] := \text{Evaluate}[\text{y}[\text{x}]/.\text{sol}]$$
$$\text{y}[\text{x}]$$

The semicolon at the end of line of input in *Mathematica* is quite different than in *Maple*, where it is required at the end of every line of input. In *Mathematica* it simply suppresses the display of the output. For example, typing **a=2** in *Mathematica* will set a equal to 2 and display 2 on the monitor, whereas typing **a=2;** will set a equal to 2, but will not display the 2.

```
sol:=dsolve({x^2*diff(y(x),x)-x*y(x)=y(x)^2,     (Maple)
    y(1)=2}, y(x));
y:=unapply(eval(y(x),sol), x);
```

Special Functions: As discussed in the text, a number of nonelementary functions are defined by integrals. Others, such as the Bessel functions discussed in Chapter 6 of the text, can be defined using infinite series. Virtually all special functions that have been tabulated in standard handbooks are built into *Mathematica* and *Maple*. Below is a table listing all of the special functions used in the text, together with the syntax for each function in both *Mathematica* and *Maple*.

Name	Notation	*Mathematica*	*Maple*
Bessel function of the first kind	$J_n(x)$	**BesselJ[n,x]**	BesselJ(n,x)
Bessel function of the second kind	$Y_n(x)$	**BesselY[n,x]**	BesselY(n,x)
Chebyshev polynomial	$T_n(x)$	**ChebyshevT[n,x]**	ChebyshevT(n,x)
Complementary error function	$\text{erfc}(x)$	**Erfc[x]**	erfc(x)
Dirac delta function	$\delta(t-a)$	**DiracDelta[t-a]**	Dirac(t-a)
Error function	$\text{erf}(x)$	**Erf[x]**	erf(x)
Fresnel sine integral	$S(x)$	**FresnelS[x]**	FresnelS(x)
Gamma	$\Gamma(x)$	**Gamma[x]**	GAMMA(x)
Heaviside (unit step function)	$\mathcal{U}(t-a)$	**UnitStep[t-a]**	Heaviside(x)
Hermite polynomial	$H_n(x)$	**HermiteH[n,x]**	HermiteH(n,x)
Imaginary error function	$\text{erfi}(x)$	**Erfi[x]**	erfi(x)
Laguerre polynomial	$L_n(x)$	**LaguerreL[n,x]**	orthopoly[L](n,x)
Legendre polynomial	$P_n(x)$	**LegendreP[n,x]**	orthopoly[P](n,x)
Modified Bessel function of the first kind	$I(x)$	**BesselI[n,x]**	BesselI(n,x)
Modified Bessel function of the second kind	$K(x)$	**BesselK[n,x]**	BesselK(n,x)
Sine integral	$Si(x)$	**SinIntegral[x]**	Si(x)

Section 2.3   Linear Equations

See Section 4.6 in this manual for a brief discussion of the integrals of $\tan x$, $\cot x$, $\sec x$, and $\csc x$. These will be used in several of the problems below.

3. For $y' + y = e^{3x}$ an integrating factor is $e^{\int dx} = e^x$ so that $\dfrac{d}{dx}[e^x y] = e^{4x}$ and $y = \frac{1}{4}e^{3x} + ce^{-x}$ for $-\infty < x < \infty$. The transient term is ce^{-x}.

6. For $y' + 2xy = x^3$ an integrating factor is $e^{\int 2x\,dx} = e^{x^2}$ so that $\dfrac{d}{dx}\left[e^{x^2}y\right] = x^3 e^{x^2}$ and $y = \frac{1}{2}x^2 - \frac{1}{2} + ce^{-x^2}$ for $-\infty < x < \infty$. The transient term is ce^{-x^2}.

9. For $y' - \dfrac{1}{x}y = x\sin x$ an integrating factor is $e^{-\int (1/x)dx} = \dfrac{1}{x}$ so that $\dfrac{d}{dx}\left[\dfrac{1}{x}y\right] = \sin x$ and $y = cx - x\cos x$ for $0 < x < \infty$.

12. For $y' - \dfrac{x}{(1+x)}y = x$ an integrating factor is $e^{-\int [x/(1+x)]dx} = (x+1)e^{-x}$ so that $\dfrac{d}{dx}\left[(x+1)e^{-x}y\right] = x(x+1)e^{-x}$ and $y = -x - \dfrac{2x+3}{x+1} + \dfrac{ce^x}{x+1}$ for $-1 < x < \infty$.

14. (*Hint*) To compute the integrating factor, $e^{\int [(1+x)/x]dx}$, begin by writing
$$\frac{1+x}{x} = \frac{1}{x} + 1.$$

15. For $\dfrac{dx}{dy} - \dfrac{4}{y}x = 4y^5$ an integrating factor is $e^{-\int (4/y)dy} = e^{\ln y^{-4}} = y^{-4}$ so that $\dfrac{d}{dy}\left[y^{-4}x\right] = 4y$ and $x = 2y^6 + cy^4$ for $0 < y < \infty$.

18. For $y' + (\cot x)y = \sec^2 x \csc x$ an integrating factor is $e^{\int \cot x\,dx} = e^{\ln|\sin x|} = \sin x$ so that $\dfrac{d}{dx}[(\sin x)\,y] = \sec^2 x$ and $y = \sec x + c\csc x$ for $0 < x < \pi/2$.

19. (*Hint*) To compute the integrating factor, $e^{\int [(x+2)/(x+1)]dx}$, begin by writing
$$\frac{x+2}{x+1} = \frac{x+1+1}{x+1} = 1 + \frac{1}{x+1},$$
or obtain this result using long division.

21. For $\dfrac{dr}{d\theta} + r\sec\theta = \cos\theta$ an integrating factor is $e^{\int \sec\theta\,d\theta} = e^{\ln|\sec x + \tan x|} = \sec\theta + \tan\theta$ so that $\dfrac{d}{d\theta}[(\sec\theta + \tan\theta)r] = 1 + \sin\theta$ and $(\sec\theta + \tan\theta)r = \theta - \cos\theta + c$ for $-\pi/2 < \theta < \pi/2$.

24. For $y' + \dfrac{2}{x^2-1}y = \dfrac{x+1}{x-1}$ an integrating factor is $e^{\int [2/(x^2-1)]dx} = \dfrac{x-1}{x+1}$ so that $\dfrac{d}{dx}\left[\dfrac{x-1}{x+1}y\right] = 1$ and $(x-1)y = x(x+1) + c(x+1)$ for $-1 < x < 1$.

27. For $\dfrac{di}{dt} + \dfrac{R}{L} i = \dfrac{E}{L}$ an integrating factor is $e^{\int (R/L)\, dt} = e^{Rt/L}$ so that $\dfrac{d}{dt}\left[e^{Rt/L}\, i \right] = \dfrac{E}{L} e^{Rt/L}$ and

$i = \dfrac{E}{R} + ce^{-Rt/L}$ for $-\infty < t < \infty$. If $i(0) = i_0$ then $c = i_0 - E/R$ and $i = \dfrac{E}{R} + \left(i_0 - \dfrac{E}{R} \right) e^{-Rt/L}$.

30. For $y' + (\tan x)y = \cos^2 x$ an integrating factor is $e^{\int \tan x\, dx} = e^{\ln |\sec x|} = \sec x$ so that $\dfrac{d}{dx}\left[(\sec x)\, y \right] = \cos x$ and $y = \sin x \cos x + c\cos x$ for $-\pi/2 < x < \pi/2$. If $y(0) = -1$ then $c = -1$ and $y = \sin x \cos x - \cos x$.

31-34. (*Suggestion*) See page 4 in this manual for the syntax used to specify piecewise-defined functions.

33. For $y' + 2xy = f(x)$ an integrating factor is e^{x^2} so that

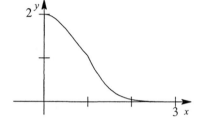

$$ y e^{x^2} = \begin{cases} \frac{1}{2} e^{x^2} + c_1, & 0 \le x \le 1 \\ c_2, & x > 1. \end{cases} $$

If $y(0) = 2$ then $c_1 = 3/2$ and for continuity we must have $c_2 = \frac{1}{2}e + \frac{3}{2}$ so that

$$ y = \begin{cases} \frac{1}{2} + \frac{3}{2} e^{-x^2}, & 0 \le x \le 1 \\ \left(\frac{1}{2}e + \frac{3}{2} \right) e^{-x^2}, & x > 1. \end{cases} $$

36. For $y' + e^x y = 1$ an integrating factor is e^{e^x}. Thus, $\dfrac{d}{dx}\left[e^{e^x} y \right] = e^{e^x}$ and $e^{e^x} y = \displaystyle\int_0^x e^{e^t}\, dt + c$. From $y(0) = 1$ we get $c = e$, so $y = e^{-e^x} \displaystyle\int_0^x e^{e^t}\, dt + e^{1-e^x}$.

When $y' + e^x y = 0$ we can separate variables and integrate:

$$ \frac{dy}{y} = -e^x\, dx \quad \text{and} \quad \ln |y| = -e^x + c. $$

Thus $y = c_1 e^{-e^x}$. From $y(0) = 1$ we get $c_1 = e$, so $y = e^{1-e^x}$.

When $y' + e^x y = e^x$ we can see by inspection that $y = 1$ is a solution.

Section 2.4   Exact Equations

The terminology and concepts listed below provide an outline of the main ideas encountered in this section. These can be useful when preparing for a quiz or test.

Terminology and Concepts

- differential of a function f of two variables

- exact DE

- criterion for exactness of $M(x,y)\,dx + N(x,y)\,dy = 0$

- use of an integrating factor to change a nonexact DE into one that is exact

The basic skills listed below summarize the more mechanical types of problems encountered in the exercise set for this section.

Basic Skills

- determine whether a DE is exact

- solve an exact DE

- find and use an integrating factor to make a DE exact

Partial Differentiation

In this section you will have to use **partial differentiation** and a reverse process which could be called **partial integration** or **partial antidifferentiation**.

Suppose $f(x,y)$ is a function of two variables defined in a region R of the plane. The mechanics of computing the partial derivative of f with respect to x, written $\partial f/\partial x$, can be summarized simply as:

To compute $\dfrac{\partial f}{\partial x}$ *, use the rules of ordinary differentiation with respect to x while treating y as a constant.*

For example, if $f(x,y) = 3x^4 y^2 + e^{5y} + 4\cos xy$, then by treating y as a constant we obtain

$$\frac{\partial f}{\partial x} = y^2 \frac{\partial}{\partial x}\, 3x^4 + \frac{\partial}{\partial x}\, e^{5y} + 4\,\frac{\partial}{\partial x}\, \cos xy^2 = y^2 \cdot 12x^3 + 0 + 4(-y^2 \sin xy^2)$$

$$= 12x^3 y^2 - 4y^2 \sin xy^2.$$

On the other hand, to compute the partial derivative of f with respect to y, written $\partial f/\partial y$, use the rules of ordinary differentiation except treat x as a constant. For example, for the function f in the preceding example, we have

$$\frac{\partial f}{\partial y} = 3x^4 \frac{\partial}{\partial y}\, y^2 + \frac{\partial}{\partial y}\, e^{5y} + 4\,\frac{\partial}{\partial y}\, \cos xy^2 = 3x^4 \cdot 2y + 5e^{5y} + 4(-x \cdot 2y \sin xy^2)$$

$$= 6x^4 y + 5e^{5y} - 8xy \sin xy^2.$$

Now suppose that we are given the partial derivative of a function f, say $\partial f/\partial x = M(x,y)$. We can recover f up to an "additive constant" by partial integration with respect to x:

$$f(x,y) = \int M(x,y)\,dx.$$

42

In the foregoing integral we use the rules of ordinary indefinite integration with respect to x while treating y as a constant. If $\partial f/\partial y = N(x, y)$, then by partial integration with respect to y we have

$$f(x, y) = \int N(x, y)\, dy,$$

where x is treated as a constant. For example, if $\partial f/\partial x = x + y$, then

$$f(x, y) = \int (x + y)\, dx = \frac{1}{2}x^2 + xy + g(y).$$

Here any function $g(y)$ is a "constant" of integration since

$$\frac{\partial f}{\partial x} = \frac{1}{2}\frac{\partial}{\partial x}x^2 + y\frac{\partial}{\partial x}x + \frac{\partial}{\partial x}g(y) = \frac{1}{2}\cdot 2x + y\cdot 1 + 0 = x + y.$$

As another example, suppose that $\partial f/\partial y = 10e^{-x^2 y}$, then, with x held constant,

$$f(x, y) = \int 10e^{-x^2 y}dy = \frac{10}{-x^2}\int e^{-x^2 y}(-x^2\, dy) = -\frac{10}{x^2}e^{-x^2 y} + h(x),$$

where this time the "constant" of integration is $h(x)$. You are encouraged to verify the last result by partial differentiation with respect to y.

Use of Computers   To obtain the partial derivative of $f(x, y) = xe^y + x^2 y^3$ with respect to y in a CAS use

```
Clear[f]                          (Mathematica)
f[x_, y_]:= x Exp[y] + x^2 y^3
f[x, y]
D[f[x, y], y]
```

```
f:=(x,y)->x*exp(y)+x^2*y^3;        (Maple)
diff(f(x,y),y);    or   D[2](f);
```

In *Maple*, `diff(f(x,y),y);` returns the expression $xe^y + 3x^2 y^2$, which cannot be directly evaluated at a point. On the other hand, `D[2](f);` returns the function $(x, y) \to xe^y + 3x^2 y^2$. In this case, the 2 in `D[2]` signifies the partial derivative with respect to the second variable, y, in the definition of f.

Exercises 2.4   *Hints, Suggestions, Solutions, and Examples*

3. Let $M = 5x + 4y$ and $N = 4x - 8y^3$ so that $M_y = 4 = N_x$. From $f_x = 5x + 4y$ we obtain $f = \frac{5}{2}x^2 + 4xy + h(y)$, $h'(y) = -8y^3$, and $h(y) = -2y^4$. A solution is $\frac{5}{2}x^2 + 4xy - 2y^4 = c$.

6. Let $M = 4x^3 - 3y \sin 3x - y/x^2$ and $N = 2y - 1/x + \cos 3x$ so that $M_y = -3 \sin 3x - 1/x^2$ and $N_x = 1/x^2 - 3 \sin 3x$. The equation is not exact.

9. Let $M = y^3 - y^2 \sin x - x$ and $N = 3xy^2 + 2y \cos x$ so that $M_y = 3y^2 - 2y \sin x = N_x$. From $f_x = y^3 - y^2 \sin x - x$ we obtain $f = xy^3 + y^2 \cos x - \frac{1}{2}x^2 + h(y)$, $h'(y) = 0$, and $h(y) = 0$. A solution is $xy^3 + y^2 \cos x - \frac{1}{2}x^2 = c$.

12. Let $M = 3x^2y + e^y$ and $N = x^3 + xe^y - 2y$ so that $M_y = 3x^2 + e^y = N_x$. From $f_x = 3x^2y + e^y$ we obtain $f = x^3y + xe^y + h(y)$, $h'(y) = -2y$, and $h(y) = -y^2$. The solution is $x^3y + xe^y - y^2 = c$.

15. Let $M = x^2y^3 - 1/\left(1 + 9x^2\right)$ and $N = x^3y^2$ so that $M_y = 3x^2y^2 = N_x$. From $f_x = x^2y^3 - 1/\left(1 + 9x^2\right)$ we obtain $f = \frac{1}{3}x^3y^3 - \frac{1}{3}\arctan(3x) + h(y)$, $h'(y) = 0$, and $h(y) = 0$. A solution is $x^3y^3 - \arctan(3x) = c$.

18. Let $M = 2y \sin x \cos x - y + 2y^2 e^{xy^2}$ and $N = -x + \sin^2 x + 4xye^{xy^2}$ so that

$$M_y = 2 \sin x \cos x - 1 + 4xy^3 e^{xy^2} + 4ye^{xy^2} = N_x.$$

From $f_x = 2y \sin x \cos x - y + 2y^2 e^{xy^2}$ we obtain $f = y \sin^2 x - xy + 2e^{xy^2} + h(y)$, $h'(y) = 0$, and $h(y) = 0$. A solution is $y \sin^2 x - xy + 2e^{xy^2} = c$.

21. Let $M = x^2 + 2xy + y^2$ and $N = 2xy + x^2 - 1$ so that $M_y = 2(x+y) = N_x$. From $f_x = x^2 + 2xy + y^2$ we obtain $f = \frac{1}{3}x^3 + x^2y + xy^2 + h(y)$, $h'(y) = -1$, and $h(y) = -y$. The general solution is $\frac{1}{3}x^3 + x^2y + xy^2 - y = c$. If $y(1) = 1$ then $c = 4/3$ and a solution of the IVP is $\frac{1}{3}x^3 + x^2y + xy^2 - y = \frac{4}{3}$.

24. Let $M = t/2y^4$ and $N = \left(3y^2 - t^2\right)/y^5$ so that $M_y = -2t/y^5 = N_t$. From $f_t = t/2y^4$ we obtain $f = \dfrac{t^2}{4y^4} + h(y)$, $h'(y) = \dfrac{3}{y^3}$, and $h(y) = -\dfrac{3}{2y^2}$. The general solution is $\dfrac{t^2}{4y^4} - \dfrac{3}{2y^2} = c$. If $y(1) = 1$ then $c = -5/4$ and a solution of the IVP is $\dfrac{t^2}{4y^4} - \dfrac{3}{2y^2} = -\dfrac{5}{4}$.

27. Equating $M_y = 3y^2 + 4kxy^3$ and $N_x = 3y^2 + 40xy^3$ we obtain $k = 10$.

30. Let $M = \left(x^2 + 2xy - y^2\right)/\left(x^2 + 2xy + y^2\right)$ and $N = \left(y^2 + 2xy - x^2\right)/\left(y^2 + 2xy + x^2\right)$ so that $M_y = -4xy/(x+y)^3 = N_x$. From $f_x = \left(x^2 + 2xy + y^2 - 2y^2\right)/(x+y)^2$ we obtain $f = x + \dfrac{2y^2}{x+y} + h(y)$, $h'(y) = -1$, and $h(y) = -y$. A solution of the DE is $x^2 + y^2 = c(x+y)$.

33. We note that $(N_x - M_y)/M = 2/y$, so an integrating factor is $e^{\int 2dy/y} = y^2$. Let $M = 6xy^3$ and $N = 4y^3 + 9x^2y^2$ so that $M_y = 18xy^2 = N_x$. From $f_x = 6xy^3$ we obtain $f = 3x^2y^3 + h(y)$, $h'(y) = 4y^3$, and $h(y) = y^4$. A solution of the DE is $3x^2y^3 + y^4 = c$.

36. We note that $(N_x - M_y)/M = -3/y$, so an integrating factor is $e^{-3\int dy/y} = 1/y^3$. Let

$$M = (y^2 + xy^3)/y^3 = 1/y + x$$

and

$$N = (5y^2 - xy + y^3 \sin y)/y^3 = 5/y - x/y^2 + \sin y,$$

44

so that $M_y = -1/y^2 = N_x$. From $f_x = 1/y + x$ we obtain $f = x/y + \frac{1}{2}x^2 + h(y)$, $h'(y) = 5/y + \sin y$, and $h(y) = 5\ln|y| - \cos y$. The solution of the DE is $x/y + \frac{1}{2}x^2 + 5\ln|y| - \cos y = c$.

39. (a) Implicitly differentiating $x^3 + 2x^2y + y^2 = c$ and solving for dy/dx we obtain

$$3x^2 + 2x^2\frac{dy}{dx} + 4xy + 2y\frac{dy}{dx} = 0 \quad \text{and} \quad \frac{dy}{dx} = -\frac{3x^2 + 4xy}{2x^2 + 2y}.$$

By writing the last equation in a different form we get $(4xy + 3x^2)dx + (2y + 2x^2)dy = 0$.

(b) Setting $x = 0$ and $y = -2$ in $x^3 + 2x^2y + y^2 = c$ we find $c = 4$, and setting $x = y = 1$ we also find $c = 4$. Thus, both initial conditions determine the same implicit solution.

(c) Solving $x^3 + 2x^2y + y^2 = 4$ for y we get

$$y_1(x) = -x^2 - \sqrt{4 - x^3 + x^4}$$

and

$$y_2(x) = -x^2 + \sqrt{4 - x^3 + x^4}.$$

Observe in the figure that $y_1(0) = -2$ and $y_2(1) = 1$.

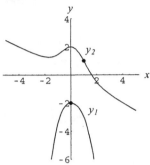

44. (*Hint*) Rewrite the DE as $-g(x)\,dx + (1/h(y))\,dy = 0$.

46. (*Suggestion*) See the discussion in this section of the *SRSM* for a way to graph several functions in a CAS.

Section 2.5 Solutions by Substitutions

The terminology and concepts listed below provide an outline of the main ideas encountered in this section. These can be useful when preparing for a quiz or test.

Terminology and Concepts

- homogeneous DE
- Bernoulli's equation
- use of a substitution to reduce a DE of the form $dy/dx = f(Ax + By + C)$ to a separable DE

Section 2.5 Solutions by Substitutions

The basic skills listed below summarize the more mechanical types of problems encountered in the exercise set for this section.

Basic Skills

- recognize and solve a homogeneous DE using an appropriate substitution
- recognize a Bernoulli equation and solve it by reducing it to a linear first-order DE
- solve a DE of the form $dy/dx = f(Ax + By + C)$ using an appropriate substitution

Homogeneous DEs   As mentioned in the text, either substitution $y = ux$ or $x = vy$ will reduce a homogeneous DE $M(x, y)\,dx + N(x, y)\,dy = 0$ to one with separable variables. As a rule of thumb, try the substitution $y = ux$ whenever the coefficient N is simpler than M. On the other hand, try the substitution $x = vy$ whenever the coefficient M is simpler than N.

Exercises 2.5   *Hints, Suggestions, Solutions, and Examples*

3. The coefficient $M = x$ is slightly simpler than the coefficient $N = y - 2x$. By the rule of thumb given above, we try the substitution $x = vy$. This gives

$$vy(v\,dy + y\,dv) + (y - 2vy)\,dy = 0$$

$$vy^2\,dv + y\left(v^2 - 2v + 1\right)dy = 0$$

$$\frac{v\,dv}{(v-1)^2} + \frac{dy}{y} = 0$$

$$\ln|v - 1| - \frac{1}{v-1} + \ln|y| = c$$

$$\ln\left|\frac{x}{y} - 1\right| - \frac{1}{x/y - 1} + \ln y = c$$

$$(x - y)\ln|x - y| - y = c(x - y).$$

6. Letting $y = ux$ and using partial fractions, we have

$$\left(u^2 x^2 + ux^2\right)dx + x^2(u\,dx + x\,du) = 0$$

$$x^2\left(u^2 + 2u\right)dx + x^3\,du = 0$$

$$\frac{dx}{x} + \frac{du}{u(u+2)} = 0$$

$$\ln|x| + \frac{1}{2}\ln|u| - \frac{1}{2}\ln|u + 2| = c$$

46

$$\frac{x^2 u}{u+2} = c_1$$

$$x^2 \frac{y}{x} = c_1 \left(\frac{y}{x} + 2 \right)$$

$$x^2 y = c_1(y + 2x).$$

9. Letting $y = ux$ we have

$$-ux \, dx + (x + \sqrt{u}\, x)(u \, dx + x \, du) = 0$$

$$(x^2 + x^2 \sqrt{u}\,) \, du + xu^{3/2} \, dx = 0$$

$$\left(u^{-3/2} + \frac{1}{u} \right) du + \frac{dx}{x} = 0$$

$$-2u^{-1/2} + \ln |u| + \ln |x| = c$$

$$\ln |y/x| + \ln |x| = 2\sqrt{x/y} + c$$

$$y(\ln |y| - c)^2 = 4x.$$

12. Letting $y = ux$ we have

$$(x^2 + 2u^2 x^2)dx - ux^2(u \, dx + x \, du) = 0$$

$$x^2(1 + u^2)dx - ux^3 \, du = 0$$

$$\frac{dx}{x} - \frac{u \, du}{1 + u^2} = 0$$

$$\ln |x| - \frac{1}{2} \ln(1 + u^2) = c$$

$$\frac{x^2}{1 + u^2} = c_1$$

$$x^4 = c_1(x^2 + y^2).$$

Using $y(-1) = 1$ we find $c_1 = 1/2$. The solution of the IVP is $2x^4 = y^2 + x^2$.

15. From $y' + \frac{1}{x} y = \frac{1}{x} y^{-2}$ and $w = y^3$ we obtain $\frac{dw}{dx} + \frac{3}{x} w = \frac{3}{x}$. An integrating factor is x^3 so that $x^3 w = x^3 + c$ or $y^3 = 1 + cx^{-3}$.

18. From $y' - \left(1 + \frac{1}{x}\right) y = y^2$ and $w = y^{-1}$ we obtain $\frac{dw}{dx} + \left(1 + \frac{1}{x}\right) w = -1$. An integrating factor is xe^x so that $xe^x w = -xe^x + e^x + c$ or $y^{-1} = -1 + \frac{1}{x} + \frac{c}{x} e^{-x}$.

21. From $y' - \dfrac{2}{x}y = \dfrac{3}{x^2}y^4$ and $w = y^{-3}$ we obtain $\dfrac{dw}{dx} + \dfrac{6}{x}w = -\dfrac{9}{x^2}$. An integrating factor is x^6 so that

$x^6 w = -\dfrac{9}{5}x^5 + c$ or $y^{-3} = -\dfrac{9}{5}x^{-1} + cx^{-6}$. If $y(1) = \dfrac{1}{2}$ then $c = \dfrac{49}{5}$ and $y^{-3} = -\dfrac{9}{5}x^{-1} + \dfrac{49}{5}x^{-6}$.

24. Let $u = x + y$ so that $du/dx = 1 + dy/dx$. Then $\dfrac{du}{dx} - 1 = \dfrac{1-u}{u}$ or $u\,du = dx$. Thus $\dfrac{1}{2}u^2 = x + c$ or $u^2 = 2x + c_1$, and $(x+y)^2 = 2x + c_1$.

27. Let $u = y - 2x + 3$ so that $du/dx = dy/dx - 2$. Then $\dfrac{du}{dx} + 2 = 2 + \sqrt{u}$ or $\dfrac{1}{\sqrt{u}}\,du = dx$. Thus $2\sqrt{u} = x + c$ and $2\sqrt{y - 2x + 3} = x + c$.

30. Let $u = 3x + 2y$ so that $du/dx = 3 + 2\,dy/dx$. Then $\dfrac{du}{dx} = 3 + \dfrac{2u}{u+2} = \dfrac{5u+6}{u+2}$ and $\dfrac{u+2}{5u+6}\,du = dx$. Now by long division

$$\frac{u+2}{5u+6} = \frac{1}{5} + \frac{4}{25u+30}$$

so we have

$$\int\left(\frac{1}{5} + \frac{4}{25u+30}\right)du = dx$$

and $\dfrac{1}{5}u + \dfrac{4}{25}\ln|25u + 30| = x + c$. Thus

$$\frac{1}{5}(3x + 2y) + \frac{4}{25}\ln|75x + 50y + 30| = x + c.$$

Setting $x = -1$ and $y = -1$ we obtain $c = \dfrac{4}{25}\ln 95$. The solution is

$$\frac{1}{5}(3x + 2y) + \frac{4}{25}\ln|75x + 50y + 30| = x + \frac{4}{25}\ln 95$$

or

$$5y - 5x + 2\ln|75x + 50y + 30| = 2\ln 95.$$

33. (*Hint*) To find a singular solution, try functions $y(x)$ that have a simplifying effect on the DE.

Section 2.6   A Numerical Method

The terminology and concepts listed below provide an outline of the main ideas encountered in this section. These can be useful when preparing for a quiz or test.

Terminology and Concepts

- direction field
- linearization of a function at a point
- Euler's method

- absolute error

- percentage relative error

- numerical solver

The basic skills listed below summarize the more mechanical types of problems encountered in the exercise set for this section.

Basic Skills

- use Euler's method to approximate the solution of a first-order DE

Use of Computers

Both *Mathematica* and *Maple* contain sophisticated programming languages. Below are examples of how each CAS can be used to implement Euler's method and display a table similar to the ones shown in Example 2 of the text. Here we will consider the IVP $y' = f(x, y) = 0.2xy$, $y(1) = 1$, and use Euler's method to approximate $y(1.5)$ using the step size $h = 0.1$. To compute the absolute error at each step we will use the fact that the analytic solution is $y = g(x) = e^{0.1(x^2 - 1)}$.

```
Clear[f, g, u, v, t, err]                                    (Mathematica)
f[x_, y_]:= 0.2x y          (* Input line *)
Print["y'= ", f[x, y]]
g[x_]:= Exp[0.1(x^2 - 1)]        (* Input line *)
Print["y(x) = ", g[x]]
h = 0.1; a = 1; b = 1.5; y0 = 1;      (* Input line *)
u[0] = a; v[0] = y0;
u[n_]:= u[n] = u[n - 1] + h
v[n_]:= v[n] = v[n - 1] + h f[u[n - 1], v[n - 1]]
t[n_]:= t[n] = g[u[n]]
err[n_]:= err[n] = Abs[t[n] - v[n]]
approxtab = Table[{u[n], v[n], t[n], err[n]}, {n, 0, Round[(b - a)/h]}];
tableheader = {"x", "y", "actual value", "abs. error"};
Prepend[approxtab, tableheader]//TableForm
```

The use of the form u[n_]:= u[n] = u[n - 1] + h in the *Mathematica* code above is a way to define a function that remembers a value once it has been computed. This greatly speeds up the running of a program that involves a recursively defined function. To make it easier to reuse the routine for subsequent problems, the lines containing inputs to the routine can be commented, as shown above, or color coded.

```
f:=(x,y)->0.2*x*y;           #Input line           (Maple)
g:=x->exp(0.1*(x^2-1));        #Input line
h:=0.1; a:=1; b:=1.5; y0:=1;      #Input line
u:=n->u(n-1)+h; u(0):=a;
v:=n->v(n-1)+h*f(u(n-1),v(n-1)); v(0):=y0;
t:=n->g(u(n));
err:=n->abs(t(n)-v(n));
u(-1):='x'; v(-1):='y'; t(-1):='actual_value';
err(-1):='abs._error';
with(linalg)
matrix(2+round((b-a)/h),4,(i,j)->piecewise(j=1,u(i-2),
    j=2,v(i-2),j=3,t(i-2),j=4,err(i-2)));
```

Exercises 2.6   *Hints, Suggestions, Solutions, and Examples*

3. Separating variables and integrating, we have

$$\frac{dy}{y} = dx \quad \text{and} \quad \ln|y| = x + c.$$

Thus $y = c_1 e^x$ and, using $y(0) = 1$, we find $c = 1$, so $y = e^x$ is the solution of the IVP.

$h=0.1$

x_n	y_n	Actual Value	Abs. Error	% Rel. Error
0.00	1.0000	1.0000	0.0000	0.00
0.10	1.1000	1.1052	0.0052	0.47
0.20	1.2100	1.2214	0.0114	0.93
0.30	1.3310	1.3499	0.0189	1.40
0.40	1.4641	1.4918	0.0277	1.86
0.50	1.6105	1.6487	0.0382	2.32
0.60	1.7716	1.8221	0.0506	2.77
0.70	1.9487	2.0138	0.0650	3.23
0.80	2.1436	2.2255	0.0820	3.68
0.90	2.3579	2.4596	0.1017	4.13
1.00	2.5937	2.7183	0.1245	4.58

$h=0.05$

x_n	y_n	Actual Value	Abs. Error	% Rel. Error
0.00	1.0000	1.0000	0.0000	0.00
0.05	1.0500	1.0513	0.0013	0.12
0.10	1.1025	1.1052	0.0027	0.24
0.15	1.1576	1.1618	0.0042	0.36
0.20	1.2155	1.2214	0.0059	0.48
0.25	1.2763	1.2840	0.0077	0.60
0.30	1.3401	1.3499	0.0098	0.72
0.35	1.4071	1.4191	0.0120	0.84
0.40	1.4775	1.4918	0.0144	0.96
0.45	1.5513	1.5683	0.0170	1.08
0.50	1.6289	1.6487	0.0198	1.20
0.55	1.7103	1.7333	0.0229	1.32
0.60	1.7959	1.8221	0.0263	1.44
0.65	1.8856	1.9155	0.0299	1.56
0.70	1.9799	2.0138	0.0338	1.68
0.75	2.0789	2.1170	0.0381	1.80
0.80	2.1829	2.2255	0.0427	1.92
0.85	2.2920	2.3396	0.0476	2.04
0.90	2.4066	2.4596	0.0530	2.15
0.95	2.5270	2.5857	0.0588	2.27
1.00	2.6533	2.7183	0.0650	2.39

6.

$h=0.1$	
x_n	y_n
0.00	1.0000
0.10	1.1000
0.20	1.2220
0.30	1.3753
0.40	1.5735
0.50	1.8371

$h=0.05$	
x_n	y_n
0.00	1.0000
0.05	1.0500
0.10	1.1053
0.15	1.1668
0.20	1.2360
0.25	1.3144
0.30	1.4039
0.35	1.5070
0.40	1.6267
0.45	1.7670
0.50	1.9332

9.

$h=0.1$	
x_n	y_n
1.00	1.0000
1.10	1.0000
1.20	1.0191
1.30	1.0588
1.40	1.1231
1.50	1.2194

$h=0.05$	
x_n	y_n
1.00	1.0000
1.05	1.0000
1.10	1.0049
1.15	1.0147
1.20	1.0298
1.25	1.0506
1.30	1.0775
1.35	1.1115
1.40	1.1538
1.45	1.2057
1.50	1.2696

12. Tables of values were computed using the Euler and RK4 methods. The resulting points were plotted and joined using **ListPlot** in *Mathematica*. A simplified version of the code used to do this is given in *Use of Computers* in this section. See the following graphs.

$h = 0.25$

$h = 0.1$

$h = 0.05$

Chapter 2 in Review

Hints, Suggestions, Solutions, and Examples

3. Start with $dy/dx = (y-1)(y-3)$. This is positive for $y < 1$ as shown in the text, but negative for $1 < y < 3$. Start multiplying by $y - 1$ and $y - 3$ and checking the signs of dy/dx. We see that if $dy/dx = (y-1)^2(y-3)$, then $dy/dx < 0$ for $y < 1$, so this does not work. Next, try $dy/dx = (y-1)^2(y-3)^2$. Since this is positive for all y except $y = 1$ and $y = 3$, this is a DE with the given phase portrait.

4. (*Suggestion*) Read the solution above for Problem 3.

6. Using a CAS we find that the zero of f occurs at approximately $P = 1.3214$. From the graph we observe that $dP/dt > 0$ for $P < 1.3214$ and $dP/dt < 0$ for $P > 1.3214$, so $P = 1.3214$ is an asymptotically stable critical point. Thus, $\lim_{t\to\infty} P(t) = 1.3214$.

Chapter 2 in Review

9. Separating variables and using the identity $\cos^2 x = \frac{1}{2}(1 + \cos 2x)$, we have

$$\cos^2 x \, dx = \frac{y}{y^2 + 1} \, dy$$

$$\int \left(\frac{1}{2} + \frac{1}{2} \cos 2x \right) dx = \frac{1}{2} \int \frac{2y}{y^2 + 1} \, dy$$

$$\frac{1}{2}x + \frac{1}{4} \sin 2x = \frac{1}{2} \ln(y^2 + 1) + c$$

$$2x + \sin 2x = 2 \ln(y^2 + 1) + c_1.$$

12. Write the DE in the form $(3y^2 + 2x)dx + (4y^2 + 6xy)dy = 0$. Letting $M = 3y^2 + 2x$ and $N = 4y^2 + 6xy$ we see that $M_y = 6y = N_x$, so the DE is exact. From $f_x = 3y^2 + 2x$ we obtain $f = 3xy^2 + x^2 + h(y)$. Then $f_y = 6xy + h'(y) = 4y^2 + 6xy$ and $h'(y) = 4y^2$, so $h(y) = \frac{4}{3}y^3$. A one-parameter family of solutions is

$$3xy^2 + x^2 + \frac{4}{3}y^3 = c.$$

15. Writing the DE in the form

$$\frac{dy}{dx} + \frac{8x}{x^2 + 4} y = \frac{2x}{x^2 + 4},$$

we see that it is linear with integrating factor $e^{\int 8x \, dx/(x^2 + 4)} = e^{4 \ln(x^2 + 4)} = (x^2 + 4)^4$. Then

$$\frac{d}{dx}[(x^2 + 4)^4 y] = \frac{2x(x^2 + 4)^4}{x^2 + 4} = 2x(x^2 + 4)^3$$

and

$$(x^2 + 4)^4 y = \frac{1}{4}(x^2 + 4)^4 + c$$

$$y = \frac{1}{4} + c(x^2 + 4)^{-4}.$$

18. Separating variables and integrating we have

$$\frac{dy}{y^2} = -2(t + 1) \, dt$$

$$-\frac{1}{y} = -(t + 1)^2 + c$$

$$y = \frac{1}{(t + 1)^2 + c_1}.$$

The initial condition $y(0) = -\frac{1}{8}$ implies $c_1 = -9$, so a solution of the IVP is

$$y = \frac{1}{(t + 1)^2 - 9} \qquad \text{or} \qquad y = \frac{1}{t^2 + 2t - 8},$$

where $-4 < t < 2$.

21. Since the solution of an IVP must be a function, we choose the portion of the black graph shown in the text that passes through the point specified by the initial condition and satisfying the vertical line test for a function.

The graph of $y_1(x)$ is the portion of the closed black curve lying in the fourth quadrant. Its interval of definition is approximately $(0.7, 4.3)$. The graph of $y_2(x)$ is the portion of the left-hand black curve lying in the third quadrant. Its interval of definition is $(-\infty, 0)$.

23. (*Suggestion*) Read the solution of Problem 24, below.

24. Since the DE is autonomous, all lineal elements on a given horizontal line have the same slope. The direction field is then as shown in the figure below. It appears from the figure that the DE has no critical points.

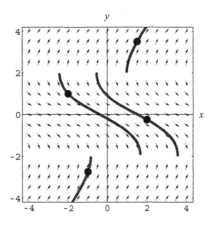

3 Modeling with First-Order Differential Equations

Linear Models

The terminology and concepts listed below provide an outline of the main ideas encountered in this section. These can be useful when preparing for a quiz or test.

Terminology and Concepts

- growth and decay
- half-life
- carbon dating
- Newton's law of cooling/warming
- mixtures
- series circuits
- steady-state part of a solution
- transient term in solution

The basic skills listed below summarize the more mechanical types of problems encountered in the exercise set for this section.

Basic Skills

- set up and solve DEs modeling populations that exhibit exponential growth or decay
- set up and solve DEs involving Newton's law of cooling/warming
- set up and solve DEs modeling mixtures of fluids
- set up and solve DEs modeling series circuits
- set up and solve DEs modeling falling objects both with and without air resistance

54

Half-life   To drive the point home that "the usual carbon-14 technique is limited to about nine half-lives of the isotope," we will estimate the amount of C-14 that would remain in a piece of wood from a tree about 50,000 years after the tree is cut down. We use the fact that the half-life of C-14 is approximately 5,600 years. Then, in $2(5,600) = 12,200$ years, $\frac{1}{4}$ of the original amount of C-14 remains; in $3(5,600) = 16,800$ years, $\frac{1}{8}$ of the original amount remains. Continuing in this fashion, in $9(5,600) = 50,400$ years, $\frac{1}{512}$ (or only about 0.2%) of the original amount of C-14 remains.

Exercises 3.1   *Hints, Suggestions, Solutions, and Examples*

2-3. (*Hint*) Remember that some questions about how fast a solution of a DE changes can be answered by simply considering the DE, that is, without explicitly solving it.

3. Let $P = P(t)$ be the population at time t. Then $dP/dt = kP$ and $P = ce^{kt}$. From $P(0) = c = 500$ we see that $P = 500e^{kt}$. Since 15% of 500 is 75, we have $P(10) = 500e^{10k} = 575$. Solving for k, we get $k = \frac{1}{10} \ln \frac{575}{500} = \frac{1}{10} \ln 1.15$. When $t = 30$,

$$P(30) = 500e^{(1/10)(\ln 1.15)30} = 500e^{3 \ln 1.15} = 760 \text{ years}$$

and

$$P'(30) = kP(30) = \frac{1}{10}(\ln 1.15)760 = 10.62 \text{ persons/year.}$$

6. Let $A = A(t)$ be the amount present at time t. From $dA/dt = kA$ and $A(0) = 100$ we obtain $A = 100e^{kt}$. Using $A(6) = 97$ we find $k = \frac{1}{6} \ln 0.97$. Then $A(24) = 100e^{(1/6)(\ln 0.97)24} = 100(0.97)^4 \approx$ 88.5 mg.

9. Let $I = I(t)$ be the intensity, t the thickness, and $I(0) = I_0$. If $dI/dt = kI$ and $I(3) = 0.25I_0$, then $I = I_0 e^{kt}$, $k = \frac{1}{3} \ln 0.25$, and $I(15) = 0.00098I_0$.

12. From Example 3, the amount of carbon present at time t is $A(t) = A_0 e^{-0.00012378t}$. Letting $t = 660$ and solving for A_0 we have $A(660) = A_0 e^{-0.0001237(660)} = 0.921553A_0$. Thus, approximately 92% of the original amount of C-14 remained in the cloth as of 1988.

15. Assume that $dT/dt = k(T - 100)$ so that $T = 100 + ce^{kt}$. If $T(0) = 20°$ and $T(1) = 22°$, then $c = -80$ and $k = \ln(39/40)$ so that $T(t) = 90°$, which implies $t = 82.1$ seconds. If $T(t) = 98°$ then $t = 145.7$ seconds.

17. The question posed at the end of this problem refers to the ambient temperature T_m of the oven.

18. (a) The initial temperature of the bath is $T_m(0) = 60°$, so in the short term the temperature of the chemical, which starts at $80°$, should decrease or cool. Over time, the temperature of the bath will increase toward $100°$ since $e^{-0.1t}$ decreases from 1 toward 0 as t increases from 0. Thus, in the long term, the temperature of the chemical should increase or warm toward $100°$.

(b) Adapting the model for Newton's law of cooling, we have

$$\frac{dT}{dt} = -0.1(T - 100 + 40e^{-0.1t}), \quad T(0) = 80.$$

Writing the DE in the form

$$\frac{dT}{dt} + 0.1T = 10 - 4e^{-0.1t}$$

we see that it is linear with integrating factor $e^{\int 0.1\, dt} = e^{0.1t}$. Thus

$$\frac{d}{dt}[e^{0.1t}T] = 10e^{0.1t} - 4$$

$$e^{0.1t}T = 100e^{0.1t} - 4t + c$$

and

$$T(t) = 100 - 4te^{-0.1t} + ce^{-0.1t}.$$

Now $T(0) = 80$ so $100 + c = 80$, $c = -20$ and

$$T(t) = 100 - 4te^{-0.1t} - 20e^{-0.1t} = 100 - (4t + 20)e^{-0.1t}.$$

The thinner curve verifies the prediction of cooling followed by warming toward $100°$. The wider curve shows the temperature T_m of the liquid bath.

21. From $dA/dt = 10 - A/100$ we obtain $A = 1000 + ce^{-t/100}$. If $A(0) = 0$ then $c = -1000$ and $A(t) = 1000 - 1000e^{-t/100}$.

22. (*Hint*) The concentration $c(t)$ of salt in the tank at time t, is the number of pounds $A(t)$ of salt divided by the number of gallons of liquid in the tank at that time.

24. With $c_{in}(t) = 2 + \sin(t/4)\,\text{lb/gal}$, the IVP is

$$\frac{dA}{dt} + \frac{1}{100}A = 6 + 3\sin\frac{t}{4}, \quad A(0) = 50.$$

The DE is linear with integrating factor $e^{\int dt/100} = e^{t/100}$, so

$$\frac{d}{dt}[e^{t/100}A(t)] = \left(6 + 3\sin\frac{t}{4}\right)e^{t/100}$$

$$e^{t/100}A(t) = 600e^{t/100} + \frac{150}{313}e^{t/100}\sin\frac{t}{4} - \frac{3750}{313}e^{t/100}\cos\frac{t}{4} + c,$$

and

$$A(t) = 600 + \frac{150}{313}\sin\frac{t}{4} - \frac{3750}{313}\cos\frac{t}{4} + ce^{-t/100}.$$

(See Section 2.2 of this manual for a discussion on how to evaluate an integral of the form $\int e^{\alpha x}\sin\beta x\, dx$.)

56

Letting $t = 0$ and $A = 50$ we have $600 - 3750/313 + c = 50$ and $c = -168400/313$. Then

$$A(t) = 600 + \frac{150}{313} \sin \frac{t}{4} - \frac{3750}{313} \cos \frac{t}{4} - \frac{168400}{313} e^{-t/100}.$$

The graphs on $[0, 300]$ and $[0, 600]$ below show the effect of the sine function in the input when compared with the graph in Figure 3.4(a) in the text.

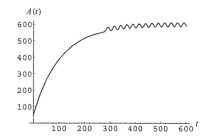

27. Assume $L\,di/dt + Ri = E(t)$, $L = 0.1$, $R = 50$, and $E(t) = 50$ so that $i = \frac{3}{5} + ce^{-500t}$. If $i(0) = 0$ then $c = -3/5$ and $\lim_{t \to \infty} i(t) = 3/5$.

30. Assume $R\,dq/dt + (1/C)q = E(t)$, $R = 1000$, $C = 5 \times 10^{-6}$, and $E(t) = 200$. Then $q = \frac{1}{1000} + ce^{-200t}$ and $i = -200ce^{-200t}$. If $i(0) = 0.4$ then $c = -\frac{1}{500}$, $q(0.005) = 0.003$ coulombs, and $i(0.005) = 0.1472$ amps. We have $q \to \frac{1}{1000}$ as $t \to \infty$.

33. **(a)** From $m\,dv/dt = mg - kv$ we obtain $v = mg/k + ce^{-kt/m}$. If $v(0) = v_0$ then $c = v_0 - mg/k$ and the solution of the initial-value problem is

$$v(t) = \frac{mg}{k} + \left(v_0 - \frac{mg}{k} \right) e^{-kt/m}.$$

(b) As $t \to \infty$ the limiting velocity is mg/k.

(c) From $ds/dt = v$ and $s(0) = 0$ we obtain

$$s(t) = \frac{mg}{k}t - \frac{m}{k}\left(v_0 - \frac{mg}{k} \right) e^{-kt/m} + \frac{m}{k}\left(v_0 - \frac{mg}{k} \right).$$

36. Assuming that the air resistance is proportional to velocity and the positive direction is downward with $s(0) = 0$, the model for the velocity is $m\,dv/dt = mg - kv$. Using separation of variables to solve this DE, we obtain $v(t) = mg/k + ce^{-kt/m}$. From $v(0) = 0$ we get $v(t) = (mg/k)(1 - e^{-kt/m})$. Letting $k = 0.5$, $m = (125 + 35)/32 = 5$, and $g = 32$, we have $v(t) = 320(1 - e^{-0.1t})$. Integrating, we find $s(t) = 320t + 3200e^{-0.1t} + c_1$. Solving $s(0) = 0$ for c_1 we find $c_1 = -3200$, therefore $s(t) = 320t + 3200e^{-0.1t} - 3200$. At $t = 15$, when the parachute opens, $v(15) = 248.598$ and $s(15) = 2314.02$. At this time the value of k changes to $k = 10$ and the new initial velocity is $v_0 = 248.598$. With the parachute open, the skydiver's velocity is $v_p(t) = mg/k + c_2 e^{-kt/m}$, where t is reset to 0 when the parachute opens. Letting $m = 5$, $g = 32$, and $k = 10$, this gives $v_p(t) = 16 + c_2 e^{-2t}$. From $v(0) = 248.598$ we find $c_2 = 232.598$, so $v_p(t) = 16 + 232.598e^{-2t}$.

Integrating, we get $s_p(t) = 16t - 116.299e^{-2t} + c_3$. Solving $s_p(0) = 0$ for c_3, we find $c_3 = 116.299$, so $s_p(t) = 16t - 116.299e^{-2t} + 116.299$. Twenty seconds after leaving the plane is five seconds after the parachute opens. The skydiver's velocity at this time is $v_p(5) = 16.0106$ ft/s and she has fallen a total of $s(15) + s_p(5) = 2314.02 + 196.294 = 2510.31$ ft. Her terminal velocity is $\lim_{t\to\infty} v_p(t) = 16$, so she has very nearly reached her terminal velocity five seconds after the parachute opens. When the parachute opens, the distance to the ground is $15{,}000 - s(15) = 15{,}000 - 2{,}314 = 12{,}686$ ft. Solving $s_p(t) = 12{,}686$ we get $t = 785.6$ s $= 13.1$ min. Thus, it will take her approximately 13.1 minutes to reach the ground after her parachute has opened and a total of $(785.6 + 15)/60 = 13.34$ minutes after she exits the plane.

39. (a) From $dP/dt = (k_1 - k_2)P$ we obtain $P = P_0 e^{(k_1 - k_2)t}$ where $P_0 = P(0)$.

(b) If $k_1 > k_2$ then $P \to \infty$ as $t \to \infty$. If $k_1 = k_2$ then $P = P_0$ for every t. If $k_1 < k_2$ then $P \to 0$ as $t \to \infty$.

45. (*Suggestion*) In part (b) of this problem, rewrite the DE

$$m\frac{dv}{dt} = mg\sin\theta - \mu mg\cos\theta$$

as

$$m\frac{dv}{dt} = mg\cos\theta(\tan\theta - \mu).$$

Solve this DE subject to the initial condition that the box starts from rest in the two cases: $\tan\theta = \mu$ and $\tan\theta < \mu$.

Section 3.2 Nonlinear Models

The terminology and concepts listed below provide an outline of the main ideas encountered in this section. These can be useful when preparing for a quiz or test.

Terminology and Concepts

- logistic DE
- logistic growth
- Gompertz DE
- second-order chemical reaction

The basic skills listed below summarize the more mechanical types of problems encountered in the exercise set for this section.

Basic Skills

- set up and solve DEs involving the logistic model

- set up and solve DEs involving chemical reactions

- set up and solve DEs involving falling bodies subject to air resistance that is proportional to the square of the instantaneous velocity

Use of Computers

A scatter plot in two dimensions is a set of points in the plane. Both *Mathematica* and *Maple* have commands for plotting a set of points. In the examples below we plot the points $(-1, -2.7)$, $(-0.5, -2.5)$, $(0, -1.8)$, $(0.5, -1.6)$, $(1, -2.3)$, $(1.5, 0)$, $(2, 1.3)$, $(2.5, 1.9)$, $(3, 3.1)$.

```
data = {{-1, -2.7}, {-0.5, -2.5}, {0, -1.8}, {0.5, -1.6},        (Mathematica)
    {1, -2.3}, {1.5, 0}, {2, 1.3}, {2.5, 1.9}, {3, 3.1}}
grdata = ListPlot[data]
```

The most commonly used options for **ListPlot** cause the dots to be enlarged or the dots to be connected with straight line segments. Although these two commands cannot be combined in a single call to **PlotStyle**, the same effect can be achieved as shown below.

```
gr0 = ListPlot[data, PlotStyle->PointSize[0.02]]        (Mathematica)
gr1 = ListPlot[data, PlotJoined->True]
Show[gr0, gr1]
```

In **Maple** a scatter plot is obtained using the `plots` package.

```
data:=[[-1,-2.7],[-0.5,-2.5],[0,-1.8],        (Maple)
    [0.5,-1.6],[1,-2.3],[1.5,0],[2,1.3],
    [2.5,1.9],[3,3.1]];
with(plots);
pointplot(data);
```

In **Maple** it is possible to show the points in a variety of symbols and symbol sizes. You can also connect the points with straight line segments. You cannot, however, do both in a single plot. The routine below shows how to accomplish this. It is assumed that the input data is as above and that the `plots` package has been activated.

```
gr0:=pointplot(data,style=point,               (Maple)
    symbol=circle);
gr1:=pointplot(data,style=line);
display([gr0,gr1]);
```

Algorithms that fit relatively simple functions to a set of data points are not difficult to develop but they tend to involve a large number of routine computations. This means that a CAS is particularly well suited to implementing a curve-fitting algorithm. The following routines use a least squares method to fit a line to the data set given above over the interval $[-1, 3]$. The resulting line is then plotted over the data points.

```
linfit = fit[data, {1, x}, x]                  (Mathematica)
gr2 = Plot[linfit, {x, -1, 3}]
Show[gr0, gr2]
```

The format of data in the routines above is as a set of points with two coordinates. In **Maple**, to apply a least squares fit of a line to the set of points, we need to reformat the data as a set of x-coordinates, which we will call xdata, and a set of y-coordinates, which we will call ydata. To accomplish this we first use the nops command to count the number of data points. Then we activate the stats pacakage which contains commands for fitting a line to a set of points in the plane.

```
n:=nops(data);                                 (Maple)
xdata:=[seq(data[i,1],i=1..n)];
ydata:=[seq(data[i,2],i=1..n)];
with(stats);
linfit:=fit[leastsquare[[x,y],y=a*x+b,{a,b}]]([xdata,ydata]);
gr3:=implicitplot(linfit,x=-1..3,y=-5..5);
display([gr0,gr3]);
```

In the above *Maple* routine, implicitplot, which is part of the plots package, is used because the result of the fit command is an equation rather than an expression or function. The range of the equation, in this case y=-5..5, is specified to be large enough to include all of the data points.

60

Exercises 3.2 *Hints, Suggestions, Solutions, and Examples*

3. From $dP/dt = P\left(10^{-1} - 10^{-7}P\right)$ and $P(0) = 5000$ we obtain $P = 500/(0.0005 + 0.0995e^{-0.1t})$ so that $P \to 1{,}000{,}000$ as $t \to \infty$. If $P(t) = 500{,}000$ then $t = 52.9\,\text{months}$.

5. (*Suggestion*) Notice that no initial population P_0 is specified in the questions on whether the fish population becomes extinct in finite time $t = T$. If there exists such a time, determine which interval

$$0 < P_0 < 1, \quad 1 < P_0 < 4, \quad \text{and} \quad P_0 > 4$$

leads to extinction and then express T in terms of P_0.

6. Solving $P(5 - P) - \frac{25}{4} = 0$ for P we obtain the equilibrium solution $P = \frac{5}{2}$. For $P \neq \frac{5}{2}$, $dP/dt < 0$. Thus, if $P_0 < \frac{5}{2}$, the population becomes extinct (otherwise there would be another equilibrium solution.) Using separation of variables to solve the IVP, we get

$$P(t) = [4P_0 + (10P_0 - 25)t]/[4 + (4P_0 - 10)t].$$

To find when the population becomes extinct for $P_0 < \frac{5}{2}$ we solve $P(t) = 0$ for t. We see that the time of extinction is $t = 4P_0/5(5 - 2P_0)$.

7. Solve the DE by separation of variables. Use completion of the square to carry out the integration:

$$\int \frac{1}{P^2 - 5P + 7}\, dP.$$

8. Solve the DE by separation of variables. The integral $\int dP/P(a - b\ln P)$ may not look familiar, but after rewriting as

$$\int \frac{dP/P}{a - b\ln P},$$

it is of the form $\int du/u$ where $u = a - b\ln P$. You will have to fix up a constant multiple to carry out the integration.

9. Let $X = X(t)$ be the amount of C at time t and $dX/dt = k(120 - 2X)(150 - X)$. If $X(0) = 0$ and $X(5) = 10$, then

$$X(t) = \frac{150 - 150e^{180kt}}{1 - 2.5e^{180kt}},$$

where $k = .0001259$ and $X(20) = 29.3\,\text{grams}$. Now by L'Hôpital's rule, $X \to 60$ as $t \to \infty$, so that the amount of $A \to 0$ and the amount of $B \to 30$ as $t \to \infty$.

10-11. (*Hint*) After separating variables, integration will involve the form

$$\int u^n\, du = \frac{u^{n+1}}{n + 1} + c, \quad n \neq -1.$$

61

12. To obtain the solution of this DE we use $h(t)$ from Problem 13 in Exercises 1.3 with A_h replaced by cA_h. Then $h(t) = (A_w\sqrt{H} - 4cA_h t)^2/A_w^2$. Solving $h(t) = 0$ with $c = 0.6$ and the values from Problem 11 we see that the tank empties in 3035.79 seconds or 50.6 minutes.

13. (*Suggestion*) Separation of variables leads to

$$\int \frac{dv}{mg - v^2} .$$

This integral is of the form $\int du/(a^2 - u^2)$. In (9) and (10) of Section 2.2 of this manual, we saw that this integral can be evaluated in two different ways:

$$\int \frac{du}{a^2 - u^2} = \begin{cases} \dfrac{1}{2a} \ln \left| \dfrac{u+a}{u-a} \right| + c \\[2mm] \dfrac{1}{2a} \tanh^{-1} \left(\dfrac{u}{a} \right) + c. \end{cases}$$

15. (a) After separating variables we obtain

$$\frac{m\,dv}{mg - kv^2} = dt$$

$$\frac{1}{g} \frac{dv}{1 - (\sqrt{k}\,v/\sqrt{mg})^2} = dt$$

$$\frac{\sqrt{mg}}{\sqrt{k}\,g} \frac{\sqrt{k/mg}\,dv}{1 - (\sqrt{k}\,v/\sqrt{mg})^2} = dt$$

$$\sqrt{\frac{m}{kg}} \tanh^{-1} \frac{\sqrt{k}\,v}{\sqrt{mg}} = t + c$$

$$\tanh^{-1} \frac{\sqrt{k}\,v}{\sqrt{mg}} = \sqrt{\frac{kg}{m}}\,t + c_1.$$

Thus the velocity at time t is

$$v(t) = \sqrt{\frac{mg}{k}} \tanh \left(\sqrt{\frac{kg}{m}}\,t + c_1 \right).$$

Setting $t = 0$ and $v = v_0$ we find $c_1 = \tanh^{-1}(\sqrt{k}\,v_0/\sqrt{mg})$.

(b) Since $\tanh t \to 1$ as $t \to \infty$, we have $v \to \sqrt{mg/k}$ as $t \to \infty$.

(c) Integrating the expression for $v(t)$ in part (a) we obtain an integral of the form $\int du/u$:

$$s(t) = \sqrt{\frac{mg}{k}} \int \tanh \left(\sqrt{\frac{kg}{m}}\,t + c_1 \right) dt = \frac{m}{k} \ln \left[\cosh \left(\sqrt{\frac{kg}{m}}\,t + c_1 \right) \right] + c_2.$$

Setting $t = 0$ and $s = 0$ we find $c_2 = -(m/k)\ln(\cosh c_1)$, where c_1 is given in part (a).

17. (*Hint*) In this manual, see the discussion on Archimedes' principle in Problem 18 of Exercises 1.3. Also, in separating variables, write

$$\frac{m\,dv}{(mg - \rho V) - (\sqrt{k}\,v)^2} = dt.$$

18. (a) Writing the equation in the form $(x - \sqrt{x^2 + y^2})dx + y\,dy = 0$ we identify $M = x - \sqrt{x^2 + y^2}$ and $N = y$. Since M and N are both homogeneous functions of degree 1 we use the substitution $y = ux$. It follows that

$$\left(x - \sqrt{x^2 + u^2 x^2}\right)dx + ux(u\,dx + x\,du) = 0$$

$$x\left[\left(1 - \sqrt{1 + u^2}\right) + u^2\right]dx + x^2 u\,du = 0$$

$$-\frac{u\,du}{1 + u^2 - \sqrt{1 + u^2}} = \frac{dx}{x}$$

$$\frac{u\,du}{\sqrt{1 + u^2}\,(1 - \sqrt{1 + u^2})} = \frac{dx}{x}.$$

Letting $w = 1 - \sqrt{1 + u^2}$ we have $dw = -u\,du/\sqrt{1 + u^2}$ so that

$$-\ln\left|1 - \sqrt{1 + u^2}\right| = \ln|x| + c$$

$$\frac{1}{1 - \sqrt{1 + u^2}} = c_1 x$$

$$1 - \sqrt{1 + u^2} = -\frac{c_2}{x} \qquad (-c_2 = 1/c_1)$$

$$1 + \frac{c_2}{x} = \sqrt{1 + \frac{y^2}{x^2}}$$

$$1 + \frac{2c_2}{x} + \frac{c_2^2}{x^2} = 1 + \frac{y^2}{x^2}.$$

Solving for y^2 we have

$$y^2 = 2c_2 x + c_2^2 = 4\left(\frac{c_2}{2}\right)\left(x + \frac{c_2}{2}\right)$$

which is a family of parabolas symmetric with respect to the x-axis with vertex at $(-c_2/2, 0)$ and focus at the origin.

(b) Let $u = x^2 + y^2$ so that

$$\frac{du}{dx} = 2x + 2y\frac{dy}{dx}.$$

Then

$$y\frac{dy}{dx} = \frac{1}{2}\frac{du}{dx} - x$$

and the DE can be written in the form

$$\frac{1}{2}\frac{du}{dx} - x = -x + \sqrt{u} \quad \text{or} \quad \frac{1}{2}\frac{du}{dx} = \sqrt{u}.$$

Separating variables and integrating gives

$$\frac{du}{2\sqrt{u}} = dx$$

$$\sqrt{u} = x + c$$

$$u = x^2 + 2cx + c^2$$

$$x^2 + y^2 = x^2 + 2cx + c^2$$

$$y^2 = 2cx + c^2.$$

20d. (*Hint*) Examine $\lim_{t\to\infty} h(t)$ and remember how to rationalize a numerator.

21. (*Suggestion*) Read **Use of Computers** in this section for information on how to use a CAS to draw scatter plots and fit a line to data points.

25. Unlike Problem 13, here it is more useful to use

$$\int \frac{du}{a^2 - u^2} = \frac{1}{2a}\ln\left|\frac{u+a}{u-a}\right| + c$$

when evaluating $\int dv/(mg - v^2)$. Also, do not be discouraged by messy algebra.

Section 3.3 Modeling with Systems of Differential Equations

The terminology and concepts listed below provide an outline of the main ideas encountered in this section. These can be useful when preparing for a quiz or test.

Terminology and Concepts

- systems of first-order DEs – linear and nonlinear
- radioactive decay series
- mixture models involving a system of DEs
- Lotka-Volterra predator-prey model
- competition models
- electrical networks involving a system of first-order DEs

The basic skills listed below summarize the more mechanical types of problems encountered in the exercise set for this section.

Basic Skills

- solve a system of DEs where one of the equations involves only a single unknown

- construct a mathematical model for mixtures involving multiple tanks

- use a computer program to numerically analyze solutions of a system of DEs

Use of Computers

There are a number of computer programs that can solve a DE or system of DEs numerically and graph the resulting numerical solution curves. We show below how *Mathematica* and *Maple* can produce graphs of numerical solutions.

In *Mathematica* **NDSolve** is used to obtain a numerical solution of an IVP which can then be graphed.

```
Clear[y]                                         (Mathematica)
sol=Flatten[NDSolve[{y'[x]==Exp[x/2]Sin[x^2], y[0]==0}, y, {x, -5, 5}]];
y[x_]:=Evaluate[y[x]/.sol]
y[x]
Plot[y[x], {x, -5, 5}]
```

Note that the interval $[-5, 5]$ specified in the **Plot** command can be no larger than the interval used in the **NDSolve** command. Following is an example of how *Mathematica* can be used to obtain a numerical solution of a system of DEs with initial conditions.

```
Clear[y]
sol=Flatten[NDSolve[{x'[t]== -0.3x[t] + 0.1x[t]y[t],
y'[t]==0.2y[t] - 0.1x[t]y[t], x[0]==6, y[0]==4}, {x, y}, {t, -2, 2}]];
x[t_]:=Evaluate[x[t]/.sol]
y[t_]:=Evaluate[y[t]/.sol]
x[t]
y[t]
Plot[{x[t], y[t]}, {t, -2, 2}]
```

In *Maple* there are two ways to plot the numerical solution curve of an IVP. The first way shown below returns a procedure that can be used with the `plots` package to graph the solution. It is also possible to define a function that can evaluate the solution at any point.

```
sol:=dsolve({diff(y(x),x)=exp(x/t)*sin(x^2),     (Maple)
```

65

```
y(0)=0},{y(x)}, type=numeric);
with(plots);
odeplots(sol,[x,y(x)],-5..5);
y1:=x1->eval(y(x),sol(x1));
y1(2.67);
```

The second way to plot a numerical solution of the DE uses the `DEtools` package.

```
with(DEtools);                                          (Maple)
DEplot(diff(y(x),x)=exp(x/2)*sin(x^2),y(x),-5..5,
   [[y(0)=0]],arrows=NONE,stepsize=0.05);
```

The `arrows` option suppresses a direction field in the background of the graph and the `stepsize` option smooths the graph out.

Maple can also graph the numerical solutions of a system of DEs with initial conditions. In this case we first specify the DEs and initial conditions before calling the `DEtools` package.

```
eq1:=diff(x(t),t)=-0.3*x(t)+0.1*x(t)*y(t);      (Maple)
eq2:=diff(y(t),t)=0.2*y(t)-0.1*x(t)*y(t);
with(DEtools);
gr1:=DEplot({eq1,eq2},[x(t),y(t)],-5..5,[[x(0)=6,y(0)=4]],
   scene=[t,x]):
gr2:=DEplot({eq1,eq2},[x(t),y(t)],-5..5,[[x(0)=6,y(0)=4]],
   scene=[t,y]):
display([gr1,gr2]);
```

Note the use of colons, rather than the usual `semicolons`, in the lines above defining `gr1` and `gr2`. This prevents the display of a very long list of uninteresting expressions.

Exercises 3.3 *Hints, Suggestions, Solutions, and Examples*

3. The amounts x and y are the same at about $t = 5$ days. The amounts x and z are the same at about $t = 20$ days. The amounts y and z are the same at about $t = 147$ days. The time when y and z are the same makes sense because most of A and half of B are gone, so half of C should have been formed.

6. Let x_1, x_2, and x_3 be the amounts of salt in tanks A, B, and C, respectively, so that

$$x_1' = \frac{1}{100}x_2 \cdot 2 - \frac{1}{100}x_1 \cdot 6 = \frac{1}{50}x_2 - \frac{3}{50}x_1$$

$$x_2' = \frac{1}{100}x_1 \cdot 6 + \frac{1}{100}x_3 - \frac{1}{100}x_2 \cdot 2 - \frac{1}{100}x_2 \cdot 5 = \frac{3}{50}x_1 - \frac{7}{100}x_2 + \frac{1}{100}x_3$$

$$x_3' = \frac{1}{100}x_2 \cdot 5 - \frac{1}{100}x_3 - \frac{1}{100}x_3 \cdot 4 = \frac{1}{20}x_2 - \frac{1}{20}x_3.$$

9. Zooming in on the graph it can be seen that the populations are first equal at about $t = 5.6$. The approximate periods of x and y are both 45.

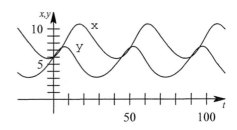

10-11. (*Suggestion*) See **Use of Computers** in this section for *Mathematica* and *Maple* routines that can be used to graph numerical solutions of a system of DEs.

12. By Kirchhoff's first law we have $i_1 = i_2 + i_3$. By Kirchhoff's second law, on each loop we have $E(t) = Li_1' + R_1 i_2$ and $E(t) = Li_1' + R_2 i_3 + q/C$ so that $q = CR_1 i_2 - CR_2 i_3$. Then $i_3 = q' = CR_1 i_2' - CR_2 i_3$ so that the system is

$$Li_2' + Li_3' + R_1 i_2 = E(t)$$

$$-R_1 i_2' + R_2 i_3' + \frac{1}{C}i_3 = 0.$$

15. We first note that $s(t) + i(t) + r(t) = n$. Now the rate of change of the number of susceptible persons, $s(t)$, is proportional to the number of contacts between the number of people infected and the number who are susceptible; that is, $ds/dt = -k_1 si$. We use $-k_1$ because $s(t)$ is decreasing. Next, the rate of change of the number of persons who have recovered is proportional to the number infected; that is, $dr/dt = k_2 i$ where k_2 is positive since r is increasing. Finally, to obtain di/dt we use

$$\frac{d}{dt}(s + i + r) = \frac{d}{dt}n = 0.$$

This gives

$$\frac{di}{dt} = -\frac{dr}{dt} - \frac{ds}{dt} = -k_2 i + k_1 si.$$

The system of equations is then

$$\frac{ds}{dt} = -k_1 s i$$

$$\frac{di}{dt} = -k_2 i + k_1 s i$$

$$\frac{dr}{dt} = k_2 i.$$

A reasonable set of initial conditions is $i(0) = i_0$, the number of infected people at time 0, $s(0) = n - i_0$, and $r(0) = 0$.

16. (*Suggestion*) See **Use of Computers** in this section for *Mathematica* and *Maple* routines that can be used to graph numerical solutions of a system of DEs.

18. (*Hint*) Let $z(t) = x(t) - y(t)$ and solve the resulting equation.

19. (*Suggestion*) See **Use of Computers** in this section for *Mathematica* and *Maple* routines that can be used to graph numerical solutions of a system of DEs.

20. (*Hint*) Assume that the metal bar is sufficiently small that it does not affect the temperature of the medium in container A. Thus, the temperature of the medium in container A is affected only by the temperature of the medium in container B.

Chapter 3 in Review

Hints, Suggestions, Solutions, and Examples

3. From $\dfrac{dP}{dt} = 0.018P$ and $P(0) = 4$ billion we obtain $P = 4e^{0.018t}$ so that $P(45) = 8.99$ billion.

6. From $V\, dC/dt = kA(C_s - C)$ and $C(0) = C_0$ we obtain $C = C_s + (C_0 - C_s)e^{-kAt/V}$.

8. (*Hint*) In part (a) use separation of variables and in part (b) write the DE as

$$\frac{dT}{dt} = [(T_m + (T - T_m))^4 - T_m^4] = kT_m^4 \left[\left(1 + \frac{T - T_m}{T_m}\right)^4 - 1 \right].$$

9. We first solve $(1 - t/10)di/dt + 0.2i = 4$. Separating variables we obtain $di/(40 - 2i) = dt/(10 - t)$. Then

$$-\frac{1}{2}\ln|40 - 2i| = -\ln|10 - t| + c \quad \text{or} \quad \sqrt{40 - 2i} = c_1(10 - t).$$

Since $i(0) = 0$ we must have $c_1 = 2/\sqrt{10}$. Solving for i we get $i(t) = 4t - \frac{1}{5}t^2$, $0 \le t < 10$. For $t \ge 10$ the equation for the current becomes $0.2i = 4$ or $i = 20$. Thus

$$i(t) = \begin{cases} 4t - \frac{1}{5}t^2, & 0 \le t < 10 \\ 20, & t \ge 10. \end{cases}$$

The graph of $i(t)$ is given in the figure.

11. (*Hint*) Be sure to express A_h in terms of feet rather than inches.

12. One hour is 3,600 seconds, so the hour mark should be placed at

$$h(3600) = [\sqrt{2} - 0.00001628(3600)]^2 \approx 1.838\,\text{ft} \approx 22.0525\,\text{in}.$$

up from the bottom of the tank. The remaining marks corresponding to the passage of 2, 3, 4, ..., 12 hours are placed at the values shown in the table. The marks are not evenly spaced because the water is not draining out at a uniform rate; that is, $h(t)$ is not a linear function of time.

time (seconds)	height (inches)
0	24.0000
1	22.0520
2	20.1864
3	18.4033
4	16.7026
5	15.0844
6	13.5485
7	12.0952
8	10.7242
9	9.4357
10	8.2297
11	7.1060
12	6.0648

15. From $dx/dt = k_1 x(\alpha - x)$ we obtain

$$\left(\frac{1/\alpha}{x} + \frac{1/\alpha}{\alpha - x} \right) dx = k_1\, dt$$

so that $x = \alpha c_1 e^{\alpha k_1 t}/(1 + c_1 e^{\alpha k_1 t})$. From $dy/dt = k_2 xy$ we obtain

$$\ln|y| = \frac{k_2}{k_1} \ln\left|1 + c_1 e^{\alpha k_1 t}\right| + c \quad \text{or} \quad y = c_2 \left(1 + c_1 e^{\alpha k_1 t}\right)^{k_2/k_1}.$$

18. Differentiating the family of curves, we have

$$y' = -\frac{1}{(x + c_1)^2} = -\frac{1}{y^2}.$$

The DE for the family of orthogonal trajectories is then $y' = y^2$. Separating variables and integrating we get

$$\frac{dy}{y^2} = dx$$

$$-\frac{1}{y} = x + c_1$$

$$y = -\frac{1}{x + c_1}.$$

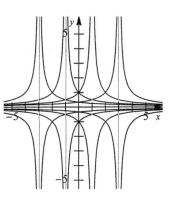

4 Higher-Order Differential Equations

Linear Differential Equations: Basic Theory

The terminology and concepts listed below provide an outline of the main ideas encountered in this section. These can be useful when preparing for a quiz or test.

Terminology and Concepts

- linear nth-order initial-value problem

- existence-uniqueness theorem for linear nth-order IVPs

- boundary-value problems for linear nth-order DEs

- no existence-uniqueness theorem for linear nth-order boundary-value problems

- homogeneous linear nth-order DEs

- representation of a DE using differential operators

- superposition principles

- linear dependence and linear independence of a set of functions

- Wronskian

- fundamental set of solutions

- general solution of a linear nth-order DE

- particular solution of a nonhomogeneous linear nth-order DE

- complementary function for a nonhomogeneous linear nth-order DE

The basic skills listed below summarize the more mechanical types of problems encountered in the exercise set for this section.

Basic Skills

- given the general solution of a linear nth-order DE, find a solution satisfying n initial conditions

70

Section 4.1 Linear Differential Equations: Basic Theory

- given a two-parameter family of solutions of a linear second-order DE, determine a member or members of the family satisfying specified boundary conditions (or show that no solution exists)

- determine whether a given set of functions is linearly independent on an interval

- show that a given set of functions forms a fundamental set of solutions on an interval

Exercises 4.1   *Hints, Suggestions, Solutions, and Examples*

3. From $y = c_1 x + c_2 x \ln x$ we find $y' = c_1 + c_2(1 + \ln x)$. Then $y(1) = c_1 = 3$, $y'(1) = c_1 + c_2 = -1$ so that $c_1 = 3$ and $c_2 = -4$. The solution is $y = 3x - 4x \ln x$.

6. In this case we have $y(0) = c_1 = 0$, $y'(0) = 2c_2 \cdot 0 = 0$ so $c_1 = 0$ and c_2 is arbitrary. Two solutions are $y = x^2$ and $y = 2x^2$.

9. Since $a_2(x) = x - 2$ and $x_0 = 0$ the problem has a unique solution for $-\infty < x < 2$.

12. In this case we have $y(0) = c_1 = 1$, $y'(1) = 2c_2 = 6$ so that $c_1 = 1$ and $c_2 = 3$. The solution is $y = 1 + 3x^2$.

15. Since $(-4)x + (3)x^2 + (1)(4x - 3x^2) = 0$ the set of functions is linearly dependent.

18. Since $(1)\cos 2x + (1)1 + (-2)\cos^2 x = 0$ the set of functions is linearly dependent.

21. The set of functions is linearly independent since

$$W\left(1 + x, x, x^2\right) = \begin{vmatrix} 1 + x & x & x^2 \\ 1 & 1 & 2x \\ 0 & 0 & 2 \end{vmatrix} = 2 \neq 0.$$

23-30. (*Hint*) Remember that in order for a set of n functions to be a fundamental set of solutions of a linear nth-order DE they must *both* satisfy the DE and be linearly independent. Furthermore, if they satisfy the DE, then linear independence can be verified using the Wronskian.

24. The functions satisfy the differential equation and are linearly independent since

$$W(\cosh 2x, \sinh 2x) = 2$$

for $-\infty < x < \infty$. The general solution is

$$y = c_1 \cosh 2x + c_2 \sinh 2x.$$

27. The functions satisfy the differential equation and are linearly independent since

$$W\left(x^3, x^4\right) = x^6 \neq 0$$

for $0 < x < \infty$. The general solution is

$$y = c_1 x^3 + c_2 x^4.$$

30. The functions satisfy the differential equation and are linearly independent since

$$W(1, x, \cos x, \sin x) = 1$$

for $-\infty < x < \infty$. The general solution is

$$y = c_1 + c_2 x + c_3 \cos x + c_4 \sin x.$$

33. The functions $y_1 = e^{2x}$ and $y_2 = xe^{2x}$ form a fundamental set of solutions of the homogeneous equation, and $y_p = x^2 e^{2x} + x - 2$ is a particular solution of the nonhomogeneous equation.

36. (a) $y_{p_1} = 5$ **(c)** $y_p = y_{p_1} + y_{p_2} = 5 - 2x$

Section 4.2 Reduction of Order

The terminology and concepts listed below provide an outline of the main ideas encountered in this section. These can be useful when preparing for a quiz or test.

Terminology and Concepts

- reduction of order

The basic skills listed below summarize the more mechanical types of problems encountered in the exercise set for this section.

Basic Skills

- given a solution y_1 of a linear second-order homogeneous DE, find a second solution y_2

Exercises 4.2 *Hints, Suggestions, Solutions, and Examples*

Reduction of order is used below in Problems 3 and 6 to find a second solution. Formula (5) from the text is used in Problems 9, 12, and 15.

3. Define $y = u(x) \cos 4x$ so

$$y' = -4u \sin 4x + u' \cos 4x, \quad y'' = u'' \cos 4x - 8u' \sin 4x - 16u \cos 4x$$

and

$$y'' + 16y = (\cos 4x)u'' - 8(\sin 4x)u' = 0 \quad \text{or} \quad u'' - 8(\tan 4x)u' = 0.$$

If $w = u'$ we obtain the first-order equation $w' - 8(\tan 4x)w = 0$ which has the integrating factor $e^{-8\int \tan 4x \, dx} = \cos^2 4x$. Now

$$\frac{d}{dx}[(\cos^2 4x)w] = 0 \quad \text{gives} \quad (\cos^2 4x)w = c.$$

Therefore $w = u' = c \sec^2 4x$ and $u = c_1 \tan 4x$. A second solution is $y_2 = \tan 4x \cos 4x = \sin 4x$.

6. Define $y = u(x)e^{5x}$ so

$$y' = 5e^{5x}u + e^{5x}u', \quad y'' = e^{5x}u'' + 10e^{5x}u' + 25e^{5x}u$$

and

$$y'' - 25y = e^{5x}(u'' + 10u') = 0 \quad \text{or} \quad u'' + 10u' = 0.$$

If $w = u'$ we obtain the first-order equation $w' + 10w = 0$ which has the integrating factor $e^{10\int dx} = e^{10x}$. Now

$$\frac{d}{dx}[e^{10x}w] = 0 \quad \text{gives} \quad e^{10x}w = c.$$

Therefore $w = u' = ce^{-10x}$ and $u = c_1 e^{-10x}$. A second solution is $y_2 = e^{-10x}e^{5x} = e^{-5x}$.

9. Identifying $P(x) = -7/x$ we have

$$y_2 = x^4 \int \frac{e^{-\int(-7/x)\,dx}}{x^8}\,dx = x^4 \int \frac{1}{x}\,dx = x^4 \ln|x|.$$

A second solution is $y_2 = x^4 \ln|x|$.

12. Identifying $P(x) = 0$ we have

$$y_2 = x^{1/2} \ln x \int \frac{e^{-\int 0\,dx}}{x(\ln x)^2}\,dx = x^{1/2} \ln x \left(-\frac{1}{\ln x}\right) = -x^{1/2}.$$

A second solution is $y_2 = x^{1/2}$.

15. Identifying $P(x) = 2(1+x)/\left(1 - 2x - x^2\right)$ we have

$$y_2 = (x+1) \int \frac{e^{-\int 2(1+x)\,dx/\left(1-2x-x^2\right)}}{(x+1)^2}\,dx = (x+1) \int \frac{e^{\ln\left(1-2x-x^2\right)}}{(x+1)^2}\,dx$$

$$= (x+1) \int \frac{1 - 2x - x^2}{(x+1)^2}\,dx = (x+1) \int \left[\frac{2}{(x+1)^2} - 1\right]dx$$

$$= (x+1)\left[-\frac{2}{x+1} - x\right] = -2 - x^2 - x.$$

73

A second solution is $y_2 = x^2 + x + 2$.

17. (*Hint*) Look for a particular solution having the form $y_p = A$.

18. Define $y = u(x) \cdot 1$ so

$$y' = u', \quad y'' = u'' \quad \text{and} \quad y'' + y' = u'' + u' = 1.$$

If $w = u'$ we obtain the first order equation $w' + w = 1$ which has the integrating factor $e^{\int dx} = e^x$. Now

$$\frac{d}{dx}[e^x w] = e^x \quad \text{gives} \quad e^x w = e^x + c.$$

Therefore $w = u' = 1 + ce^{-x}$ and $u = x + c_1 e^{-x} + c_2$. The general solution is

$$y = u = x + c_1 e^{-x} + c_2.$$

19. (*Hint*) Look for a particular solution having the form $y_p = Ae^{3x}$.

20. (*Hint*) Look for a particular solution having the form $y_p = Ax + B$.

22. (*Hint*) Recall the infinite series reperesentation of the exponential function: $e^x = \sum_{n=0}^{\infty}(1/n!)x^n$.

Section 4.3   Homogeneous Linear Equations with Constant Coefficients

The terminology and concepts listed below provide an outline of the main ideas encountered in this section. These can be useful when preparing for a quiz or test.

Terminology and Concepts

- auxiliary equation

The basic skills listed below summarize the more mechanical types of problems encountered in the exercise set for this section.

Basic Skills

- find the general solution of a homogeneous linear DE with constant coefficients when the roots of the auxiliary equation are real and distinct, real and repeated, conjugate complex, and repeated conjugate complex

- find the solution of an nth-order homogeneous linear DE with constant coefficients and n initial conditions

Factorization of Polynomials In Section 4.3 you will see that the solution method for homogeneous linear DEs with constant coefficents relies on your ability to solve polynomial equations. For example, you know that a quadratic polynomial equation

$$am^2 + bm + c = 0, \quad a \neq 0, \ a, b, c \text{ real constants,}$$

can always be solved using the quadratic formula. The two roots m_1 and m_2 are

$$m_1 = \frac{-b + \sqrt{b^2 - 4ac}}{2a} \quad \text{and} \quad m_2 = \frac{-b - \sqrt{b^2 - 4ac}}{2a}.$$

The roots m_1 and m_2 are also called the **zeros** of the polynomial $am^2 + bm + c$. Note that m_1 and m_2 are real and distinct ($m_1 \neq m_2$) if $b^2 - 4ac > 0$, real and equal ($m_1 = m_2$) if $b^2 - 4ac = 0$, and finally, m_1 and m_2 are complex conjugates ($m_1 = \alpha + i\beta$, $m_1 = \alpha - i\beta$, α and β are positive real numbers) if $b^2 - 4ac < 0$. The Factor Theorem of Algebra states that finding zeros of a polynomial and factoring the polynomial are equivalent problems. Thus, in the case of a quadratic polynomial $am^2 + bm + c$ we can write

$$am^2 + bm + c = a(m - m_1)(m - m_2).$$

A polynomial of degree n, $a_n m^n + a_{n-1} m^{n-1} + \cdots + a_1 m + a_0$ with real coefficients has n zeros (counting multiplicities) and hence can be factored as

$$a_n m^n + a_{n-1} m^{n-1} + \cdots + a_1 m + a_0 = a_n(m - m_1)(m - m_2) \cdots (m - m_n).$$

Finding the zeros requires that we find roots of a polynomial equation of degree > 2, and *that* is the fundamental problem here. Analogous to quadratic equations, cubic and quartic polynomial equations can be solved by (complicated) formulas involving radicals, but there exist no such formulas for solving polynomial equations of degree $n > 4$. But if the polynomial has *integer* coefficients and if p/q is a rational zero (p and q are integers and $q \neq 0$), then it can be proved that

$$p \text{ must be an integer factor of } a_0,$$

and

$$q \text{ must be an integer factor of } a_n.$$

By forming all possible quotients of each factor of a_0 to each factor of a_n, we can construct a list of all *possible* rational zeros. For example, consider the fourth-degree polynomial equation $3m^4 - 10m^3 - 3m^2 + 8m - 2 = 0$. We identify $a_0 = -2$ and $a_n = 3$ and list all integer factors of a_0 and a_n, respectively:

$$p : \pm 1, \pm 2 \quad \text{and} \quad q : \pm 1, \pm 3.$$

The eight possible rational roots of the equation are then

$$\frac{p}{q} : -1, 1, -\frac{1}{3}, \frac{1}{3}, -2, 2, -\frac{2}{3}, \frac{2}{3}.$$

75

To determine which, if any, of these numbers are roots, we could use brute force substitution of a number p/q into the polynomial to see whether it satisfies the equation. We could also use long division. If the term $m - p/q$ divides the polynomial evenly, that is, the remainder r is zero, then $m - p/q$ is a factor and hence p/q is a root. A short-hand version of long division, called **synthetic division**, is an efficient way of proceeding. We begin by testing -1 by synthetically dividing by $m - (-1) = m + 1$:

$$
\begin{array}{r|rrrrr}
-1 & 3 & -10 & -3 & 8 & -2 \\
   &   & -3 & 13 & -10 & 2 \\
\hline
   & 3 & -13 & 10 & -2 & \boxed{0 = r}
\end{array}
$$

Since the remainder upon division is $r = 0$, we can conclude $m + 1$ is a factor of the polynomial and so -1 is a root. Now the division shows that

$$3m^4 - 10m^3 - 3m^2 + 8m - 2 = (m + 1)(3m^3 - 13m^2 + 10m - 2).$$

Observe that if the original polynomial has additional rational zeros they must be zeros of the second, or cubic, factor $3m^3 - 13m^2 + 10m - 2$. Thus we continue the testing by dividing this factor by $m - 1$:

$$
\begin{array}{r|rrrr}
1 & 3 & -13 & 10 & -2 \\
  &   & 3 & -10 & 0 \\
\hline
  & 3 & -10 & 0 & \boxed{-2 = r}
\end{array}
$$

The foregoing division shows that $m - 1$ is not a factor since the remainder r is not zero, and so 1 is not a rational root. Checking $\frac{1}{3}$ we have

$$
\begin{array}{r|rrrr}
\frac{1}{3} & 3 & -13 & 10 & -2 \\
            &   & 1 & -4 & 2 \\
\hline
            & 3 & -12 & 6 & \boxed{0 = r}
\end{array}
$$

Thus, $\frac{1}{3}$ is a root, and we now have the factorization

$$3m^4 - 10m^3 - 3m^2 + 8m - 2 = (m + 1)\left(m - \frac{1}{3}\right)(3m^2 - 12m + 6)$$

$$= (m + 1)\left(m - \frac{1}{3}\right)(3)(m^2 - 4m + 2)$$

$$= (m + 1)(3m - 1)(m^2 - 4m + 2).$$

Since the factor $m^2 - 4m + 2$ discovered by the division is quadratic, the remaining roots, $2 + \sqrt{2}$ and $2 - \sqrt{2}$, can be determined from the quadratic formula. Therefore, the original polynomial equation has four real roots: two rational and two irrational.

Exercises 4.3   *Hints, Suggestions, Solutions, and Examples*

3. From $m^2 - m - 6 = 0$ we obtain $m = 3$ and $m = -2$ so that $y = c_1 e^{3x} + c_2 e^{-2x}$.

6. From $m^2 - 10m + 25 = 0$ we obtain $m = 5$ and $m = 5$ so that $y = c_1 e^{5x} + c_2 x e^{5x}$.

9. From $m^2 + 9 = 0$ we obtain $\underline{m = 3i}$ and $\underline{m = -3i}$ so that $y = c_1 \cos 3x + c_2 \sin 3x$.

12. From $2m^2 + 2m + 1 = 0$ we obtain $m = -1/2 \pm i/2$ so that
$$y = e^{-x/2}[c_1 \cos(x/2) + c_2 \sin(x/2)].$$

15. From $m^3 - 4m^2 - 5m = 0$ we obtain $m = 0$, $m = 5$, and $m = -1$ so that
$$y = c_1 + c_2 e^{5x} + c_3 e^{-x}.$$

18. From $m^3 + 3m^2 - 4m - 12 = 0$ we obtain $m = -2$, $m = 2$, and $m = -3$ so that
$$y = c_1 e^{-2x} + c_2 e^{2x} + c_3 e^{-3x}.$$

21. From $m^3 + 3m^2 + 3m + 1 = 0$ we obtain $m = -1$, $m = -1$, and $m = -1$ so that
$$y = c_1 e^{-x} + c_2 x e^{-x} + c_3 x^2 e^{-x}.$$

24. From $m^4 - 2m^2 + 1 = 0$ we obtain $m = 1$, $m = 1$, $m = -1$, and $m = -1$ so that
$$y = c_1 e^{x} + c_2 x e^{x} + c_3 e^{-x} + c_4 x e^{-x}.$$

27. From $m^5 + 5m^4 - 2m^3 - 10m^2 + m + 5 = 0$ we obtain $m = -1$, $m = -1$, $m = 1$, and $m = 1$, and $m = -5$ so that
$$u = c_1 e^{-r} + c_2 r e^{-r} + c_3 e^{r} + c_4 r e^{r} + c_5 e^{-5r}.$$

30. From $m^2 + 1 = 0$ we obtain $m = \pm i$ so that $y = c_1 \cos \theta + c_2 \sin \theta$. If $y(\pi/3) = 0$ and $y'(\pi/3) = 2$ then
$$\frac{1}{2}c_1 + \frac{\sqrt{3}}{2}c_2 = 0$$
$$-\frac{\sqrt{3}}{2}c_1 + \frac{1}{2}c_2 = 2,$$
so $c_1 = -\sqrt{3}$, $c_2 = 1$, and $y = -\sqrt{3} \cos \theta + \sin \theta$.

33. From $m^2 + m + 2 = 0$ we obtain $m = -1/2 \pm \sqrt{7}\, i/2$ so that $y = e^{-x/2}[c_1 \cos(\sqrt{7}\, x/2) + c_2 \sin(\sqrt{7}\, x/2)]$. If $y(0) = 0$ and $y'(0) = 0$ then $c_1 = 0$ and $c_2 = 0$ so that $y = 0$.

36. From $m^3 + 2m^2 - 5m - 6 = 0$ we obtain $m = -1$, $m = 2$, and $m = -3$ so that

$$y = c_1 e^{-x} + c_2 e^{2x} + c_3 e^{-3x}.$$

If $y(0) = 0$, $y'(0) = 0$, and $y''(0) = 1$ then

$$c_1 + c_2 + c_3 = 0, \quad -c_1 + 2c_2 - 3c_3 = 0, \quad c_1 + 4c_2 + 9c_3 = 1,$$

so $c_1 = -1/6$, $c_2 = 1/15$, $c_3 = 1/10$, and

$$y = -\frac{1}{6}e^{-x} + \frac{1}{15}e^{2x} + \frac{1}{10}e^{-3x}.$$

39. From $m^2 + 1 = 0$ we obtain $m = \pm i$ so that $y = c_1 \cos x + c_2 \sin x$ and $y' = -c_1 \sin x + c_2 \cos x$. From $y'(0) = c_1(0) + c_2(1) = c_2 = 0$ and $y'(\pi/2) = -c_1(1) = 0$ we find $c_1 = c_2 = 0$. A solution of the BVP is $y = 0$.

42. The auxiliary equation is $m^2 - 1 = 0$ which has roots -1 and 1. By (10) the general solution is $y = c_1 e^x + c_2 e^{-x}$. By (11) the general solution is $y = c_1 \cosh x + c_2 \sinh x$. For $y = c_1 e^x + c_2 e^{-x}$ the boundary conditions imply $c_1 + c_2 = 1$, $c_1 e - c_2 e^{-1} = 0$. Solving for c_1 and c_2 we find $c_1 = 1/(1 + e^2)$ and $c_2 = e^2/(1 + e^2)$ so $y = e^x/(1 + e^2) + e^2 e^{-x}/(1 + e^2)$. For $y = c_1 \cosh x + c_2 \sinh x$ the boundary conditions imply $c_1 = 1$, $c_2 = -\tanh 1$, so $y = \cosh x - (\tanh 1) \sinh x$.

45. The auxiliary equation should have a pair of complex roots $\alpha \pm \beta i$ where $a < 0$, so that the solution has the form $e^{\alpha x}(c_1 \cos \beta x + c_2 \sin \beta x)$. Thus, the DE is (e).

48. The DE should have the form $y'' + k^2 y = 0$ where $k = 2$ so that the period of the solution is π. Thus, the DE is (b).

49. (*Hint*) Read the discussion above in this section about the relationship between finding zeros of a polynomial and factoring the polynomial.

Section 4.4
Undetermined Coefficients – Superposition Approach

The terminology and concepts listed below provide an outline of the main ideas encountered in this section. These can be useful when preparing for a quiz or test.

Terminology and Concepts

- undetermined coefficients
- superposition principle
- multiplication rule

The basic skills listed below summarize the more mechanical types of problems encountered in the exercise set for this section.

Basic Skills

- use undetermined coefficients with the superposition approach to find a particular solution of a nonhomogeneous linear DE with constant coefficients when (1) no function in the assumed particular solution is a solution of the associated homogeneous DE, and (2) a function in the assumed particular solution is also a solution of the associated homogeneous DE

- recognize when the input function (right-hand side of the DE) is appropriate for the method of undetermined coefficients.

Exercises 4.4   *Hints, Suggestions, Solutions, and Examples*

3. From $m^2 - 10m + 25 = 0$ we find $m_1 = m_2 = 5$. Then $y_c = c_1 e^{5x} + c_2 x e^{5x}$ and we assume $y_p = Ax + B$. Substituting into the DE we obtain $25A = 30$ and $-10A + 25B = 3$. Then $A = \frac{6}{5}$, $B = \frac{6}{5}$, $y_p = \frac{6}{5}x + \frac{6}{5}$, and

$$y = c_1 e^{5x} + c_2 x e^{5x} + \frac{6}{5}x + \frac{6}{5}.$$

6. From $m^2 - 8m + 20 = 0$ we find $m_1 = 4 + 2i$ and $m_2 = 4 - 2i$. Then $y_c = e^{4x}(c_1 \cos 2x + c_2 \sin 2x)$ and we assume $y_p = Ax^2 + Bx + C + (Dx + E)e^x$. Substituting into the DE we obtain

$$2A - 8B + 20C = 0$$

$$-6D + 13E = 0$$

$$-16A + 20B = 0$$

$$13D = -26$$

$$20A = 100.$$

Then $A = 5$, $B = 4$, $C = \frac{11}{10}$, $D = -2$, $E = -\frac{12}{13}$, $y_p = 5x^2 + 4x + \frac{11}{10} + \left(-2x - \frac{12}{13}\right)e^x$ and

$$y = e^{4x}(c_1 \cos 2x + c_2 \sin 2x) + 5x^2 + 4x + \frac{11}{10} + \left(-2x - \frac{12}{13}\right)e^x.$$

9. From $m^2 - m = 0$ we find $m_1 = 1$ and $m_2 = 0$. Then $y_c = c_1 e^x + c_2$ and we assume $y_p = Ax$. Substituting into the DE we obtain $-A = -3$. Then $A = 3$, $y_p = 3x$ and $y = c_1 e^x + c_2 + 3x$.

12. From $m^2 - 16 = 0$ we find $m_1 = 4$ and $m_2 = -4$. Then $y_c = c_1 e^{4x} + c_2 e^{-4x}$ and we assume $y_p = Axe^{4x}$. Substituting into the DE we obtain $8A = 2$. Then $A = \frac{1}{4}$, $y_p = \frac{1}{4}xe^{4x}$ and

$$y = c_1 e^{4x} + c_2 e^{-4x} + \frac{1}{4}xe^{4x}.$$

Section 4.4 Undetermined Coefficients – Superposition Approach

15. From $m^2 + 1 = 0$ we find $m_1 = i$ and $m_2 = -i$. Then $y_c = c_1 \cos x + c_2 \sin x$ and we assume $y_p = (Ax^2 + Bx) \cos x + (Cx^2 + Dx) \sin x$. Substituting into the DE we obtain $4C = 0$, $2A + 2D = 0$, $-4A = 2$, and $-2B + 2C = 0$. Then $A = -\frac{1}{2}$, $B = 0$, $C = 0$, $D = \frac{1}{2}$, $y_p = -\frac{1}{2}x^2 \cos x + \frac{1}{2}x \sin x$, and

$$y = c_1 \cos x + c_2 \sin x - \frac{1}{2}x^2 \cos x + \frac{1}{2}x \sin x.$$

18. From $m^2 - 2m + 2 = 0$ we find $m_1 = 1 + i$ and $m_2 = 1 - i$. Then $y_c = e^x(c_1 \cos x + c_2 \sin x)$ and we assume $y_p = Ae^{2x} \cos x + Be^{2x} \sin x$. Substituting into the DE we obtain $A + 2B = 1$ and $-2A + B = -3$. Then $A = \frac{7}{5}$, $B = -\frac{1}{5}$, $y_p = \frac{7}{5}e^{2x} \cos x - \frac{1}{5}e^{2x} \sin x$ and

$$y = e^x(c_1 \cos x + c_2 \sin x) + \frac{7}{5}e^{2x} \cos x - \frac{1}{5}e^{2x} \sin x.$$

21. From $m^3 - 6m^2 = 0$ we find $m_1 = m_2 = 0$ and $m_3 = 6$. Then $y_c = c_1 + c_2 x + c_3 e^{6x}$ and we assume $y_p = Ax^2 + B \cos x + C \sin x$. Substituting into the DE we obtain $-12A = 3$, $6B - C = -1$, and $B + 6C = 0$. Then $A = -\frac{1}{4}$, $B = -\frac{6}{37}$, $C = \frac{1}{37}$, $y_p = -\frac{1}{4}x^2 - \frac{6}{37} \cos x + \frac{1}{37} \sin x$, and

$$y = c_1 + c_2 x + c_3 e^{6x} - \frac{1}{4}x^2 - \frac{6}{37} \cos x + \frac{1}{37} \sin x.$$

24. From $m^3 - m^2 - 4m + 4 = 0$ we find $m_1 = 1$, $m_2 = 2$, and $m_3 = -2$. Then $y_c = c_1 e^x + c_2 e^{2x} + c_3 e^{-2x}$ and we assume $y_p = A + Bxe^x + Cxe^{2x}$. Substituting into the DE we obtain $4A = 5$, $-3B = -1$, and $4C = 1$. Then $A = \frac{5}{4}$, $B = \frac{1}{3}$, $C = \frac{1}{4}$, $y_p = \frac{5}{4} + \frac{1}{3}xe^x + \frac{1}{4}xe^{2x}$, and

$$y = c_1 e^x + c_2 e^{2x} + c_3 e^{-2x} + \frac{5}{4} + \frac{1}{3}xe^x + \frac{1}{4}xe^{2x}.$$

27. We have $y_c = c_1 \cos 2x + c_2 \sin 2x$ and we assume $y_p = A$. Substituting into the DE we find $A = -\frac{1}{2}$. Thus $y = c_1 \cos 2x + c_2 \sin 2x - \frac{1}{2}$. From the initial conditions we obtain $c_1 = 0$ and $c_2 = \sqrt{2}$, so $y = \sqrt{2} \sin 2x - \frac{1}{2}$.

30. We have $y_c = c_1 e^{-2x} + c_2 xe^{-2x}$ and we assume $y_p = (Ax^3 + Bx^2)e^{-2x}$. Substituting into the DE we find $A = \frac{1}{6}$ and $B = \frac{3}{2}$. Thus $y = c_1 e^{-2x} + c_2 xe^{-2x} + \left(\frac{1}{6}x^3 + \frac{3}{2}x^2\right) e^{-2x}$. From the initial conditions we obtain $c_1 = 2$ and $c_2 = 9$, so

$$y = 2e^{-2x} + 9xe^{-2x} + \left(\frac{1}{6}x^3 + \frac{3}{2}x^2\right) e^{-2x}.$$

33. We have $x_c = c_1 \cos \omega t + c_2 \sin \omega t$ and we assume $x_p = At \cos \omega t + Bt \sin \omega t$. Substituting into the DE we find $A = -F_0/2\omega$ and $B = 0$. Thus $x = c_1 \cos \omega t + c_2 \sin \omega t - (F_0/2\omega)t \cos \omega t$. From the initial conditions we obtain $c_1 = 0$ and $c_2 = F_0/2\omega^2$, so

$$x = (F_0/2\omega^2) \sin \omega t - (F_0/2\omega)t \cos \omega t.$$

34. (*Suggestion*) Assume $\gamma \neq \omega$. (In Problem 33, $\gamma = \omega$.)

36. We have $y_c = c_1 e^{-2x} + e^x(c_2 \cos \sqrt{3}\,x + c_3 \sin \sqrt{3}\,x)$ and we assume $y_p = Ax + B + Cxe^{-2x}$. Substituting into the DE we find $A = \frac{1}{4}$, $B = -\frac{5}{8}$, and $C = \frac{2}{3}$. Thus

$$y = c_1 e^{-2x} + e^x(c_2 \cos \sqrt{3}\,x + c_3 \sin \sqrt{3}\,x) + \frac{1}{4}x - \frac{5}{8} + \frac{2}{3}xe^{-2x}.$$

From the initial conditions we obtain $c_1 = -\frac{23}{12}$, $c_2 = -\frac{59}{24}$, and $c_3 = \frac{17}{72}\sqrt{3}$, so

$$y = -\frac{23}{12}e^{-2x} + e^x\left(-\frac{59}{24}\cos \sqrt{3}\,x + \frac{17}{72}\sqrt{3}\sin \sqrt{3}\,x\right) + \frac{1}{4}x - \frac{5}{8} + \frac{2}{3}xe^{-2x}.$$

39. The general solution of the DE $y'' + 3y = 6x$ is $y = c_1 \cos \sqrt{3}x + c_2 \sin \sqrt{3}x + 2x$. The condition $y(0) = 0$ implies $c_1 = 0$ and so $y = c_2 \sin \sqrt{3}x + 2x$. The condition $y(1) + y'(1) = 0$ implies $c_2 \sin \sqrt{3} + 2 + c_2\sqrt{3}\cos \sqrt{3} + 2 = 0$ so $c_2 = -4/(\sin \sqrt{3} + \sqrt{3}\cos \sqrt{3})$. The solution is

$$y = \frac{-4\sin \sqrt{3}x}{\sin \sqrt{3} + \sqrt{3}\cos \sqrt{3}} + 2x.$$

42. We have $y_c = e^x(c_1 \cos 3x + c_2 \sin 3x)$ and we assume $y_p = A$ on $[0, \pi]$. Substituting into the DE we find $A = 2$. Thus, $y = e^x(c_1 \cos 3x + c_2 \sin 3x) + 2$ on $[0, \pi]$. On (π, ∞) we have $y = e^x(c_3 \cos 3x + c_4 \sin 3x)$. From $y(0) = 0$ and $y'(0) = 0$ we obtain

$$c_1 = -2, \qquad c_1 + 3c_2 = 0.$$

Solving this system, we find $c_1 = -2$ and $c_2 = \frac{2}{3}$. Thus $y = e^x(-2\cos 3x + \frac{2}{3}\sin 3x) + 2$ on $[0, \pi]$. Now, continuity of y at $x = \pi$ implies

$$e^\pi\left(-2\cos 3\pi + \frac{2}{3}\sin 3\pi\right) + 2 = e^\pi(c_3 \cos 3\pi + c_4 \sin 3\pi)$$

or $2 + 2e^\pi = -c_3 e^\pi$ or $c_3 = -2e^{-\pi}(1 + e^\pi)$. Continuity of y' at π implies

$$\frac{20}{3}e^\pi \sin 3\pi = e^\pi[(c_3 + 3c_4)\cos 3\pi + (-3c_3 + c_4)\sin 3\pi]$$

or $-c_3 e^\pi - 3c_4 e^\pi = 0$. Since $c_3 = -2e^{-\pi}(1 + e^\pi)$ we have $c_4 = \frac{2}{3}e^{-\pi}(1 + e^\pi)$. The solution of the IVP is

$$y(x) = \begin{cases} e^x(-2\cos 3x + \frac{2}{3}\sin 3x) + 2, & 0 \le x \le \pi \\ (1 + e^\pi)e^{x-\pi}(-2\cos 3x + \frac{2}{3}\sin 3x), & x > \pi. \end{cases}$$

Section 4.5

Undetermined Coefficients – Annihilator Approach

The terminology and concepts listed below provide an outline of the main ideas encountered in this section. These can be useful when preparing for a quiz or test.

Terminology and Concepts

- annihilator operator
- method of undetermined coefficients

The basic skills listed below summarize the more mechanical types of problems encountered in the exercise set for this section.

Basic Skills

- write a DE using differential operator notation
- find a differential operator that annihilates a given function
- find linearly independent functions that are annihilated by a given differential operator
- solve a nonhomogeneous linear DE with constant coefficients using annihilators and undetermined coefficients

Exercises 4.5   *Hints, Suggestions, Solutions, and Examples*

3. $(D^2 - 4D - 12)y = (D - 6)(D + 2)y = x - 6$

6. $(D^3 + 4D)y = D(D^2 + 4)y = e^x \cos 2x$

9. $(D^4 + 8D)y = D(D + 2)(D^2 - 2D + 4)y = 4$

12. $(2D - 1)y = (2D - 1)4e^{x/2} = 8De^{x/2} - 4e^{x/2} = 4e^{x/2} - 4e^{x/2} = 0$

15. D^4 because of x^3

18. $D^2(D - 6)^2$ because of x and xe^{6x}

21. $D^3(D^2 + 16)$ because of x^2 and $\sin 4x$

24. $D(D - 1)(D - 2)$ because of 1, e^x, and e^{2x}

27. $1, x, x^2, x^3, x^4$

30. $D^2 - 9D - 36 = (D - 12)(D + 3)$; e^{12x}, e^{-3x}

33. $D^3 - 10D^2 + 25D = D(D-5)^2$; $\quad 1, e^{5x}, xe^{5x}$

36. Applying D to the DE we obtain

$$D(2D^2 - 7D + 5)y = 0.$$

Then

$$y = \underbrace{c_1 e^{5x/2} + c_2 e^x}_{y_c} + c_3$$

and $y_p = A$. Substituting y_p into the DE yields $5A = -29$ or $A = -29/5$. The general solution is

$$y = c_1 e^{5x/2} + c_2 e^x - \frac{29}{5}.$$

39. Applying D^2 to the DE we obtain

$$D^2(D^2 + 4D + 4)y = D^2(D+2)^2 y = 0.$$

Then

$$y = \underbrace{c_1 e^{-2x} + c_2 xe^{-2x}}_{y_c} + c_3 + c_4 x$$

and $y_p = Ax + B$. Substituting y_p into the DE yields $4Ax + (4A + 4B) = 2x + 6$. Equating coefficients gives

$$4A = 2$$

$$4A + 4B = 6.$$

Then $A = 1/2$, $B = 1$, and the general solution is

$$y = c_1 e^{-2x} + c_2 xe^{-2x} + \frac{1}{2}x + 1.$$

42. Applying D^4 to the DE we obtain

$$D^4(D^2 - 2D + 1)y = D^4(D-1)^2 y = 0.$$

Then

$$y = \underbrace{c_1 e^x + c_2 xe^x}_{y_c} + c_3 x^3 + c_4 x^2 + c_5 x + c_6$$

and $y_p = Ax^3 + Bx^2 + Cx + D$. Substituting y_p into the DE yields

$$Ax^3 + (B - 6A)x^2 + (6A - 4B + C)x + (2B - 2C + D) = x^3 + 4x.$$

Equating coefficients gives

$$A = 1$$

$$B - 6A = 0$$

$$6A - 4B + C = 4$$

$$2B - 2C + D = 0.$$

83

Then $A = 1$, $B = 6$, $C = 22$, $D = 32$, and the general solution is

$$y = c_1 e^x + c_2 x e^x + x^3 + 6x^2 + 22x + 32.$$

45. Applying $D(D-1)$ to the DE we obtain

$$D(D-1)(D^2 - 2D - 3)y = D(D-1)(D+1)(D-3)y = 0.$$

Then

$$y = \underbrace{c_1 e^{3x} + c_2 e^{-x}}_{y_c} + c_3 e^x + c_4$$

and $y_p = A e^x + B$. Substituting y_p into the DE yields $-4Ae^x - 3B = 4e^x - 9$. Equating coefficients gives $A = -1$ and $B = 3$. The general solution is

$$y = c_1 e^{3x} + c_2 e^{-x} - e^x + 3.$$

48. Applying $D(D^2 + 1)$ to the DE we obtain

$$D(D^2 + 1)(D^2 + 4)y = 0.$$

Then

$$y = \underbrace{c_1 \cos 2x + c_2 \sin 2x}_{y_c} + c_3 \cos x + c_4 \sin x + c_5$$

and $y_p = A \cos x + B \sin x + C$. Substituting y_p into the DE yields

$$3A \cos x + 3B \sin x + 4C = 4 \cos x + 3 \sin x - 8.$$

Equating coefficients gives $A = 4/3$, $B = 1$, and $C = -2$. The general solution is

$$y = c_1 \cos 2x + c_2 \sin 2x + \frac{4}{3} \cos x + \sin x - 2.$$

51. Applying $D(D-1)^3$ to the DE we obtain

$$D(D-1)^3(D^2 - 1)y = D(D-1)^4(D+1)y = 0.$$

Then

$$y = \underbrace{c_1 e^x + c_2 e^{-x}}_{y_c} + c_3 x^3 e^x + c_4 x^2 e^x + c_5 x e^x + c_6$$

and $y_p = A x^3 e^x + B x^2 e^x + C x e^x + D$. Substituting y_p into the DE yields

$$6Ax^2 e^x + (6A + 4B)x e^x + (2B + 2C)e^x - D = x^2 e^x + 5.$$

Equating coefficients gives

$$6A = 1$$

$$6A + 4B = 0$$

$$2B + 2C = 0$$

$$-D = 5.$$

84

Then $A = 1/6$, $B = -1/4$, $C = 1/4$, $D = -5$, and the general solution is

$$y = c_1 e^x + c_2 e^{-x} + \frac{1}{6}x^3 e^x - \frac{1}{4}x^2 e^x + \frac{1}{4}xe^x - 5.$$

54. Applying $D^2 - 2D + 10$ to the DE we obtain

$$(D^2 - 2D + 10)\left(D^2 + D + \frac{1}{4}\right)y = (D^2 - 2D + 10)\left(D + \frac{1}{2}\right)^2 y = 0.$$

Then

$$y = \underbrace{c_1 e^{-x/2} + c_2 x e^{-x/2}}_{y_c} + c_3 e^x \cos 3x + c_4 e^x \sin 3x$$

and $y_p = Ae^x \cos 3x + Be^x \sin 3x$. Substituting y_p into the DE yields

$$(9B - 27A/4)e^x \cos 3x - (9A + 27B/4)e^x \sin 3x = -e^x \cos 3x + e^x \sin 3x.$$

Equating coefficients gives

$$-\frac{27}{4}A + 9B = -1$$

$$-9A - \frac{27}{4}B = 1.$$

Then $A = -4/225$, $B = -28/225$, and the general solution is

$$y = c_1 e^{-x/2} + c_2 x e^{-x/2} - \frac{4}{225}e^x \cos 3x - \frac{28}{225}e^x \sin 3x.$$

57. Applying $(D^2 + 1)^2$ to the DE we obtain

$$(D^2 + 1)^2(D^2 + D + 1) = 0.$$

Then

$$y = \underbrace{e^{-x/2}\left[c_1 \cos \frac{\sqrt{3}}{2}x + c_2 \sin \frac{\sqrt{3}}{2}x\right]}_{y_c} + c_3 \cos x + c_4 \sin x + c_5 x \cos x + c_6 x \sin x$$

and $y_p = A\cos x + B\sin x + Cx\cos x + Dx\sin x$. Substituting y_p into the DE yields

$$(B + C + 2D)\cos x + Dx\cos x + (-A - 2C + D)\sin x - Cx\sin x = x\sin x.$$

Equating coefficients gives

$$B + C + 2D = 0$$

$$D = 0$$

$$-A - 2C + D = 0$$

$$-C = 1.$$

85

Then $A = 2$, $B = 1$, $C = -1$, and $D = 0$, and the general solution is

$$y = e^{-x/2}\left[c_1 \cos \frac{\sqrt{3}}{2}x + c_2 \sin \frac{\sqrt{3}}{2}x\right] + 2\cos x + \sin x - x\cos x.$$

60. Applying $D(D-1)^2(D+1)$ to the DE we obtain

$$D(D-1)^2(D+1)(D^3 - D^2 + D - 1) = D(D-1)^3(D+1)(D^2+1) = 0.$$

Then

$$y = \underbrace{c_1 e^x + c_2 \cos x + c_3 \sin x}_{y_c} + c_4 + c_5 e^{-x} + c_6 x e^x + c_7 x^2 e^x$$

and $y_p = A + Be^{-x} + Cxe^x + Dx^2 e^x$. Substituting y_p into the DE yields

$$4Dxe^x + (2C+4D)e^x - 4Be^{-x} - A = xe^x - e^{-x} + 7.$$

Equating coefficients gives

$$4D = 1$$

$$2C + 4D = 0$$

$$-4B = -1$$

$$-A = 7.$$

Then $A = -7$, $B = 1/4$, $C = -1/2$, and $D = 1/4$, and the general solution is

$$y = c_1 e^x + c_2 \cos x + c_3 \sin x - 7 + \frac{1}{4}e^{-x} - \frac{1}{2}xe^x + \frac{1}{4}x^2 e^x.$$

63. Applying $D(D-1)$ to the DE we obtain

$$D(D-1)(D^4 - 2D^3 + D^2) = D^3(D-1)^3 = 0.$$

Then

$$y = \underbrace{c_1 + c_2 x + c_3 e^x + c_4 x e^x}_{y_c} + c_5 x^2 + c_6 x^2 e^x$$

and $y_p = Ax^2 + Bx^2 e^x$. Substituting y_p into the DE yields $2A + 2Be^x = 1 + e^x$. Equating coefficients gives $A = 1/2$ and $B = 1/2$. The general solution is

$$y = c_1 + c_2 x + c_3 e^x + c_4 x e^x + \frac{1}{2}x^2 + \frac{1}{2}x^2 e^x.$$

66. The complementary function is $y_c = c_1 + c_2 e^{-x}$. Using D^2 to annihilate x we find $y_p = Ax + Bx^2$. Substituting y_p into the DE we obtain $(A + 2B) + 2Bx = x$. Thus $A = -1$ and $B = 1/2$, and

$$y = c_1 + c_2 e^{-x} - x + \frac{1}{2}x^2$$

$$y' = -c_2 e^{-x} - 1 + x.$$

The initial conditions imply

$$c_1 + c_2 = 1$$

$$-c_2 = 1.$$

Thus $c_1 = 2$ and $c_2 = -1$, and

$$y = 2 - e^{-x} - x + \frac{1}{2}x^2.$$

69. The complementary function is $y_c = c_1 \cos x + c_2 \sin x$. Using $(D^2 + 1)(D^2 + 4)$ to annihilate $8 \cos 2x - 4 \sin x$ we find $y_p = Ax \cos x + Bx \sin x + C \cos 2x + D \sin 2x$. Substituting y_p into the DE we obtain $2B \cos x - 3C \cos 2x - 2A \sin x - 3D \sin 2x = 8 \cos 2x - 4 \sin x$. Thus $A = 2$, $B = 0$, $C = -8/3$, and $D = 0$, and

$$y = c_1 \cos x + c_2 \sin x + 2x \cos x - \frac{8}{3} \cos 2x$$

$$y' = -c_1 \sin x + c_2 \cos x + 2 \cos x - 2x \sin x + \frac{16}{3} \sin 2x.$$

The initial conditions imply

$$c_2 + \frac{8}{3} = -1$$

$$-c_1 - \pi = 0.$$

Thus $c_1 = -\pi$ and $c_2 = -11/3$, and

$$y = -\pi \cos x - \frac{11}{3} \sin x + 2x \cos x - \frac{8}{3} \cos 2x.$$

72. The complementary function is $y_c = c_1 + c_2 x + c_3 x^2 + c_4 e^x$. Using $D^2(D-1)$ to annihilate $x + e^x$ we find $y_p = Ax^3 + Bx^4 + Cxe^x$. Substituting y_p into the DE we obtain $(-6A + 24B) - 24Bx + Ce^x = x + e^x$. Thus $A = -1/6$, $B = -1/24$, and $C = 1$, and

$$y = c_1 + c_2 x + c_3 x^2 + c_4 e^x - \frac{1}{6}x^3 - \frac{1}{24}x^4 + xe^x$$

$$y' = c_2 + 2c_3 x + c_4 e^x - \frac{1}{2}x^2 - \frac{1}{6}x^3 + e^x + xe^x$$

$$y'' = 2c_3 + c_4 e^x - x - \frac{1}{2}x^2 + 2e^x + xe^x.$$

$$y''' = c_4 e^x - 1 - x + 3e^x + xe^x$$

The initial conditions imply

$$c_1 + c_4 = 0$$

$$c_2 + c_4 + 1 = 0$$

$$2c_3 + c_4 + 2 = 0$$

$$2 + c_4 = 0.$$

Thus $c_1 = 2$, $c_2 = 1$, $c_3 = 0$, and $c_4 = -2$, and

$$y = 2 + x - 2e^x - \frac{1}{6}x^3 - \frac{1}{24}x^4 + xe^x.$$

Section 4.6   Variation of Parameters

The terminology and concepts listed below provide an outline of the main ideas encountered in this section. These can be useful when preparing for a quiz or test.

Terminology and Concepts

- variation of parameters

The basic skills listed below summarize the more mechanical types of problems encountered in the exercise set for this section.

Basic Skills

- solve a nonhomogeneous linear second-order DE with constant coefficients using variation of parameters

Integrals of Trig Functions   In Example 2 in the text we had to integrate

$$\int \cot 3x \, dx = \int \frac{\cos 3x}{\sin 3x} \, dx.$$

If we identify $u = \sin 3x$, then $du = 3\cos 3x \, dx$. The integral is recognized as one of the form $\int du/u = \ln|u| + c$:

$$\frac{1}{3} \int \frac{\overbrace{3\cos 3x \, dx}^{du}}{\underbrace{\sin 3x}_{u}} = \frac{1}{3} \ln|\sin 3x| + c.$$

In general, we can integrate $\tan u$ and $\cot u$ in terms of logarithms:

$$\int \tan u \, du = -\ln|\cos u| + c \tag{1}$$

$$\int \cot u \, du = \ln|\sin u| + c. \tag{2}$$

The integrals in (1) and (2), along with

$$\int \sec u \, du = \ln|\sec u + \tan u| + c \tag{3}$$

$$\int \csc u \, du = \ln|\csc u - \cot u| + c \tag{4}$$

occur frequently in the solution of differential equations by variation of parameters. Watch for them in Exercises 4.6.

Exercises 4.6	*Hints, Suggestions, Solutions, and Examples*

The particular solution, $y_p = u_1 y_1 + u_2 y_2$, in the following problems can take on a variety of forms, especially where trigonometric functions are involved. The validity of a particular form can best be checked by substituting it back into the differential equation.

3. The auxiliary equation is $m^2 + 1 = 0$, so $y_c = c_1 \cos x + c_2 \sin x$ and

$$W = \begin{vmatrix} \cos x & \sin x \\ -\sin x & \cos x \end{vmatrix} = 1.$$

Identifying $f(x) = \sin x$ we obtain

$$u_1' = -\sin^2 x$$

$$u_2' = \cos x \sin x.$$

Then

$$u_1 = \frac{1}{4} \sin 2x - \frac{1}{2}x = \frac{1}{2} \sin x \cos x - \frac{1}{2}x$$

$$u_2 = -\frac{1}{2}\cos^2 x.$$

and

$$y = c_1 \cos x + c_2 \sin x + \frac{1}{2} \sin x \cos^2 x - \frac{1}{2} x \cos x - \frac{1}{2} \cos^2 x \sin x$$

$$= c_1 \cos x + c_2 \sin x - \frac{1}{2} x \cos x.$$

5. (*Hint*) To integrate $\sin x \cos^2 x$ think of $\int u^2 \, du$ with $u = \cos x$ and $du = -\sin x \, dx$.

Section 4.6 Variation of Parameters

6. The auxiliary equation is $m^2 + 1 = 0$, so $y_c = c_1 \cos x + c_2 \sin x$ and

$$W = \begin{vmatrix} \cos x & \sin x \\ -\sin x & \cos x \end{vmatrix} = 1.$$

Identifying $f(x) = \sec^2 x$ we obtain

$$u_1' = -\frac{\sin x}{\cos^2 x}$$

$$u_2' = \sec x.$$

Then

$$u_1 = -\frac{1}{\cos x} = -\sec x$$

$$u_2 = \ln|\sec x + \tan x|$$

and

$$y = c_1 \cos x + c_2 \sin x - \cos x \sec x + \sin x \ln|\sec x + \tan x|$$

$$= c_1 \cos x + c_2 \sin x - 1 + \sin x \ln|\sec x + \tan x|.$$

9. The auxiliary equation is $m^2 - 4 = 0$, so $y_c = c_1 e^{2x} + c_2 e^{-2x}$ and

$$W = \begin{vmatrix} e^{2x} & e^{-2x} \\ 2e^{2x} & -2e^{-2x} \end{vmatrix} = -4.$$

Identifying $f(x) = e^{2x}/x$ we obtain $u_1' = 1/4x$ and $u_2' = -e^{4x}/4x$. Then

$$u_1 = \frac{1}{4}\ln|x|,$$

$$u_2 = -\frac{1}{4}\int_{x_0}^{x} \frac{e^{4t}}{t}\, dt$$

and

$$y = c_1 e^{2x} + c_2 e^{-2x} + \frac{1}{4}\left(e^{2x}\ln|x| - e^{-2x}\int_{x_0}^{x}\frac{e^{4t}}{t}\,dt\right), \qquad x_0 > 0.$$

12. The auxiliary equation is $m^2 - 2m + 1 = (m-1)^2 = 0$, so $y_c = c_1 e^x + c_2 x e^x$ and

$$W = \begin{vmatrix} e^x & xe^x \\ e^x & xe^x + e^x \end{vmatrix} = e^{2x}.$$

Identifying $f(x) = e^x/\left(1 + x^2\right)$ we obtain

$$u_1' = -\frac{xe^x e^x}{e^{2x}(1+x^2)} = -\frac{x}{1+x^2}$$

$$u_2' = \frac{e^x e^x}{e^{2x}(1+x^2)} = \frac{1}{1+x^2}.$$

Then $u_1 = -\frac{1}{2}\ln\left(1 + x^2\right)$, $u_2 = \tan^{-1} x$, and

$$y = c_1 e^x + c_2 x e^x - \frac{1}{2}e^x \ln\left(1 + x^2\right) + x e^x \tan^{-1} x.$$

13. (*Hint*) To integrate $e^{2x}\sin e^x$ think of the form $\int u\sin u\,du$, with $u = e^x$, $du = e^x\,dx$, and then use integration by parts.

15. The auxiliary equation is $m^2 + 2m + 1 = (m+1)^2 = 0$, so $y_c = c_1 e^{-t} + c_2 t e^{-t}$ and

$$W = \begin{vmatrix} e^{-t} & te^{-t} \\ -e^{-t} & -te^{-t} + e^{-t} \end{vmatrix} = e^{-2t}.$$

Identifying $f(t) = e^{-t}\ln t$ we obtain

$$u_1' = -\frac{te^{-t}e^{-t}\ln t}{e^{-2t}} = -t\ln t$$

$$u_2' = \frac{e^{-t}e^{-t}\ln t}{e^{-2t}} = \ln t.$$

Then

$$u_1 = -\frac{1}{2}t^2 \ln t + \frac{1}{4}t^2$$

$$u_2 = t\ln t - t$$

and

$$y = c_1 e^{-t} + c_2 t e^{-t} - \frac{1}{2}t^2 e^{-t}\ln t + \frac{1}{4}t^2 e^{-t} + t^2 e^{-t}\ln t - t^2 e^{-t}$$

$$= c_1 e^{-t} + c_2 t e^{-t} + \frac{1}{2}t^2 e^{-t}\ln t - \frac{3}{4}t^2 e^{-t}.$$

18. The auxiliary equation is $4m^2 - 4m + 1 = (2m-1)^2 = 0$, so $y_c = c_1 e^{x/2} + c_2 x e^{x/2}$ and

$$W = \begin{vmatrix} e^{x/2} & xe^{x/2} \\ \frac{1}{2}e^{x/2} & \frac{1}{2}xe^{x/2} + e^{x/2} \end{vmatrix} = e^x.$$

Identifying $f(x) = \frac{1}{4}e^{x/2}\sqrt{1-x^2}$ we obtain

$$u_1' = -\frac{xe^{x/2}e^{x/2}\sqrt{1-x^2}}{4e^x} = -\frac{1}{4}x\sqrt{1-x^2}$$

$$u_2' = \frac{e^{x/2}e^{x/2}\sqrt{1-x^2}}{4e^x} = \frac{1}{4}\sqrt{1-x^2}.$$

To find u_1 and u_2 we use the substitution $v = 1 - x^2$ and the trig substitution $x = \sin\theta$, respectively:

$$u_1 = \frac{1}{12}\left(1 - x^2\right)^{3/2}$$

$$u_2 = \frac{x}{8}\sqrt{1-x^2} + \frac{1}{8}\sin^{-1} x.$$

91

Section 4.6 Variation of Parameters

Thus

$$y = c_1 e^{x/2} + c_2 x e^{x/2} + \frac{1}{12} e^{x/2} \left(1 - x^2\right)^{3/2} + \frac{1}{8} x^2 e^{x/2} \sqrt{1 - x^2} + \frac{1}{8} x e^{x/2} \sin^{-1} x.$$

21. The auxiliary equation is $m^2 + 2m - 8 = (m - 2)(m + 4) = 0$, so $y_c = c_1 e^{2x} + c_2 e^{-4x}$ and

$$W = \begin{vmatrix} e^{2x} & e^{-4x} \\ 2e^{2x} & -4e^{-4x} \end{vmatrix} = -6e^{-2x}.$$

Identifying $f(x) = 2e^{-2x} - e^{-x}$ we obtain

$$u_1' = \frac{1}{3} e^{-4x} - \frac{1}{6} e^{-3x}$$

$$u_2' = \frac{1}{6} e^{3x} - \frac{1}{3} e^{2x}.$$

Then

$$u_1 = -\frac{1}{12} e^{-4x} + \frac{1}{18} e^{-3x}$$

$$u_2 = \frac{1}{18} e^{3x} - \frac{1}{6} e^{2x}.$$

Thus

$$y = c_1 e^{2x} + c_2 e^{-4x} - \frac{1}{12} e^{-2x} + \frac{1}{18} e^{-x} + \frac{1}{18} e^{-x} - \frac{1}{6} e^{-2x}$$

$$= c_1 e^{2x} + c_2 e^{-4x} - \frac{1}{4} e^{-2x} + \frac{1}{9} e^{-x}$$

and

$$y' = 2c_1 e^{2x} - 4c_2 e^{-4x} + \frac{1}{2} e^{-2x} - \frac{1}{9} e^{-x}.$$

The initial conditions imply

$$c_1 + c_2 - \frac{5}{36} = 1$$

$$2c_1 - 4c_2 + \frac{7}{18} = 0.$$

Thus $c_1 = 25/36$ and $c_2 = 4/9$, and

$$y = \frac{25}{36} e^{2x} + \frac{4}{9} e^{-4x} - \frac{1}{4} e^{-2x} + \frac{1}{9} e^{-x}.$$

24. Write the equation in the form

$$y'' + \frac{1}{x} y' + \frac{1}{x^2} y = \frac{\sec(\ln x)}{x^2}$$

and identify $f(x) = \sec(\ln x)/x^2$. From $y_1 = \cos(\ln x)$ and $y_2 = \sin(\ln x)$ we compute

$$W = \begin{vmatrix} \cos(\ln x) & \sin(\ln x) \\ -\frac{\sin(\ln x)}{x} & \frac{\cos(\ln x)}{x} \end{vmatrix} = \frac{1}{x}.$$

92

Now

$$u_1' = -\frac{\tan(\ln x)}{x} \quad \text{so} \quad u_1 = \ln|\cos(\ln x)|,$$

and

$$u_2' = \frac{1}{x} \quad \text{so} \quad u_2 = \ln x.$$

Thus, a particular solution is

$$y_p = \cos(\ln x)\ln|\cos(\ln x)| + (\ln x)\sin(\ln x).$$

25-26. (*Suggestion*) In the text, see the paragraph following Example 6 in Appendix II.

Section 4.7 Cauchy-Euler Equation

The terminology and concepts listed below provide an outline of the main ideas encountered in this section. These can be useful when preparing for a quiz or test.

Terminology and Concepts

- Cauchy-Euler equation
- auxiliary equation
- variation of parameters

The basic skills listed below summarize the more mechanical types of problems encountered in the exercise set for this section.

Basic Skills

- find the general solution of a homogeneous Cauchy-Euler equation
- use variation of parameters to find a particular solution of a nonhomogeneous Cauchy-Euler equation
- use the substitution $x = e^t$ to transform a Cauchy-Euler equation into a linear DE with constant coefficients

---

Exercises 4.7 *Hints, Suggestions, Solutions, and Examples*

3. The auxiliary equation is $m^2 = 0$ so that $y = c_1 + c_2 \ln x$.

93

6. The auxiliary equation is $m^2 + 4m + 3 = (m+1)(m+3) = 0$ so that $y = c_1 x^{-1} + c_2 x^{-3}$.

9. The auxiliary equation is $25m^2 + 1 = 0$ so that $y = c_1 \cos\left(\frac{1}{5}\ln x\right) + c_2 \sin\left(\frac{1}{5}\ln x\right)$.

12. The auxiliary equation is $m^2 + 7m + 6 = (m+1)(m+6) = 0$ so that $y = c_1 x^{-1} + c_2 x^{-6}$.

15. Assuming that $y = x^m$ and substituting into the DE we obtain

$$m(m-1)(m-2) - 6 = m^3 - 3m^2 + 2m - 6 = (m-3)(m^2 + 2) = 0.$$

Thus

$$y = c_1 x^3 + c_2 \cos\left(\sqrt{2}\ln x\right) + c_3 \sin\left(\sqrt{2}\ln x\right).$$

18. Assuming that $y = x^m$ and substituting into the DE we obtain

$$m(m-1)(m-2)(m-3) + 6m(m-1)(m-2) + 9m(m-1) + 3m + 1 = m^4 + 2m^2 + 1 = (m^2 + 1)^2 = 0.$$

Thus

$$y = c_1 \cos(\ln x) + c_2 \sin(\ln x) + c_3(\ln x)\cos(\ln x) + c_4(\ln x)\sin(\ln x).$$

21. The auxiliary equation is $m^2 - 2m + 1 = (m-1)^2 = 0$ so that $y_c = c_1 x + c_2 x \ln x$ and

$$W(x, x\ln x) = \begin{vmatrix} x & x\ln x \\ 1 & 1 + \ln x \end{vmatrix} = x.$$

Identifying $f(x) = 2/x$ we obtain $u_1' = -2\ln x/x$ and $u_2' = 2/x$. Then $u_1 = -(\ln x)^2$, $u_2 = 2\ln x$, and

$$y = c_1 x + c_2 x \ln x - x(\ln x)^2 + 2x(\ln x)^2$$

$$= c_1 x + c_2 x \ln x + x(\ln x)^2.$$

24. The auxiliary equation $m(m-1) + m - 1 = m^2 - 1 = 0$ has roots $m_1 = -1$, $m_2 = 1$, so $y_c = c_1 x^{-1} + c_2 x$. With $y_1 = x^{-1}$, $y_2 = x$, and the identification $f(x) = 1/x^2(x+1)$, we get from (6) and (5) of Section 4.6 in the text

$$W = 2x^{-1}, \qquad W_1 = -1/x(x+1), \qquad \text{and} \qquad W_2 = 1/x^3(x+1).$$

Then $u_1' = W_1/W = -1/2(x+1)$, $u_2' = W_2/W = 1/2x^2(x+1)$, and integration (by partial fractions for u_2') gives

$$u_1 = -\frac{1}{2}\ln(x+1)$$

$$u_2 = -\frac{1}{2}x^{-1} - \frac{1}{2}\ln x + \frac{1}{2}\ln(x+1),$$

so

$$y_p = u_1 y_1 + u_2 y_2 = \left[-\frac{1}{2}\ln(x+1)\right]x^{-1} + \left[-\frac{1}{2}x^{-1} - \frac{1}{2}\ln x + \frac{1}{2}\ln(x+1)\right]x$$

$$= -\frac{1}{2} - \frac{1}{2}x\ln x + \frac{1}{2}x\ln(x+1) - \frac{\ln(x+1)}{2x} = -\frac{1}{2} + \frac{1}{2}x\ln\left(1 + \frac{1}{x}\right) - \frac{\ln(x+1)}{2x}$$

and

$$y = y_c + y_p = c_1 x^{-1} + c_2 x - \frac{1}{2} + \frac{1}{2} x \ln\left(1 + \frac{1}{x}\right) - \frac{\ln(x+1)}{2x}, \qquad x > 0.$$

27. The auxiliary equation is $m^2 + 1 = 0$, so that

$$y = c_1 \cos(\ln x) + c_2 \sin(\ln x)$$

and

$$y' = -c_1 \frac{1}{x} \sin(\ln x) + c_2 \frac{1}{x} \cos(\ln x).$$

The initial conditions imply $c_1 = 1$ and $c_2 = 2$. Thus
$y = \cos(\ln x) + 2\sin(\ln x)$.

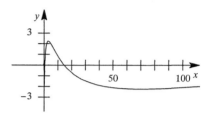

30. The auxiliary equation is $m^2 - 6m + 8 = (m-2)(m-4) = 0$,
so that $y_c = c_1 x^2 + c_2 x^4$ and

$$W = \begin{vmatrix} x^2 & x^4 \\ 2x & 4x^3 \end{vmatrix} = 2x^5.$$

Identifying $f(x) = 8x^4$ we obtain $u_1' = -4x^3$ and $u_2' = 4x$. Then
$u_1 = -x^4$, $u_2 = 2x^2$, and $y = c_1 x^2 + c_2 x^4 + x^6$. The initial conditions imply

$$\frac{1}{4} c_1 + \frac{1}{16} c_2 = -\frac{1}{64}$$

$$c_1 + \frac{1}{2} c_2 = -\frac{3}{16}.$$

Thus $c_1 = \frac{1}{16}$, $c_2 = -\frac{1}{2}$, and $y = \frac{1}{16} x^2 - \frac{1}{2} x^4 + x^6$.

33. Substituting $x = e^t$ into the DE we obtain

$$\frac{d^2 y}{dt^2} + 9\frac{dy}{dt} + 8y = e^{2t}.$$

The auxiliary equation is $m^2 + 9m + 8 = (m+1)(m+8) = 0$ so that $y_c = c_1 e^{-t} + c_2 e^{-8t}$. Using undetermined coefficients we try $y_p = Ae^{2t}$. This leads to $30Ae^{2t} = e^{2t}$, so that $A = 1/30$ and

$$y = c_1 e^{-t} + c_2 e^{-8t} + \frac{1}{30} e^{2t} = c_1 x^{-1} + c_2 x^{-8} + \frac{1}{30} x^2.$$

36. From

$$\frac{d^2 y}{dx^2} = \frac{1}{x^2}\left(\frac{d^2 y}{dt^2} - \frac{dy}{dt}\right)$$

95

it follows that

$$\frac{d^3y}{dx^3} = \frac{1}{x^2}\frac{d}{dx}\left(\frac{d^2y}{dt^2} - \frac{dy}{dt}\right) - \frac{2}{x^3}\left(\frac{d^2y}{dt^2} - \frac{dy}{dt}\right)$$

$$= \frac{1}{x^2}\frac{d}{dx}\left(\frac{d^2y}{dt^2}\right) - \frac{1}{x^2}\frac{d}{dx}\left(\frac{dy}{dt}\right) - \frac{2}{x^3}\frac{d^2y}{dt^2} + \frac{2}{x^3}\frac{dy}{dt}$$

$$= \frac{1}{x^2}\frac{d^3y}{dt^3}\left(\frac{1}{x}\right) - \frac{1}{x^2}\frac{d^2y}{dt^2}\left(\frac{1}{x}\right) - \frac{2}{x^3}\frac{d^2y}{dt^2} + \frac{2}{x^3}\frac{dy}{dt}$$

$$= \frac{1}{x^3}\left(\frac{d^3y}{dt^3} - 3\frac{d^2y}{dt^2} + 2\frac{dy}{dt}\right).$$

Substituting into the DE we obtain

$$\frac{d^3y}{dt^3} - 3\frac{d^2y}{dt^2} + 2\frac{dy}{dt} - 3\left(\frac{d^2y}{dt^2} - \frac{dy}{dt}\right) + 6\frac{dy}{dt} - 6y = 3 + 3t$$

or

$$\frac{d^3y}{dt^3} - 6\frac{d^2y}{dt^2} + 11\frac{dy}{dt} - 6y = 3 + 3t.$$

The auxiliary equation is $m^3 - 6m^2 + 11m - 6 = (m-1)(m-2)(m-3) = 0$ so that $y_c = c_1e^t + c_2e^{2t} + c_3e^{3t}$. Using undetermined coefficients we try $y_p = A + Bt$. This leads to $(11B - 6A) - 6Bt = 3 + 3t$, so that $A = -17/12$, $B = -1/2$, and

$$y = c_1e^t + c_2e^{2t} + c_3e^{3t} - \frac{17}{12} - \frac{1}{2}t = c_1x + c_2x^2 + c_3x^3 - \frac{17}{12} - \frac{1}{2}\ln x.$$

39. (*Hint*) Think substitution.

Section 4.8   Solving Systems of Linear Equations by Elimination

The terminology and concepts listed below provide an outline of the main ideas encountered in this section. These can be useful when preparing for a quiz or test.

Terminology and Concepts

- system of linear DEs
- systematic elimination to find the solution of a system of linear DEs

The basic skills listed below summarize the more mechanical types of problems encountered in the exercise set for this section.

Basic Skills

- use systematic elimnation to solve a system of linear DEs

Exercises 4.8 *Hints, Suggestions, Solutions, and Examples*

3. From $Dx = -y + t$ and $Dy = x - t$ we obtain $y = t - Dx$, $Dy = 1 - D^2 x$, and $(D^2 + 1)x = 1 + t$. The solution is

$$x = c_1 \cos t + c_2 \sin t + 1 + t$$

$$y = c_1 \sin t - c_2 \cos t + t - 1.$$

6. From $(D + 1)x + (D - 1)y = 2$ and $3x + (D + 2)y = -1$ we obtain $x = -\frac{1}{3} - \frac{1}{3}(D + 2)y$, $Dx = -\frac{1}{3}(D^2 + 2D)y$, and $(D^2 + 5)y = -7$. The solution is

$$y = c_1 \cos \sqrt{5}\,t + c_2 \sin \sqrt{5}\,t - \frac{7}{5}$$

$$x = \left(-\frac{2}{3}c_1 - \frac{\sqrt{5}}{3}c_2\right) \cos \sqrt{5}\,t + \left(\frac{\sqrt{5}}{3}c_1 - \frac{2}{3}c_2\right) \sin \sqrt{5}\,t + \frac{3}{5}.$$

9. From $Dx + D^2 y = e^{3t}$ and $(D + 1)x + (D - 1)y = 4e^{3t}$ we obtain $D(D^2 + 1)x = 34e^{3t}$ and $D(D^2 + 1)y = -8e^{3t}$. The solution is

$$y = c_1 + c_2 \sin t + c_3 \cos t - \frac{4}{15}e^{3t}$$

$$x = c_4 + c_5 \sin t + c_6 \cos t + \frac{17}{15}e^{3t}.$$

Substituting into $(D + 1)x + (D - 1)y = 4e^{3t}$ gives

$$(c_4 - c_1) + (c_5 - c_6 - c_3 - c_2) \sin t + (c_6 + c_5 + c_2 - c_3) \cos t = 0$$

so that $c_4 = c_1$, $c_5 = c_3$, $c_6 = -c_2$, and

$$x = c_1 - c_2 \cos t + c_3 \sin t + \frac{17}{15}e^{3t}.$$

12. From $(2D^2 - D - 1)x - (2D + 1)y = 1$ and $(D - 1)x + Dy = -1$ we obtain $(2D + 1)(D - 1)(D + 1)x = -1$ and $(2D + 1)(D + 1)y = -2$. The solution is

$$x = c_1 e^{-t/2} + c_2 e^{-t} + c_3 e^t + 1$$

$$y = c_4 e^{-t/2} + c_5 e^{-t} - 2.$$

Substituting into $(D-1)x + Dy = -1$ gives

$$\left(-\frac{3}{2}c_1 - \frac{1}{2}c_4\right)e^{-t/2} + (-2c_2 - c_5)e^{-t} = 0$$

so that $c_4 = -3c_1$, $c_5 = -2c_2$, and

$$y = -3c_1 e^{-t/2} - 2c_2 e^{-t} - 2.$$

15. Multiplying the first equation by $D+1$ and the second equation by $D^2 + 1$ and subtracting we obtain $(D^4 - D^2)x = 1$. Then

$$x = c_1 + c_2 t + c_3 e^t + c_4 e^{-t} - \frac{1}{2}t^2.$$

Multiplying the first equation by $D+1$ and subtracting we obtain $D^2(D+1)y = 1$. Then

$$y = c_5 + c_6 t + c_7 e^{-t} - \frac{1}{2}t^2.$$

Substituting into $(D-1)x + (D^2 + 1)y = 1$ gives

$$(-c_1 + c_2 + c_5 - 1) + (-2c_4 + 2c_7)e^{-t} + (-1 - c_2 + c_6)t = 1$$

so that $c_5 = c_1 - c_2 + 2$, $c_6 = c_2 + 1$, and $c_7 = c_4$. The solution of the system is

$$x = c_1 + c_2 t + c_3 e^t + c_4 e^{-t} - \frac{1}{2}t^2$$

$$y = (c_1 - c_2 + 2) + (c_2 + 1)t + c_4 e^{-t} - \frac{1}{2}t^2.$$

18. From $Dx + z = e^t$, $(D-1)x + Dy + Dz = 0$, and $x + 2y + Dz = e^t$ we obtain $z = -Dx + e^t$, $Dz = -D^2 x + e^t$, and the system $(-D^2 + D - 1)x + Dy = -e^t$ and $(-D^2 + 1)x + 2y = 0$. Then $y = \frac{1}{2}(D^2 - 1)x$, $Dy = \frac{1}{2}D(D^2 - 1)x$, and $(D-2)(D^2 + 1)x = -2e^t$ so that the solution is

$$x = c_1 e^{2t} + c_2 \cos t + c_3 \sin t + e^t$$

$$y = \frac{3}{2}c_1 e^{2t} - c_2 \cos t - c_3 \sin t$$

$$z = -2c_1 e^{2t} - c_3 \cos t + c_2 \sin t.$$

21. From $(D+5)x + y = 0$ and $4x - (D+1)y = 0$ we obtain $y = -(D+5)x$ so that $Dy = -(D^2 + 5D)x$. Then $4x + (D^2 + 5D)x + (D+5)x = 0$ and $(D+3)^2 x = 0$. Thus

$$x = c_1 e^{-3t} + c_2 t e^{-3t}$$

$$y = -(2c_1 + c_2)e^{-3t} - 2c_2 t e^{-3t}.$$

Using $x(1) = 0$ and $y(1) = 1$ we obtain

$$c_1 e^{-3} + c_2 e^{-3} = 0$$

$$-(2c_1 + c_2)e^{-3} - 2c_2 e^{-3} = 1$$

or

$$c_1 + c_2 = 0$$

$$2c_1 + 3c_2 = -e^3.$$

Thus $c_1 = e^3$ and $c_2 = -e^3$. The solution of the initial value problem is

$$x = e^{-3t+3} - te^{-3t+3}$$

$$y = -e^{-3t+3} + 2te^{-3t+3}.$$

24. From Newton's second law in the x-direction we have

$$m\frac{d^2x}{dt^2} = -k\cos\theta = -k\frac{1}{v}\frac{dx}{dt} = -|c|\frac{dx}{dt}.$$

In the y-direction we have

$$m\frac{d^2y}{dt^2} = -mg - k\sin\theta = -mg - k\frac{1}{v}\frac{dy}{dt} = -mg - |c|\frac{dy}{dt}.$$

From $mD^2x + |c|Dx = 0$ we have $D(mD + |c|)x = 0$ so that $(mD + |c|)x = c_1$ or $(D + |c|/m)x = c_2$. This is a linear first-order DE. An integrating factor is $e^{\int |c|dt/m} = e^{|c|t/m}$ so that

$$\frac{d}{dt}[e^{|c|t/m}x] = c_2 e^{|c|t/m}$$

and $e^{|c|t/m}x = (c_2m/|c|)e^{|c|t/m} + c_3$. The general solution of this equation is $x(t) = c_4 + c_3 e^{-|c|t/m}$.

From $(mD^2 + |c|D)y = -mg$ we have $D(mD + |c|)y = -mg$ so that $(mD + |c|)y = -mgt + c_1$ or $(D + |c|/m)y = -gt + c_2$. This is a linear first-order DE with integrating factor $e^{\int |c|dt/m} = e^{|c|t/m}$. Thus

$$\frac{d}{dt}[e^{|c|t/m}y] = (-gt + c_2)e^{|c|t/m}$$

$$e^{|c|t/m}y = -\frac{mg}{|c|}te^{|c|t/m} + \frac{m^2g}{c^2}e^{|c|t/m} + c_3 e^{|c|t/m} + c_4$$

and

$$y(t) = -\frac{mg}{|c|}t + \frac{m^2g}{c^2} + c_3 + c_4 e^{-|c|t/m}.$$

27. (*Suggestion*) Start by solving the first DE for x_1 and substituting the solution into the second DE.

| Section 4.9 | **Nonlinear Differential Equations** |

The terminology and concepts listed below provide an outline of the main ideas encountered in this section. These can be useful when preparing for a quiz or test.

Terminology and Concepts

- superposition of solutions does *not* apply to nonlinear DEs
- nonlinear DEs can possess singular solutions
- using a substitution, express a nonlinear second-order DE, $F(x, y', y'') = 0$, as a first-order DE
- using a substitution, express a nonlinear second-order DE, $F(y, y', y'') = 0$, as a first-order DE
- Taylor series solution of an IVP

The basic skills listed below summarize the more mechanical types of problems encountered in the exercise set for this section.

Basic Skills

- solve a nonlinear second-order DE using reduction of order when appropriate
- use a substitution to find a Taylor series solution of a nonlinear second-order IVP

Exercises 4.9 *Hints, Suggestions, Solutions, and Examples*

3. Let $u = y'$ so that $u' = y''$. The equation becomes $u' = -u - 1$ which is separable. Thus

$$\frac{du}{u^2 + 1} = -dx \implies \tan^{-1} u = -x + c_1 \implies y' = \tan(c_1 - x) \implies y = \ln|\cos(c_1 - x)| + c_2.$$

6. Let $u = y'$ so that $y'' = u\, du/dy$. The equation becomes $(y + 1)u\, du/dy = u^2$. Separating variables we obtain

$$\frac{du}{u} = \frac{dy}{y + 1} \implies \ln|u| = \ln|y + 1| + \ln c_1 \implies u = c_1(y + 1)$$

$$\implies \frac{dy}{dx} = c_1(y + 1) \implies \frac{dy}{y + 1} = c_1\, dx$$

$$\implies \ln|y + 1| = c_1 x + c_2 \implies y + 1 = c_3 e^{c_1 x}.$$

9. (a)

(b) Let $u = y'$ so that $y'' = u\, du/dy$. The equation becomes $u\, du/dy + yu = 0$. Separating variables we obtain

$$du = -y\, dy \implies u = -\frac{1}{2}y^2 + c_1 \implies y' = -\frac{1}{2}y^2 + c_1.$$

When $x = 0$, $y = 1$ and $y' = -1$ so $-1 = -1/2 + c_1$ and $c_1 = -1/2$. Then

$$\frac{dy}{dx} = -\frac{1}{2}y^2 - \frac{1}{2} \implies \frac{dy}{y^2 + 1} = -\frac{1}{2}\, dx \implies \tan^{-1} y = -\frac{1}{2}x + c_2$$

$$\implies y = \tan\left(-\frac{1}{2}x + c_2\right).$$

When $x = 0$, $y = 1$ so $1 = \tan c_2$ and $c_2 = \pi/4$. The solution of the initial-value problem is

$$y = \tan\left(\frac{\pi}{4} - \frac{1}{2}x\right).$$

The graph is shown in part (a).

(c) The interval of definition is $-\pi/2 < \pi/4 - x/2 < \pi/2$ or $-\pi/2 < x < 3\pi/2$.

12. Let $u = y'$ so that $u' = y''$. The equation becomes $u' - (1/x)u = u^2$, which is a Bernoulli DE. Using the substitution $w = u^{-1}$ we obtain $dw/dx + (1/x)w = -1$. An integrating factor is x, so

$$\frac{d}{dx}[xw] = -x \implies w = -\frac{1}{2}x + \frac{1}{x}c \implies \frac{1}{u} = \frac{c_1 - x^2}{2x} \implies u = \frac{2x}{c_1 - x^2} \implies y = -\ln\left|c_1 - x^2\right| + c_2.$$

15. We look for a solution of the form

$$y(x) = y(0) + y'(0) + \frac{1}{2!}y''(0) + \frac{1}{3!}y'''(0) + \frac{1}{4!}y^{(4)}(x) + \frac{1}{5!}y^{(5)}(x).$$

From $y''(x) = x^2 + y^2 - 2y'$ we compute

$$y'''(x) = 2x + 2yy' - 2y''$$

$$y^{(4)}(x) = 2 + 2(y')^2 + 2yy'' - 2y'''$$

$$y^{(5)}(x) = 6y'y'' + 2yy''' - 2y^{(4)}.$$

Using $y(0) = 1$ and $y'(0) = 1$ we find

$$y''(0) = -1, \quad y'''(0) = 4, \quad y^{(4)}(0) = -6, \quad y^{(5)}(0) = 14.$$

An approximate solution is

$$y(x) = 1 + x - \frac{1}{2}x^2 + \frac{2}{3}x^3 - \frac{1}{4}x^4 + \frac{7}{60}x^5.$$

In this problem the thinner curve is obtained using a numerical solver, while the thicker curve is the graph of the Taylor polynomial.

21. (*Suggestion*) Use a CAS to evaluate

$$t = -\frac{1}{k}\sqrt{\frac{x_0}{2}} \int \sqrt{\frac{x}{x_0 - x}}\, dx.$$

Chapter 4 in Review

Hints, Suggestions, Solutions, and Examples

3. It is not true unless the DE is homogeneous. For example, $y_1 = x$ is a solution of $y'' + y = x$, but $y_2 = 5x$ is not.

6. (a) Since $f_2(x) = 2\ln x = 2f_1(x)$, the set of functions is linearly dependent.

(b) Since x^{n+1} is not a constant multiple of x^n, the set of functions is linearly independent.

(c) Since $x + 1$ is not a constant multiple of x, the set of functions is linearly independent.

(d) Since $f_1(x) = \cos x \cos(\pi/2) - \sin x \sin(\pi/2) = -\sin x = -f_2(x)$, the set of functions is linearly dependent.

(e) Since $f_1(x) = 0 \cdot f_2(x)$, the set of functions is linearly dependent.

(f) Since $2x$ is not a constant multiple of 2, the set of functions is linearly independent.

(g) Since $3(x^2) + 2(1 - x^2) - (2 + x^2) = 0$, the set of functions is linearly dependent.

(h) Since $xe^{x+1} + 0(4x - 5)e^x - exe^x = 0$, the set of functions is linearly dependent.

9. From $m^2 - 2m - 2 = 0$ we obtain $m = 1 \pm \sqrt{3}$ so that

$$y = c_1 e^{(1+\sqrt{3})x} + c_2 e^{(1-\sqrt{3})x}.$$

12. From $2m^3 + 9m^2 + 12m + 5 = 0$ we obtain $m = -1$, $m = -1$, and $m = -5/2$ so that

$$y = c_1 e^{-5x/2} + c_2 e^{-x} + c_3 x e^{-x}.$$

15. Applying D^4 to the DE we obtain $D^4(D^2 - 3D + 5) = 0$. Then

$$y = \underbrace{e^{3x/2}\left(c_1 \cos \frac{\sqrt{11}}{2}x + c_2 \sin \frac{\sqrt{11}}{2}x\right)}_{y_c} + c_3 + c_4 x + c_5 x^2 + c_6 x^3$$

and $y_p = A + Bx + Cx^2 + Dx^3$. Substituting y_p into the DE yields

$$(5A - 3B + 2C) + (5B - 6C + 6D)x + (5C - 9D)x^2 + 5Dx^3 = -2x + 4x^3.$$

Equating coefficients gives $A = -222/625$, $B = 46/125$, $C = 36/25$, and $D = 4/5$. The general solution is

$$y = e^{3x/2}\left(c_1 \cos \frac{\sqrt{11}}{2}x + c_2 \sin \frac{\sqrt{11}}{2}x\right) - \frac{222}{625} + \frac{46}{125}x + \frac{36}{25}x^2 + \frac{4}{5}x^3.$$

18. Applying D to the DE we obtain $D(D^3 - D^2) = D^3(D - 1) = 0$. Then

$$y = \underbrace{c_1 + c_2 x + c_3 e^x}_{y_c} + c_4 x^2$$

and $y_p = Ax^2$. Substituting y_p into the DE yields $-2A = 6$. Equating coefficients gives $A = -3$. The general solution is

$$y = c_1 + c_2 x + c_3 e^x - 3x^2.$$

21. The auxiliary equation is $6m^2 - m - 1 = 0$ so that

$$y = c_1 x^{1/2} + c_2 x^{-1/3}.$$

24. The auxiliary equation is $m^2 - 2m + 1 = (m - 1)^2 = 0$ and a particular solution is $y_p = \frac{1}{4}x^3$ so that

$$y = c_1 x + c_2 x \ln x + \frac{1}{4}x^3.$$

27. (a) The auxiliary equation is $m^4 - 2m^2 + 1 = (m^2 - 1)^2 = 0$, so the general solution of the DE is

$$y = c_1 \sinh x + c_2 \cosh x + c_3 x \sinh x + c_4 x \cosh x.$$

(b) Since both $\sinh x$ and $x \sinh x$ are solutions of the associated homogeneous DE, a particular solution of $y^{(4)} - 2y'' + y = \sinh x$ has the form $y_p = Ax^2 \sinh x + Bx^2 \cosh x$.

Chapter 4 in Review

30. The auxiliary equation is $m^2 + 2m + 1 = (m+1)^2 = 0$, so that $y = c_1 e^{-x} + c_2 x e^{-x}$. Setting $y(-1) = 0$ and $y'(0) = 0$ we get $c_1 e - c_2 e = 0$ and $-c_1 + c_2 = 0$. Thus $c_1 = c_2$ and $y = c_1(e^{-x} + xe^{-x})$ is a solution of the boundary-value problem for any real number c.

33. Let $u = y'$ so that $u' = y''$ and the DE becomes $u \, du/dy = 4x$. Separating variables and integrating, we have

$$u \, du = 4x \, dx$$

$$\frac{1}{2} u^2 = 2x^2 + c_1$$

$$u^2 = 4x^2 + c_2.$$

From the initial conditions, when $x = 1$, $y' = u = 2$, so $4 = 4 + c_2$ and $c_2 = 0$. Then $u^2 = 4x^2$ or $(dy/dx)^2 = 4x^2$, so $dy/dx = \pm 2x$. Integrating, we have $y = x^2 + c_3$ or $y = -x^2 + c_4$. Now, when $x = 1$, $y = 5$, so $5 = 1 + c_3$ or $5 = -1 + c_4$. Thus $c_3 = 4$ and $c_4 = 6$. We have $y = x^2 + 4$ and $y = -x^2 + 6$. Note, however, that when $y = -x^2 + 6$, $y' = -2x$ and $y'(1) = -2 \neq 2$. Thus, the solution of the IVP is $y = x^2 + 4$.

36. Consider $xy'' + y' = 0$ and look for a solution of the form $y = x^m$. Substituting into the DE we have

$$xy'' + y' = m(m-1)x^{m-1} + mx^{m-1} = m^2 x^{m-1}.$$

Thus, the general solution of $xy'' + y' = 0$ is $y_c = c_1 + c_2 \ln x$. To find a particular solution of $xy'' + y' = -\sqrt{x}$ we use variation of parameters. The Wronskian is

$$W = \begin{vmatrix} 1 & \ln x \\ 0 & 1/x \end{vmatrix} = \frac{1}{x}.$$

Identifying $f(x) = -x^{-1/2}$ we obtain

$$u_1' = \frac{x^{-1/2} \ln x}{1/x} = \sqrt{x} \ln x \quad \text{and} \quad u_2' = \frac{-x^{-1/2}}{1/x} = -\sqrt{x},$$

so that

$$u_1 = x^{3/2}\left(\frac{2}{3} \ln x - \frac{4}{9}\right) \quad \text{and} \quad u_2 = -\frac{2}{3} x^{3/2}.$$

Then

$$y_p = x^{3/2}\left(\frac{2}{3} \ln x - \frac{4}{9}\right) - \frac{2}{3} x^{3/2} \ln x = -\frac{4}{9} x^{3/2}$$

and the general solution of the DE is

$$y = c_1 + c_2 \ln x - \frac{4}{9} x^{3/2}.$$

The initial conditions are $y(1) = 0$ and $y'(1) = 0$. These imply that $c_1 = \frac{4}{9}$ and $c_2 = \frac{2}{3}$. The solution of the IVP is

$$y = \frac{4}{9} + \frac{2}{3} \ln x - \frac{4}{9} x^{3/2}.$$

39. From $(D-2)x - y = -e^t$ and $-3x + (D-4)y = -7e^t$ we obtain $(D-1)(D-5)x = -4e^t$ so that

$$x = c_1 e^t + c_2 e^{5t} + t e^t.$$

Then

$$y = (D-2)x + e^t = -c_1 e^t + 3 c_2 e^{5t} - t e^t + 2 e^t.$$

5

Modeling with Higher-Order Differential Equations

Section 5.1
Linear Models: Initial-Value Problems

The terminology and concepts listed below provide an outline of the main ideas encountered in this section. These can be useful when preparing for a quiz or test.

Terminology and Concepts

- Hooke's law

- spring constant (k)

- Newton's second law of motion

- mass (m) versus weight $(W = mg)$

- equilibrium position

- DE of free undamped motion: $m\, d^2x/dt^2 + kx = 0$ or $d^2x/dt^2 + \omega^2 x = 0$, where $\omega^2 = k/m$

- equation of free undamped (simple harmonic) motion: $x(t) = c_1 \cos \omega t + c_2 \sin \omega t$

- period of motion: $T = 2\pi/\omega$

- cycle

- frequency of motion: $f = \omega/2\pi$

- amplitude of free vibrations (A)

- phase angle (ϕ)

- damping constant (β)

- DE of free damped motion: $m\, d^2x/dt^2 + \beta\, dx/dt + kx = 0$ or $d^2x/dt^2 + 2\lambda\, dx/dt + \omega^2 x = 0$
  where $2\lambda = \beta/m$ and $\omega^2 = k/m$

- equation of motion of an overdamped system: $x(t) = e^{-\lambda t}(c_1 e^{\sqrt{\lambda^2 - \omega^2}\, t} + c_2 e^{-\sqrt{\lambda^2 - \omega^2}\, t})$

- equation of motion of a critically damped system: $x(t) = e^{-\lambda t}(c_1 + c_2 t)$

106

- equation of motion of an underdamped system:

$$x(t) = e^{-\lambda t}(c_1 \cos \sqrt{\omega^2 - \lambda^2}\, t + c_2 \sin \sqrt{\omega^2 - \lambda^2}\, t)$$

- damped amplitude: $Ae^{-\lambda t}$
- quasi period: $2\pi/\sqrt{\omega^2 - \lambda^2}$
- DE of driven (forced) motion with damping: $m\, d^2x/dt^2 + \beta\, dx/dt + kx = f(t)$
- transient term (solution)
- steady-state term (solution)
- DE of driven motion without damping: $m\, d^2x/dt^2 + kx = f(t)$
- pure resonance
- aging spring
- Airy's differential equation
- LRC series electrical circuit
- free electrical vibrations
- reactance
- impedance

The basic skills listed below summarize the more mechanical types of problems encountered in the exercise set for this section.

Basic Skills

- set up and solve DEs modeling free undamped motion of a spring/mass system
- set up and solve DEs modeling free damped motion of a spring/mass system
- set up and solve DEs modeling driven (forced) motion of a spring/mass system
- set up and solve DEs modeling an LRC series circuit

Phase Angle   As pointed out in the text, be careful in computing the phase angle ϕ in a solution of the form $x(t) = A\sin(\omega t + \phi)$. The problem with using the inverse tangent $\tan^{-1} X$ is that all calculators are programmed for this particular inverse trigonometric function to give angular values that are either in the first or fourth quadrants. In other words, $-\pi/2 < \tan^{-1} X < \pi/2$. For example, because of periodicity there are infinitely many angles for which $\tan \phi = 1$ and for which $\tan \phi = -1/\sqrt{3}$. But there is only *one* answer when we use the notation $\tan^{-1} 1$ and $\tan^{-1}(-1/\sqrt{3})$. The exact values are:

$$\tan^{-1} 1 = \frac{\pi}{4} \quad \text{(first quadrant) because } \tan \frac{\pi}{4} = 1,$$

107

Section 5.1 Linear Models: Initial-Value Problems

$$\tan^{-1}\left(-\frac{1}{\sqrt{3}}\right) = -\frac{\pi}{6} \quad \text{(fourth quadrant) because } \tan\left(-\frac{\pi}{6}\right) = -\frac{1}{\sqrt{3}}.$$

Of course, if you use a calculator you will either get

$$\tan^{-1} 1 = 45° \quad \text{or} \quad \tan^{-1} 1 = 0.78539816\ldots$$

depending on whether your calculator is set in degree or radian mode. The decimal number $0.78539816\ldots$ is the number π divided by 4. Now in order to compute ϕ in $x(t) = A\sin(\omega t + \phi)$ we must take into consideration all three components given in (7) of Section 5.1 in the text. Since $A > 0$, the signs of $\sin\phi = c_1/A$ and $\cos\phi = c_2/A$ are determined from the algebraic signs of c_1 and c_2 and these signs, in turn, indicate the quadrant in which ϕ lies. We summarize the four possibilities:

$$\phi = \tan^{-1}\left(\frac{c_1}{c_2}\right) \quad \text{if} \quad \overbrace{c_1 > 0, \ c_2 > 0}^{\phi \text{ in 1st quadrant}} \quad \text{or if} \quad \overbrace{c_1 < 0, \ c_2 > 0}^{\phi \text{ in 4th quadrant}}$$

$$\phi = \pi + \tan^{-1}\left(\frac{c_1}{c_2}\right) \quad \text{if} \quad \overbrace{c_1 > 0, \ c_2 < 0}^{\phi \text{ in 2nd quadrant}} \quad \text{or if} \quad \overbrace{c_1 < 0, \ c_2 < 0}^{\phi \text{ in 3rd quadrant}}.$$

Damping In Examples 4 and 5 in the text, the statements

"...a damping force numerically equal to 2 times the instantaneous velocity..."

and

"...surrounding medium offers a resistance numerically equal to the instantaneous velocity..."

are to be interpreted: In Example 4, the damping constant is $\beta = 2$ in equation (10) and in Example 5, the damping constant is $\beta = 1$ in equation (10).

Exercises 5.1   *Hints, Suggestions, Solutions, and Examples*

3. If the mass weighs 24 pounds then $24 = m(32)$ and $m = \frac{24}{32} = \frac{3}{4}$. To find k we use $F = ks$ where, in this case, $F = 24$ and $s = 4$ in. $= \frac{4}{12} = \frac{1}{3}$ ft. Thus, $k = 24/(1/3) = 72$ and the DE is $\frac{3}{4}x'' + 72x = 0$ or $x'' + 96x = 0$. Since the mass is initially 3 inches ($\frac{3}{12} = \frac{1}{4}$ feet) *above* the equilibrium position, $x(0) = -\frac{1}{4}$. Since the mass is released from *rest*, $x'(0) = 0$. Identifying $\omega = \sqrt{96} = 4\sqrt{6}$, we have $x(t) = c_1\cos 4\sqrt{6}\,t + c_2\sin 4\sqrt{6}\,t$. The initial conditions give $x(0) = c_1 = -\frac{1}{4}$ and $x'(0) = 4\sqrt{6} = c_2 = 0$. The equation of motion of the spring/mass system is $x(t) = -\frac{1}{4}\cos 4\sqrt{6}\,t$.

6. To find k we identify $F = 400$ and $s = 2$ and solve $400 = k(2)$, so $k = 200$. The DE is then $50x'' + 200x = 0$ or $x'' + 4x = 0$. Since the spring is released from the equilibrium position, $x(0) = 0$. Since it starts with an *upward* velocity of 10 m/s, $x'(0) = -10$. The solution of the DE

108

is $x(t) = c_1 \cos 2t + c_2 \sin 2t$. From $x(0) = c_1 = 0$ and $x'(0) = 2c_2 = -10$ we see that the equation of motion is $x(t) = -5 \sin 2t$.

9. If the mass weighs 8 pounds then $8 = m(32)$ and $m = \frac{8}{32} = \frac{1}{4}$. The spring constant is $k = 1$ so the DE is $\frac{1}{4}x'' + x = 0$ or $x'' + 4x = 0$ and the equation of motion is $x(t) = c_1 \cos 2t + c_2 \sin 2t$. Since the mass is initially 6 inches ($\frac{6}{12} = \frac{1}{2}$ ft) *below* the equilibrium position $x(0) = c_1 = \frac{1}{2}$. Since the mass is given an initial *downward* velocity of $\frac{3}{2}$ ft/s, $x'(0) = 2c_2 = \frac{3}{2}$ and $c_2 = \frac{3}{4}$. Thus $x(t) = \frac{1}{2} \cos 2t + \frac{3}{4} \sin 2t$. To put this in form of equation (6) in the text we first note that $\omega = \sqrt{4} = 2$ and $A = \sqrt{\frac{1}{4} + \frac{9}{16}} = \sqrt{13}/4$. Next, after reading the material on **Phase Angle** above in this manual, we note that c_1 and c_2 are both positive, so $\phi = \tan^{-1}(c_1/c_2) = \tan^{-1} \frac{2}{3} \approx 0.588$. Thus, $x(t) = \frac{1}{4}\sqrt{13} \sin(2t + 0.588)$.

10-11. (*Suggestion*) Read the material above in this section on **Phase Angle**.

12. From $x'' + 9x = 0$, $x(0) = -1$, and $x'(0) = -\sqrt{3}$ we obtain

$$x = -\cos 3t - \frac{\sqrt{3}}{3}\sin 3t = \frac{2}{\sqrt{3}}\sin\left(3t + \frac{4\pi}{3}\right)$$

and $x' = 2\sqrt{3}\cos(3t + 4\pi/3)$. If $x' = 3$ then $t = -7\pi/18 + 2n\pi/3$ and $t = -\pi/2 + 2n\pi/3$ for $n = 1, 2, 3, \ldots$.

15. For large values of t, the DE is approximated by $x'' = 0$. The solution of this equation is the linear function $x = c_1 t + c_2$. Thus, for large time, the restoring force will have decayed to the point where the spring is incapable of returning the mass, and the spring will simply keep on stretching.

18. (a) Since $x(0) > 0$, the initial displacement is below the equilibrium position.

    (b) Since $x'(0) = 0$, the mass is initially released from rest.

21. Since the mass weighs 4 pounds, $4 = m(32)$ and $m = \frac{1}{8}$. Since the damping force is dx/dt, $\beta = 1$ and the DE is $\frac{1}{8}d^2x/dt^2 + dx/dt + 2x = 0$ or $d^2x/dt^2 + 2(4)dx/dt + 16x = 0$. Identifying $\lambda = 4$ and $\omega = 4$ we see that $\lambda^2 - \omega^2 = 0$ so that the equation of motion is $x = e^{-4t}(c_1 + c_2 t)$. The initial conditions are $x(0) = -1$ and $x'(0) = 8$, from which we determine that $c_1 = -1$ and $c_2 = 4$. Thus, $x(t) = -e^{-4t} + 4te^{-4t}$. Solving $x(0) = -1$, we find $t = \frac{1}{4}$ second. Now $x'(t) = 8e^{-4t} - 16te^{-4t}$, so when $x'(0) = 0$ we have $t = \frac{1}{2}$ as the time when the mass attains its extreme displacement from the equilibrium position. The position at this time is $x(\frac{1}{2}) = 1/e^2 \approx 0.135$ ft.

24. (a) In Problem 23(a), $x(t) = \frac{4}{3}e^{-2t} - \frac{1}{3}e^{-8t} = \frac{1}{3}e^{-8t}(4e^{6t} - 1)$, which is never 0 for $t \geq 0$. Since $x'(t) < 0$ for $t > 0$, the maximum displacement from the equilibrium position occurs at $t = 0$. The maximum displacement is $x(0) = \frac{4}{3} - \frac{1}{3} = 1$ meter.

    (b) In Problem 23(b), $x(t) = -\frac{2}{3}e^{-2t} + \frac{5}{3}e^{-8t} = \frac{1}{3}e^{-8t}(5 - 2e^{6t})$, which is 0 when $t = \frac{1}{6}\ln\frac{5}{2} \approx 0.153$ second. Solving $x'(t) = \frac{4}{3}e^{-8t}(e^{6t} - 10) = 0$, we see that the mass attains its maximum displacement when $t = \frac{1}{6}\ln 10 \approx 0.384$ second. The maximum displacement is $x(\frac{1}{6}\ln 10) =$

109

$-1/210^{1/3} \approx -0.232$ meter. Since this is negative, the maximum displacement is above the equilibrium position.

27. From $\frac{5}{16}x''+\beta x'+5x = 0$ we find that the roots of the auxiliary equation are $m = -\frac{8}{5}\beta \pm \frac{4}{5}\sqrt{4\beta^2 - 25}$.

    (a) If $4\beta^2 - 25 > 0$ then $\beta > 5/2$.
    (b) If $4\beta^2 - 25 = 0$ then $\beta = 5/2$.
    (c) If $4\beta^2 - 25 < 0$ then $0 < \beta < 5/2$.

30. (a) If $x'' + 2x' + 5x = 12\cos 2t + 3\sin 2t$, $x(0) = 1$, and $x'(0) = 5$ then $x_c = e^{-t}(c_1\cos 2t + c_2\sin 2t)$ and $x_p = 3\sin 2t$ so that the equation of motion is

$$x = e^{-t}\cos 2t + 3\sin 2t.$$

(b)

(c)

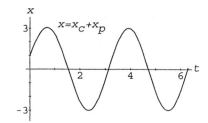

33. From $2x'' + 32x = 68e^{-2t}\cos 4t$, $x(0) = 0$, and $x'(0) = 0$ we obtain $x_c = c_1\cos 4t + c_2\sin 4t$ and $x_p = \frac{1}{2}e^{-2t}\cos 4t - 2e^{-2t}\sin 4t$ so that

$$x = -\frac{1}{2}\cos 4t + \frac{9}{4}\sin 4t + \frac{1}{2}e^{-2t}\cos 4t - 2e^{-2t}\sin 4t.$$

36. (a) From $100x'' + 1600x = 1600\sin 8t$, $x(0) = 0$, and $x'(0) = 0$ we obtain $x_c = c_1\cos 4t + c_2\sin 4t$ and $x_p = -\frac{1}{3}\sin 8t$ so that by a trig identity

$$x = \frac{2}{3}\sin 4t - \frac{1}{3}\sin 8t = \frac{2}{3}\sin 4t - \frac{2}{3}\sin 4t\cos 4t.$$

(b) If $x = \frac{1}{3}\sin 4t(2 - 2\cos 4t) = 0$ then $t = n\pi/4$ for $n = 0, 1, 2, \ldots$.

(c) If $x' = \frac{8}{3}\cos 4t - \frac{8}{3}\cos 8t = \frac{8}{3}(1-\cos 4t)(1+2\cos 4t) = 0$ then $t = \pi/3 + n\pi/2$ and $t = \pi/6 + n\pi/2$ for $n = 0, 1, 2, \ldots$ at the extreme values. *Note:* There are many other values of t for which $x' = 0$.

(d) $x(\pi/6 + n\pi/2) = \sqrt{3}/2$ cm and $x(\pi/3 + n\pi/2) = -\sqrt{3}/2$ cm

(e)

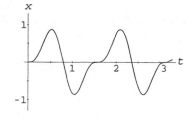

110

39. (a) From $x'' + \omega^2 x = F_0 \cos \gamma t$, $x(0) = 0$, and $x'(0) = 0$ we obtain $x_c = c_1 \cos \omega t + c_2 \sin \omega t$ and $x_p = (F_0 \cos \gamma t)/(\omega^2 - \gamma^2)$ so that

$$x = -\frac{F_0}{\omega^2 - \gamma^2} \cos \omega t + \frac{F_0}{\omega^2 - \gamma^2} \cos \gamma t.$$

(b) Using L'Hôpital's rule, we have

$$\lim_{\gamma \to \omega} \frac{F_0}{\omega^2 - \gamma^2} (\cos \gamma t - \cos \omega t) = \lim_{\gamma \to \omega} \frac{-F_0 t \sin \gamma t}{-2\gamma} = \frac{F_0}{2\omega} t \sin \omega t.$$

45. Solving $\frac{1}{20}q'' + 2q' + 100q = 0$ we obtain $q(t) = e^{-20t}(c_1 \cos 40t + c_2 \sin 40t)$. The initial conditions $q(0) = 5$ and $q'(0) = 0$ imply $c_1 = 5$ and $c_2 = 5/2$. Thus

$$q(t) = e^{-20t}\left(5 \cos 40t + \frac{5}{2} \sin 40t\right) = \sqrt{25 + 25/4}\, e^{-20t} \sin(40t + 1.1071)$$

and $q(0.01) \approx 4.5676$ coulombs. The charge is zero for the first time when $40t + 1.1071 = \pi$ or $t \approx 0.0509$ second.

48. Solving $q'' + 100q' + 2500q = 30$ we obtain $q(t) = c_1 e^{-50t} + c_2 t e^{-50t} + 0.012$. The initial conditions $q(0) = 0$ and $q'(0) = 2$ imply $c_1 = -0.012$ and $c_2 = 1.4$. Thus, using $i(t) = q'(t)$ we get

$$q(t) = -0.012 e^{-50t} + 1.4t e^{-50t} + 0.012 \quad \text{and} \quad i(t) = 2e^{-50t} - 70t e^{-50t}.$$

Solving $i(t) = 0$ we see that the maximum charge occurs when $t = 1/35$ second and $q(1/35) \approx 0.01871$ coulomb.

51. The DE is $\frac{1}{2}q'' + 20q' + 1000q = 100 \sin t$. To use Example 10 in the text we identify $E_0 = 100$ and $\gamma = 60$. Then

$$X = L\gamma - \frac{1}{c\gamma} = \frac{1}{2}(60) - \frac{1}{0.001(60)} \approx 13.3333,$$

$$Z = \sqrt{X^2 + R^2} = \sqrt{X^2 + 400} \approx 24.0370,$$

and

$$\frac{E_0}{Z} = \frac{100}{Z} \approx 4.1603.$$

In Problem 50 it is shown that

$$i_p(t) = \frac{E_0}{Z}\left(\frac{R}{Z} \sin \gamma t - \frac{X}{Z} \cos \gamma t\right) = 4.1603 \sin(60t + \phi),$$

where $\sin \phi = -X/Z$ and $\cos \phi = R/Z$. Thus $\tan \phi = -X/R \approx -0.6667$ and ϕ is a fourth quadrant angle. Now $\phi \approx -0.5880$ and

$$i_p(t) = 4.1603 \sin(60t - 0.5880).$$

54. In Problem 50 it is shown that the amplitude of the steady-state current is E_0/Z, where $Z = \sqrt{X^2 + R^2}$ and $X = L\gamma - 1/C\gamma$. Since E_0 is constant the amplitude will be a maximum when

Z is a minimum. Since R is constant, Z will be a minimum when $X = 0$. Solving $L\gamma - 1/C\gamma = 0$ for γ we obtain $\gamma = 1/\sqrt{LC}$. The maximum amplitude will be E_0/R.

57. In an LC series circuit there is no resistor, so the DE is

$$L\frac{d^2q}{dt^2} + \frac{1}{C}q = E(t).$$

Then $q(t) = c_1 \cos\left(t/\sqrt{LC}\right) + c_2 \sin\left(t/\sqrt{LC}\right) + q_p(t)$ where $q_p(t) = A\sin\gamma t + B\cos\gamma t$. Substituting $q_p(t)$ into the DE we find

$$\left(\frac{1}{C} - L\gamma^2\right)A\sin\gamma t + \left(\frac{1}{C} - L\gamma^2\right)B\cos\gamma t = E_0\cos\gamma t.$$

Equating coefficients we obtain $A = 0$ and $B = E_0C/(1 - LC\gamma^2)$. Thus, the charge is

$$q(t) = c_1\cos\frac{1}{\sqrt{LC}}t + c_2\sin\frac{1}{\sqrt{LC}}t + \frac{E_0C}{1 - LC\gamma^2}\cos\gamma t.$$

The initial conditions $q(0) = q_0$ and $q'(0) = i_0$ imply $c_1 = q_0 - E_0C/(1 - LC\gamma^2)$ and $c_2 = i_0\sqrt{LC}$. The current is $i(t) = q'(t)$ or

$$i(t) = -\frac{c_1}{\sqrt{LC}}\sin\frac{1}{\sqrt{LC}}t + \frac{c_2}{\sqrt{LC}}\cos\frac{1}{\sqrt{LC}}t - \frac{E_0C\gamma}{1 - LC\gamma^2}\sin\gamma t$$

$$= i_0\cos\frac{1}{\sqrt{LC}}t - \frac{1}{\sqrt{LC}}\left(q_0 - \frac{E_0C}{1 - LC\gamma^2}\right)\sin\frac{1}{\sqrt{LC}}t - \frac{E_0C\gamma}{1 - LC\gamma^2}\sin\gamma t.$$

Section 5.2 Linear Models: Boundary-Value Problems

The terminology and concepts listed below provide an outline of the main ideas encountered in this section. These can be useful when preparing for a quiz or test.

Terminology and Concepts

- axis of symmetry of a beam
- deflection curve or elastic curve of a beam
- cantilever beam
- eigenvalues and corresponding eigenfunctions of a boundary-value problem

The basic skills listed below summarize the more mechanical types of problems encountered in the exercise set for this section.

Basic Skills

- solve problems involving beams satisfying various end (boundary) conditions such as embedded, free, or simply supported

- find the eigenvalues and corresponding eigenfunctions for second-order boundary-value problems

Exercises 5.2   *Hints, Suggestions, Solutions, and Examples*

3. (a) The general solution is

$$y(x) = c_1 + c_2 x + c_3 x^2 + c_4 x^3 + \frac{w_0}{24EI} x^4.$$

The boundary conditions are $y(0) = 0$, $y'(0) = 0$, $y(L) = 0$, $y''(L) = 0$. The first two conditions give $c_1 = 0$ and $c_2 = 0$. The conditions at $x = L$ give the system

$$c_3 L^2 + c_4 L^3 + \frac{w_0}{24EI} L^4 = 0$$

$$2c_3 + 6c_4 L + \frac{w_0}{2EI} L^2 = 0.$$

Solving, we obtain $c_3 = w_0 L^2 / 16EI$ and $c_4 = -5w_0 L / 48EI$. The deflection is

$$y(x) = \frac{w_0}{48EI} (3L^2 x^2 - 5Lx^3 + 2x^4).$$

(b)

6. (a) $y_{\max} = y(L) = w_0 L^4 / 8EI$

(b) Replacing both L and x by $L/2$ in $y(x)$ we obtain $w_0 L^4 / 128EI$, which is $1/16$ of the maximum deflection when the length of the beam is L.

(c) $y_{\max} = y(L/2) = 5w_0 L^4 / 384EI$

(d) The maximum deflection in Example 1 is $y(L/2) = (w_0/24EI)L^4/16 = w_0L^4/384EI$, which is $1/5$ of the maximum displacement of the beam in Problem 2.

9. This is Example 2 in the text with $L = \pi$. The eigenvalues are $\lambda_n = n^2\pi^2/\pi^2 = n^2$, $n = 1, 2, 3, \ldots$ and the corresponding eigenfunctions are $y_n = \sin(n\pi x/\pi) = \sin nx$, $n = 1, 2, 3, \ldots$.

12. For $\lambda \leq 0$ the only solution of the boundary-value problem is $y = 0$. For $\lambda = \alpha^2 > 0$ we have

$$y = c_1 \cos \alpha x + c_2 \sin \alpha x.$$

Since $y(0) = 0$ implies $c_1 = 0$, $y = c_2 \sin x \, dx$. Now

$$y'\left(\frac{\pi}{2}\right) = c_2\alpha \cos \alpha \frac{\pi}{2} = 0$$

gives

$$\alpha\frac{\pi}{2} = \frac{(2n-1)\pi}{2} \quad \text{or} \quad \lambda = \alpha^2 = (2n-1)^2, \ n = 1, 2, 3, \ldots.$$

The eigenvalues $\lambda_n = (2n-1)^2$ correspond to the eigenfunctions $y_n = \sin(2n-1)x$.

15. The auxiliary equation has solutions

$$m = \frac{1}{2}\left(-2 \pm \sqrt{4 - 4(\lambda+1)}\right) = -1 \pm \alpha.$$

For $\lambda = -\alpha^2 < 0$ we have

$$y = e^{-x}\left(c_1 \cosh \alpha x + c_2 \sinh \alpha x\right).$$

The boundary conditions imply

$$y(0) = c_1 = 0$$

$$y(5) = c_2 e^{-5} \sinh 5\alpha = 0$$

so $c_1 = c_2 = 0$ and the only solution of the BVP is $y = 0$.

For $\lambda = 0$ we have

$$y = c_1 e^{-x} + c_2 x e^{-x}$$

and the only solution of the BVP is $y = 0$.

For $\lambda = \alpha^2 > 0$ we have

$$y = e^{-x}\left(c_1 \cos \alpha x + c_2 \sin \alpha x\right).$$

Now $y(0) = 0$ implies $c_1 = 0$, so

$$y(5) = c_2 e^{-5} \sin 5\alpha = 0$$

gives

$$5\alpha = n\pi \quad \text{or} \quad \lambda = \alpha^2 = \frac{n^2\pi^2}{25}, \ n = 1, 2, 3, \ldots.$$

The eigenvalues $\lambda_n = \dfrac{n^2\pi^2}{25}$ correspond to the eigenfunctions $y_n = e^{-x} \sin \dfrac{n\pi}{5}x$ for $n = 1, 2, 3, \ldots$.

16. (*Suggestion*) Let $\lambda + 1 = 0$, $\lambda + 1 = -\alpha^2$, and $\lambda + 1 = \alpha^2$.

18. For $\lambda = 0$ the general solution is $y = c_1 + c_2 \ln x$. Now $y' = c_2/x$, so $y'(e^{-1}) = c_2 e = 0$ implies $c_2 = 0$. Then $y = c_1$ and $y(1) = 0$ gives $c_1 = 0$. Thus $y(x) = 0$.

For $\lambda = -\alpha^2 < 0$, $y = c_1 x^{-\alpha} + c_2 x^{\alpha}$. The initial conditions give $c_2 = c_1 e^{2\alpha}$ and $c_1 = 0$, so that $c_2 = 0$ and $y(x) = 0$.

For $\lambda = \alpha^2 > 0$, $y = c_1 \cos(\alpha \ln x) + c_2 \sin(\alpha \ln x)$. From $y(1) = 0$ we obtain $c_1 = 0$ and $y = c_2 \sin(\alpha \ln x)$. Now $y' = c_2(\alpha/x) \cos(\alpha \ln x)$, so $y'(e^{-1}) = c_2 e \alpha \cos \alpha = 0$ implies $\cos \alpha = 0$ or $\alpha = (2n-1)\pi/2$ and $\lambda = \alpha^2 = (2n-1)^2 \pi^2/4$ for $n = 1, 2, 3, \ldots$. The corresponding eigenfunctions are

$$y_n = \sin\left(\frac{2n-1}{2} \pi \ln x\right).$$

21. If restraints are put on the column at $x = L/4$, $x = L/2$, and $x = 3L/4$, then the critical load will be P_4.

24. The boundary-value problem is

$$\frac{d^4 y}{dx^4} + \lambda \frac{d^2 y}{dx^2} = 0, \quad y(0) = 0, y''(0) = 0, \ y(L) = 0, y'(L) = 0,$$

where $\lambda = \alpha^2 = P/EI$. The solution of the DE is $y = c_1 \cos \alpha x + c_2 \sin \alpha x + c_3 x + c_4$ and the conditions $y(0) = 0$, $y''(0) = 0$ yield $c_1 = 0$ and $c_4 = 0$. Next, by applying $y(L) = 0$, $y'(L) = 0$ to $y = c_2 \sin \alpha x + c_3 x$ we get the system of equations

$$c_2 \sin \alpha L + c_3 L = 0$$

$$\alpha c_2 \cos \alpha L + c_3 = 0.$$

To obtain nontrivial solutions c_2, c_3, we must have the determinant of the coefficients equal to zero:

$$\begin{vmatrix} \sin \alpha L & L \\ \alpha \cos \alpha L & 1 \end{vmatrix} = 0 \quad \text{or} \quad \tan \beta = \beta,$$

where $\beta = \alpha L$. If β_n denotes the positive roots of the last equation, then the eigenvalues are found from $\beta_n = \alpha_n L = \sqrt{\lambda_n} L$ or $\lambda_n = (\beta_n/L)^2$. From $\lambda = P/EI$ we see that the critical loads are $P_n = \beta_n^2 EI/L^2$. With the aid of a CAS we find that the first positive root of $\tan \beta = \beta$

is (approximately) $\beta_1 = 4.4934$, and so the Euler load is (approximately) $P_1 = 20.1907EI/L^2$. Finally, if we use $c_3 = -c_2\alpha\cos\alpha L$, then the deflection curves are

$$y_n(x) = c_2\sin\alpha_n x + c_3 x = c_2\left[\sin\left(\frac{\beta_n}{L}x\right) - \left(\frac{\beta_n}{L}\cos\beta_n\right)x\right].$$

With $L = 1$ and c_2 appropriately chosen, the general shape of the first buckling mode,

$$y_1(x) = c_2\left[\sin\left(\frac{4.4934}{L}x\right) - \left(\frac{4.4934}{L}\cos(4.4934)\right)x\right],$$

is shown below.

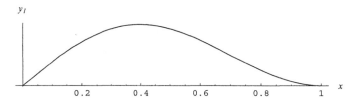

27. The auxiliary equation is $m^2 + m = m(m + 1) = 0$ so that $u(r) = c_1 r^{-1} + c_2$. The boundary conditions $u(a) = u_0$ and $u(b) = u_1$ yield the system $c_1 a^{-1} + c_2 = u_0$, $c_1 b^{-1} + c_2 = u_1$. Solving gives

$$c_1 = \left(\frac{u_0 - u_1}{b - a}\right)ab \quad \text{and} \quad c_2 = \frac{u_1 b - u_0 a}{b - a}.$$

Thus

$$u(r) = \left(\frac{u_0 - u_1}{b - a}\right)\frac{ab}{r} + \frac{u_1 b - u_0 a}{b - a}.$$

Section 5.3   Nonlinear Equations

The terminology and concepts listed below provide an outline of the main ideas encountered in this section. These can be useful when preparing for a quiz or test.

Terminology and Concepts

- nonlinear spring models
- hard spring
- soft spring
- simple pendulum
- linearization of a DE

The basic skills listed below summarize the more mechanical types of problems encountered in the exercise set for this section.

Basic Skills

- use a numerical solver to plot solution curves of nonlinear DEs

Exercises 5.3   *Hints, Suggestions, Solutions, and Examples*

3. The period corresponding to $x(0) = 1$, $x'(0) = 1$ is approximately 5.8. The second initial-value problem does not have a periodic solution.

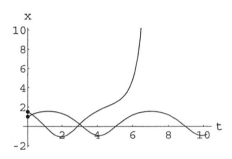

6. From the graphs we see that the interval is approximately $(-0.8, 1.1)$.

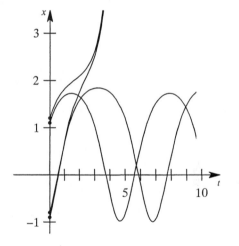

9. This is a damped hard spring, so x will approach 0 as t approaches ∞.

117

12. (a)

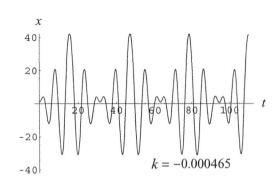

The system appears to be oscillatory for $-0.000465 \le k_1 < 0$ and nonoscillatory for $k_1 \le -0.000466$.

(b)

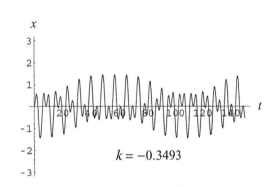

The system appears to be oscillatory for $-0.3493 \le k_1 < 0$ and nonoscillatory for $k_1 \le -0.3494$.

15. (a) Intuitively, one might expect that only half of a 10-pound chain could be lifted by a 5-pound vertical force.

(b) Since $x = 0$ when $t = 0$, and $v = dx/dt = \sqrt{160 - 64x/3}$, we have $v(0) = \sqrt{160} \approx 12.65$ ft/s.

(c) Since x should always be positive, we solve $x(t) = 0$, getting $t = 0$ and $t = \frac{3}{2}\sqrt{5/2} \approx 2.3717$. Since the graph of $x(t)$ is a parabola, the maximum value occurs at $t_m = \frac{3}{4}\sqrt{5/2}$. (This can also be obtained by solving $x'(t) = 0$.) At this time the height of the chain is $x(t_m) \approx 7.5$ ft. This is higher than predicted because of the momentum generated by the force. When the chain is 5 feet high it still has a positive velocity of about 7.3 ft/s, which keeps it going higher for a while.

17. (*Hint*) In part (c) express t in terms of a definite integral from $s = 3$ to $s = x$, then use a CAS to evaluate the definite integral from $s = 3$ to $s = 7$.

18. (a) There are two forces acting on the chain as it falls from the platform. One is the force due to gravity on the portion of the chain hanging over the edge of the platform. This is $F_1 = 2x$. The second is due to the motion of the portion of the chain stretched out on the platform. By

Newton's second law this is

$$F_2 = \frac{d}{dt}[mv] = \frac{d}{dt}\left[\frac{(8-x)2}{32}v\right] = \frac{d}{dt}\left[\frac{8-x}{16}v\right]$$

$$= \frac{8-x}{16}\frac{dv}{dt} - \frac{1}{16}v\frac{dx}{dt} = \frac{1}{16}\left[(8-x)\frac{dv}{dt} - v^2\right].$$

From $\frac{d}{dt}[mv] = F_1 - F_2$ we have

$$\frac{d}{dt}\left[\frac{2x}{32}v\right] = 2x - \frac{1}{16}\left[(8-x)\frac{dv}{dt} - v^2\right]$$

$$\frac{x}{16}\frac{dv}{dt} + \frac{1}{16}v\frac{dx}{dt} = 2x - \frac{1}{16}\left[(8-x)\frac{dv}{dt} - v^2\right]$$

$$x\frac{dv}{dt} + v^2 = 32x - (8-x)\frac{dv}{dt} + v^2$$

$$x\frac{dv}{dt} = 32x - 8\frac{dv}{dt} + x\frac{dv}{dt}$$

$$8\frac{dv}{dt} = 32x.$$

By the Chain Rule, $dv/dt = (dv/dx)(dx/dt) = v\,dv/dx$, so

$$8\frac{dv}{dt} = 8v\frac{dv}{dx} = 32x \quad \text{and} \quad v\frac{dv}{dx} = 4x.$$

(b) Integrating $v\,dv = 4x\,dx$ we get $\frac{1}{2}v^2 = 2x^2 + c$. Since $v = 0$ when $x = 3$, we have $c = -18$. Then $v^2 = 4x^2 - 36$ and $v = \sqrt{4x^2 - 36}$. Using $v = dx/dt$, separating variables, and integrating we obtain

$$\frac{dx}{\sqrt{x^2 - 9}} = 2\,dt \quad \text{and} \quad \cosh^{-1}\frac{x}{3} = 2t + c_1.$$

Solving for x we get $x(t) = 3\cosh(2t + c_1)$. Since $x = 3$ when $t = 0$, we have $\cosh c_1 = 1$ and $c_1 = 0$. Thus, $x(t) = 3\cosh 2t$. Differentiating, we find $v(t) = dx/dt = 6\sinh 2t$.

(c) To find the time when the back end of the chain leaves the platform we solve $x(t) = 3\cosh 2t = 8$. This gives $t_1 = \frac{1}{2}\cosh^{-1}\frac{8}{3} \approx 0.8184$ seconds. The velocity at this instant is

$$v(t_1) = 6\sinh\left(\cosh^{-1}\frac{8}{3}\right) = 2\sqrt{55} \approx 14.83 \text{ ft/s}.$$

(d) Replacing 8 with L and 32 with g in part (a) we have $L\,dv/dt = gx$. Then

$$L\frac{dv}{dt} = Lv\frac{dv}{dx} = gx \quad \text{and} \quad v\frac{dv}{dx} = \frac{g}{L}x.$$

Integrating we get $\frac{1}{2}v^2 = (g/2L)x^2 + c$. Setting $x = x_0$ and $v = 0$, we find $c = -(g/2L)x_0^2$. Solving for v we find

$$v(x) = \sqrt{\frac{g}{L}x^2 - \frac{g}{L}x_0^2}\,.$$

119

Then the velocity at which the end of the chain leaves the edge of the platform is

$$v(L) = \sqrt{\frac{g}{L}(L^2 - x_0^2)} \ .$$

Chapter 5 in Review

Hints, Suggestions, Solutions, and Examples

3. $5/4$ m, since $x = -\cos 4t + \frac{3}{4}\sin 4t$.

6. False; since the equation of motion in this case is $x(t) = e^{-\lambda t}(c_1 + c_2 t)$ and $x(t) = 0$ can have at most one real solution.

9. $y = 0$ because $\lambda = 8$ is not an eigenvalue.

12. (a) Solving $\frac{3}{8}x'' + 6x = 0$ subject to $x(0) = 1$ and $x'(0) = -4$ we obtain

$$x = \cos 4t - \sin 4t = \sqrt{2}\sin(4t + 3\pi/4)\ .$$

  (b) The amplitude is $\sqrt{2}$, period is $\pi/2$, and frequency is $2/\pi$.

  (c) If $x = 1$ then $t = n\pi/2$ and $t = -\pi/8 + n\pi/2$ for $n = 1, 2, 3, \ldots$.

  (d) If $x = 0$ then $t = \pi/16 + n\pi/4$ for $n = 0, 1, 2, \ldots$. The motion is upward for n even and downward for n odd.

  (e) $x'(3\pi/16) = 0$

  (f) If $x' = 0$ then $4t + 3\pi/4 = \pi/2 + n\pi$ or $t = 3\pi/16 + n\pi$.

15. From $mx'' + 4x' + 2x = 0$ we see that nonoscillatory motion results if $16 - 8m \geq 0$ or $0 < m \leq 2$.

18. By inspection we see that $x_p = A/\omega^2$ suffices.

21. From $q'' + 10^4 q = 100\sin 50t$, $q(0) = 0$, and $q'(0) = 0$ we obtain $q_c = c_1\cos 100t + c_2\sin 100t$, $q_p = \frac{1}{75}\sin 50t$, and

  (a) $q = -\frac{1}{150}\sin 100t + \frac{1}{75}\sin 50t$,

  (b) $i = -\frac{2}{3}\cos 100t + \frac{2}{3}\cos 50t$, and

  (c) $q = 0$ when $\sin 50t(1 - \cos 50t) = 0$ or $t = n\pi/50$ for $n = 0, 1, 2, \ldots$.

24. (a) The differential equation is $d^2r/dt^2 - \omega^2 r = -g\sin\omega t$. The auxiliary equation is $m^2 - \omega^2 = 0$, so $r_c = c_1 e^{\omega t} + c_2 e^{-\omega t}$. A particular solution has the form $r_p = A\sin\omega t + B\cos\omega t$. Substituting into the DE we find $-2A\omega^2\sin\omega t - 2B\omega^2\cos\omega t = -g\sin\omega t$. Thus, $B = 0$, $A = g/2\omega^2$, and $r_p = (g/2\omega^2)\sin\omega t$. The general solution of the DE is $r(t) = c_1 e^{\omega t} + c_2 e^{-\omega t} + (g/2\omega^2)\sin\omega t$.

The initial conditions imply $c_1 + c_2 = r_0$ and $g/2\omega - \omega c_1 + \omega c_2 = v_0$. Solving for c_1 and c_2 we get

$$c_1 = (2\omega^2 r_0 + 2\omega v_0 - g)/4\omega^2 \quad \text{and} \quad c_2 = (2\omega^2 r_0 - 2\omega v_0 + g)/4\omega^2,$$

so that

$$r(t) = \frac{2\omega^2 r_0 + 2\omega v_0 - g}{4\omega^2} e^{\omega t} + \frac{2\omega^2 r_0 - 2\omega v_0 + g}{4\omega^2} e^{-\omega t} + \frac{g}{2\omega^2} \sin \omega t.$$

(b) The bead will exhibit simple harmonic motion when the exponential terms are missing. Solving $c_1 = 0$, $c_2 = 0$ for r_0 and v_0 we find $r_0 = 0$ and $v_0 = g/2\omega$.

To find the minimum length of rod that will accommodate simple harmonic motion we determine the amplitude of $r(t)$ and double it. Thus $L = g/\omega^2$.

(c) As t increases, $e^{\omega t}$ approaches infinity and $e^{-\omega t}$ approaches 0. Since $\sin \omega t$ is bounded, the distance, $r(t)$, of the bead from the pivot point increases without bound and the distance of the bead from P will eventually exceed $L/2$.

(d)

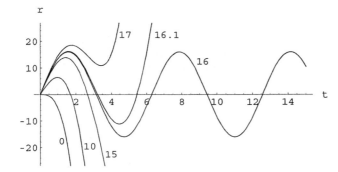

(e) For each v_0 we want to find the smallest value of t for which $r(t) = \pm 20$. Whether we look for $r(t) = -20$ or $r(t) = 20$ is determined by looking at the graphs in part (d). The total times that the bead stays on the rod is shown in the table below.

v_0	0	10	15	16.1	17
r	-20	-20	-20	20	20
t	1.55007	2.35494	3.43088	6.11627	4.22339

When $v_0 = 16$ the bead never leaves the rod.

6

Series Solutions of Linear Equations

Section 6.1

Solutions about Ordinary Points

The terminology and concepts listed below provide an outline of the main ideas encountered in this section. These can be useful when preparing for a quiz or test.

Terminology and Concepts

- power series centered at a; $\sum_{n=0}^{\infty} c_n (x-a)^n$
- convergence of a power series
- interval of convergence
- radius of convergence
- absolute convergence
- ratio test
- a power series defines a function
- a function analytic at a point
- ordinary point
- singular point
- power series solution of a DE
- recurrence relation

The basic skills listed below summarize the more mechanical types of problems encountered in the exercise set for this section.

Basic Skills

- find the radius of convergence of a power series
- convert an expression like $\sum_{n=3}^{\infty} (n+1)x^{n-2} + \sum_{n=2}^{\infty} nx^n$ to an expression involving a single infinite summation: $4x + \sum_{k=2}^{\infty} (2k+3)x^k$

- find power series solutions of linear second-order DEs

Summation Notation   Without a doubt, the key to successfully solving a linear DE with variable coefficients by the power series method depends on your ability to recast a combination of power series centered at the same point in terms of a *single* summation. In Example 1 of Section 6.1 we gave an example involving two series; here is another example of how this is done using three series. We wish to combine the three sums

$$\sum_{n=2}^{\infty} c_n n(n-1)x^{n-2} + 2\sum_{n=1}^{\infty} c_n n x^n + 5\sum_{n=0}^{\infty} c_n x^{n+1}$$

into one term. Notice that the first series starts (for $n = 2$) with x^0, the second series starts (for $n = 1$) with x^1, and the third series also starts (for $n = 0$) with x^1. To bring the first series in line with the last two, we write out the term corresponding to $n = 2$ and observe that now all three series

$$2c_2 + \sum_{n=3}^{\infty} c_n n(n-1)x^{n-2} + \sum_{n=1}^{\infty} 2c_n n x^n + \sum_{n=0}^{\infty} 5c_n x^{n+1}$$

begin with the same power x^1. The next thing that must be done is to get a common summation index for the series:

- In the first series, we let $k = n - 2$. As $n = 3, 4, 5, \ldots$, k takes on the values $1, 2, 3, \ldots$.

- In the second series, we let $k = n$. Here $n = 1, 2, 3, \ldots$ corresponds to $k = 1, 2, 3, \ldots$.

- In the third series, we let $k = n + 1$. Then $n = 0, 1, 2, \ldots$ also yields $k = 1, 2, 3, \ldots$.

Next, we substitute $n = k + 2$ in the first series, $n = k$ in the second series, and then $n = k - 1$ in the third:

$$2c_2 + \sum_{k=1}^{\infty} c_{k+2}(k+2)(k+1)x^k + \sum_{k=1}^{\infty} 2c_k k x^k + \sum_{k=1}^{\infty} 5c_{k-1}x^k.$$

At this point, when all the series start with the same power of x and have the same summation index starting with the same value (in this case, $k = 1$), we can add the series term-by-term to get

$$2c_2 + \sum_{k=1}^{\infty} [c_{k+2}(k+2)(k+1) + 2c_k k + 5c_{k-1}]x^k.$$

Alternatively, we can make the change of summation variable first as illustrated in the solution of Problem 18 in this section of the manual.

Use of Computers   Both *Mathematica* and *Maple* can express the sum of several terms in a concise way without actually listing each term. The commands below can be used to express the first four terms of the infinite sum $\sum_{n=0}^{\infty} c_n(-1)^n/(n+1)^2$.

Clear[c]   (*Mathematica*)

123

$$\text{Sum}[c[n](-1)\hat{}\,n/(n+1)\hat{}\,2, \{n, 0, 3\}]$$

```
sum(c(n)*(-1)^n/(n+1)^2, n=0..3);                    (Maple)
```

To obtain the power series representation about $x = 0$ of a function like $f(x) = e^x$ use

$$\text{Series}[\text{Exp}[x], \{x, 0, 5\}]   (Mathematica)$$

or

```
series(exp(x),x=0,5);                                (Maple)
```

The *Mathematica* output corresponding to the above command is

$$1 + x + \frac{1}{2}x^2 + \frac{1}{6}x^3 + \frac{1}{24}x^4 + \frac{1}{120}x^5 + O[x]^6,$$

whereas for *Maple* the output is

$$1 + x + \frac{1}{2}x^2 + \frac{1}{6}x^3 + \frac{1}{24}x^4 + O(x^5).$$

Note that in the *Mathematica* command the 5 tells the program to compute the series out to, and including, the term containing x^5. On the other hand, in *Maple*, the 5 tells the program to compute the series out to, but excluding, the term containing x^5.

Depending on how you want to use the series representation of a function found using a CAS, it may be inconvenient to have the order of the series included in the output. For example, neither *Mathematica* nor *Maple* will plot the graph of an expression containing $O[x]\hat{}\,6$, in the case of *Mathematica*, or $O(x^5)$ in the case of *Maple*. To get around this, both programs have commands that truncate the series, converting it to a polynomial.

$$\text{Normal}[\text{Series}[\text{Exp}[x], \{x, 0, 5\}]\ ]   (Mathematica)$$

or

$$\text{Series}[\text{Exp}[x], \{x, 0, 5\}]//\text{Normal}$$

```
convert(series(exp(x),x=0,5),polynom);               (Maple)
```

Exercises 6.1   *Hints, Suggestions, Solutions, and Examples*

3. By the ratio test,

$$\lim_{k\to\infty} \left| \frac{(x-5)^{k+1}/10^{k+1}}{(x-5)^k/10^k} \right| = \lim_{k\to\infty} \frac{1}{10}|x-5| = \frac{1}{10}|x-5|.$$

The series is absolutely convergent for $\frac{1}{10}|x-5| < 1$, $|x-5| < 10$, or on $(-5, 15)$. The radius of convergence is $R = 10$. At $x = -5$, the series $\sum_{k=1}^{\infty}(-1)^k(-10)^k/10^k = \sum_{k=1}^{\infty} 1$ diverges by the n-th term test. At $x = 15$, the series $\sum_{k=1}^{\infty}(-1)^k 10^k/10^k = \sum_{k=1}^{\infty}(-1)^k$ diverges by the n-th term test. Thus, the series converges on $(-5, 15)$.

6. $e^{-x}\cos x = \left(1 - x + \dfrac{x^2}{2} - \dfrac{x^3}{6} + \dfrac{x^4}{24} - \cdots\right)\left(1 - \dfrac{x^2}{2} + \dfrac{x^4}{24} - \cdots\right) = 1 - x + \dfrac{x^3}{3} - \dfrac{x^4}{6} + \cdots$

This problem can also be done using a CAS:

        `s1=Series[Exp[-x], {x, 0, 4}]`   *(Mathematica)*

        `s2=Series[Cos[x], {x, 0, 4}]`

        `s1 s2`

```
s1=convert(series(exp(-x),x=0,5),polynom);     (Maple)
s2=convert(series(cos(x),x=0,5),polynom);
expand(s1*s2);
```

9. Let $k = n + 2$ so that $n = k - 2$ and

$$\sum_{n=1}^{\infty} n c_n x^{n+2} = \sum_{k=3}^{\infty} (k-2)c_{k-2}x^k.$$

12. $\displaystyle\sum_{n=2}^{\infty} n(n-1)c_n x^n + 2\sum_{n=2}^{\infty} n(n-1)c_n x^{n-2} + 3\sum_{n=1}^{\infty} n c_n x^n$

$= 2 \cdot 2 \cdot 1 c_2 x^0 + 2\cdot 3\cdot 2 c_3 x^1 + 3\cdot 1\cdot c_1 x^1 + \underbrace{\sum_{n=2}^{\infty} n(n-1)c_n x^n}_{k=n} + 2\underbrace{\sum_{n=4}^{\infty} n(n-1)c_n x^{n-2}}_{k=n-2} + 3\underbrace{\sum_{n=2}^{\infty} n c_n x^n}_{k=n}$

$= 4c_2 + (3c_1 + 12c_3)x + \sum_{k=2}^{\infty} k(k-1)c_k x^k + 2\sum_{k=2}^{\infty}(k+2)(k+1)c_{k+2}x^k + 3\sum_{k=2}^{\infty} k c_k x^k$

$= 4c_2 + (3c_1 + 12c_3)x + \sum_{k=2}^{\infty}([k(k-1) + 3k]c_k + 2(k+2)(k+1)c_{k+2})x^k$

$= 4c_2 + (3c_1 + 12c_3)x + \sum_{k=2}^{\infty}[k(k+2)c_k + 2(k+1)(k+2)c_{k+2}]x^k$

15. The singular points of $(x^2 - 25)y'' + 2xy' + y = 0$ are -5 and 5. The distance from 0 to either of these points is 5. The distance from 1 to the closest of these points is 4.

125

Section 6.1 Solutions about Ordinary Points

18. Substituting $y = \sum_{n=0}^{\infty} c_n x^n$ into the DE we have

$$y'' + x^2 y = \underbrace{\sum_{n=2}^{\infty} n(n-1)c_n x^{n-2}}_{k=n-2} + \underbrace{\sum_{n=0}^{\infty} c_n x^{n+2}}_{k=n+2} = \sum_{k=0}^{\infty} (k+2)(k+1)c_{k+2} x^k + \sum_{k=2}^{\infty} c_{k-2} x^k$$

$$= 2c_2 + 6c_3 x + \sum_{k=2}^{\infty} [(k+2)(k+1)c_{k+2} + c_{k-2}]x^k = 0.$$

Thus

$$c_2 = c_3 = 0$$

$$(k+2)(k+1)c_{k+2} + c_{k-2} = 0$$

and

$$c_{k+2} = -\frac{1}{(k+2)(k+1)} c_{k-2}, \quad k = 2, 3, 4, \dots.$$

Choosing $c_0 = 1$ and $c_1 = 0$ we find

$$c_4 = -\frac{1}{12}$$

$$c_5 = c_6 = c_7 = 0$$

$$c_8 = \frac{1}{672}$$

and so on. For $c_0 = 0$ and $c_1 = 1$ we obtain

$$c_4 = 0$$

$$c_5 = -\frac{1}{20}$$

$$c_6 = c_7 = c_8 = 0$$

$$c_9 = \frac{1}{1440}$$

and so on. Thus, two solutions are

$$y_1 = 1 - \frac{1}{12}x^4 + \frac{1}{672}x^8 - \cdots \qquad \text{and} \qquad y_2 = x - \frac{1}{20}x^5 + \frac{1}{1440}x^9 - \cdots.$$

126

21. Substituting $y = \sum_{n=0}^{\infty} c_n x^n$ into the DE we have

$$y'' + x^2 y' + xy = \underbrace{\sum_{n=2}^{\infty} n(n-1)c_n x^{n-2}}_{k=n-2} + \underbrace{\sum_{n=1}^{\infty} nc_n x^{n+1}}_{k=n+1} + \underbrace{\sum_{n=0}^{\infty} c_n x^{n+1}}_{k=n+1}$$

$$= \sum_{k=0}^{\infty} (k+2)(k+1)c_{k+2}x^k + \sum_{k=2}^{\infty} (k-1)c_{k-1}x^k + \sum_{k=1}^{\infty} c_{k-1}x^k$$

$$= 2c_2 + (6c_3 + c_0)x + \sum_{k=2}^{\infty} [(k+2)(k+1)c_{k+2} + kc_{k-1}]x^k = 0.$$

Thus

$$c_2 = 0$$

$$6c_3 + c_0 = 0$$

$$(k+2)(k+1)c_{k+2} + kc_{k-1} = 0$$

and

$$c_2 = 0$$

$$c_3 = -\frac{1}{6}c_0$$

$$c_{k+2} = -\frac{k}{(k+2)(k+1)}c_{k-1}, \quad k = 2, 3, 4, \ldots.$$

Choosing $c_0 = 1$ and $c_1 = 0$ we find

$$c_3 = -\frac{1}{6}$$

$$c_4 = c_5 = 0$$

$$c_6 = \frac{1}{45}$$

and so on. For $c_0 = 0$ and $c_1 = 1$ we obtain

$$c_3 = 0$$

$$c_4 = -\frac{1}{6}$$

$$c_5 = c_6 = 0$$

$$c_7 = \frac{5}{252}$$

and so on. Thus, two solutions are

$$y_1 = 1 - \frac{1}{6}x^3 + \frac{1}{45}x^6 - \cdots \qquad \text{and} \qquad y_2 = x - \frac{1}{6}x^4 + \frac{5}{252}x^7 - \cdots.$$

127

24. Substituting $y = \sum_{n=0}^{\infty} c_n x^n$ into the DE we have

$$(x+2)y'' + xy' - y = \sum_{n=2}^{\infty} n(n-1)c_n x^{n-1} + \sum_{n=2}^{\infty} 2n(n-1)c_n x^{n-2} + \underbrace{\sum_{n=1}^{\infty} nc_n x^n}_{k=n} - \underbrace{\sum_{n=0}^{\infty} c_n x^n}_{k=n}$$
$$\underbrace{\phantom{\sum_{n=2}^{\infty} n(n-1)c_n x^{n-1}}}_{k=n-1} \underbrace{\phantom{\sum_{n=2}^{\infty} 2n(n-1)c_n x^{n-2}}}_{k=n-2}$$

$$= \sum_{k=1}^{\infty} (k+1)k c_{k+1} x^k + \sum_{k=0}^{\infty} 2(k+2)(k+1)c_{k+2} x^k + \sum_{k=1}^{\infty} kc_k x^k - \sum_{k=0}^{\infty} c_k x^k$$

$$= 4c_2 - c_0 + \sum_{k=1}^{\infty} [(k+1)k c_{k+1} + 2(k+2)(k+1)c_{k+2} + (k-1)c_k] x^k = 0.$$

Thus

$$4c_2 - c_0 = 0$$

$$(k+1)k c_{k+1} + 2(k+2)(k+1)c_{k+2} + (k-1)c_k = 0, \quad k = 1, 2, 3, \ldots$$

and

$$c_2 = \frac{1}{4}c_0$$

$$c_{k+2} = -\frac{(k+1)k c_{k+1} + (k-1)c_k}{2(k+2)(k+1)}, \quad k = 1, 2, 3, \ldots .$$

Choosing $c_0 = 1$ and $c_1 = 0$ we find

$$c_1 = 0, \qquad c_2 = \frac{1}{4}, \qquad c_3 = -\frac{1}{24}, \qquad c_4 = 0, \qquad c_5 = \frac{1}{480}$$

and so on. For $c_0 = 0$ and $c_1 = 1$ we obtain

$$c_2 = 0$$

$$c_3 = 0$$

$$c_4 = c_5 = c_6 = \cdots = 0.$$

Thus, two solutions are

$$y_1 = c_0 \left[1 + \frac{1}{4}x^2 - \frac{1}{24}x^3 + \frac{1}{480}x^5 + \cdots \right] \qquad \text{and} \qquad y_2 = c_1 x.$$

27. Substituting $y = \sum_{n=0}^{\infty} c_n x^n$ into the DE we have

$$\left(x^2 + 2\right)y'' + 3xy' - y = \underbrace{\sum_{n=2}^{\infty} n(n-1)c_n x^n}_{k=n} + 2\underbrace{\sum_{n=2}^{\infty} n(n-1)c_n x^{n-2}}_{k=n-2} + 3\underbrace{\sum_{n=1}^{\infty} nc_n x^n}_{k=n} - \underbrace{\sum_{n=0}^{\infty} c_n x^n}_{k=n}$$

$$= \sum_{k=2}^{\infty} k(k-1)c_k x^k + 2\sum_{k=0}^{\infty}(k+2)(k+1)c_{k+2}x^k + 3\sum_{k=1}^{\infty} kc_k x^k - \sum_{k=0}^{\infty} c_k x^k$$

$$= (4c_2 - c_0) + (12c_3 + 2c_1)x + \sum_{k=2}^{\infty}\left[2(k+2)(k+1)c_{k+2} + \left(k^2 + 2k - 1\right)c_k\right]x^k = 0.$$

Thus

$$4c_2 - c_0 = 0$$

$$12c_3 + 2c_1 = 0$$

$$2(k+2)(k+1)c_{k+2} + \left(k^2 + 2k - 1\right)c_k = 0$$

and

$$c_2 = \frac{1}{4}c_0$$

$$c_3 = -\frac{1}{6}c_1$$

$$c_{k+2} = -\frac{k^2 + 2k - 1}{2(k+2)(k+1)}c_k, \quad k = 2, 3, 4, \ldots.$$

Choosing $c_0 = 1$ and $c_1 = 0$ we find

$$c_2 = \frac{1}{4}$$

$$c_3 = c_5 = c_7 = \cdots = 0$$

$$c_4 = -\frac{7}{96}$$

and so on. For $c_0 = 0$ and $c_1 = 1$ we obtain

$$c_2 = c_4 = c_6 = \cdots = 0$$

$$c_3 = -\frac{1}{6}$$

$$c_5 = \frac{7}{120}$$

and so on. Thus, two solutions are

$$y_1 = 1 + \frac{1}{4}x^2 - \frac{7}{96}x^4 + \cdots \qquad \text{and} \qquad y_2 = x - \frac{1}{6}x^3 + \frac{7}{120}x^5 - \cdots.$$

30. Substituting $y = \sum_{n=0}^{\infty} c_n x^n$ into the DE we have

$$(x+1)y'' - (2-x)y' + y$$

$$= \sum_{n=2}^{\infty} \underbrace{n(n-1)c_n x^{n-1}}_{k=n-1} + \sum_{n=2}^{\infty} \underbrace{n(n-1)c_n x^{n-2}}_{k=n-2} - 2\sum_{n=1}^{\infty} \underbrace{nc_n x^{n-1}}_{k=n-1} + \sum_{n=1}^{\infty} \underbrace{nc_n x^n}_{k=n} + \sum_{n=0}^{\infty} \underbrace{c_n x^n}_{k=n}$$

$$= \sum_{k=1}^{\infty} (k+1)kc_{k+1}x^k + \sum_{k=0}^{\infty} (k+2)(k+1)c_{k+2}x^k - 2\sum_{k=0}^{\infty} (k+1)c_{k+1}x^k + \sum_{k=1}^{\infty} kc_k x^k + \sum_{k=0}^{\infty} c_k x^k$$

$$= 2c_2 - 2c_1 + c_0 + \sum_{k=1}^{\infty} [(k+2)(k+1)c_{k+2} - (k+1)c_{k+1} + (k+1)c_k]x^k = 0.$$

Thus

$$2c_2 - 2c_1 + c_0 = 0$$

$$(k+2)(k+1)c_{k+2} - (k+1)c_{k+1} + (k+1)c_k = 0$$

and

$$c_2 = c_1 - \frac{1}{2}c_0$$

$$c_{k+2} = \frac{1}{k+2}c_{k+1} - \frac{1}{k+2}c_k, \quad k = 1, 2, 3, \ldots .$$

Choosing $c_0 = 1$ and $c_1 = 0$ we find

$$c_2 = -\frac{1}{2}, \qquad c_3 = -\frac{1}{6}, \qquad c_4 = \frac{1}{12}$$

and so on. For $c_0 = 0$ and $c_1 = 1$ we obtain

$$c_2 = 1, \qquad c_3 = 0, \qquad c_4 = -\frac{1}{4}$$

and so on. Thus,

$$y = C_1 \left(1 - \frac{1}{2}x^2 - \frac{1}{6}x^3 + \frac{1}{12}x^4 + \cdots\right) + C_2 \left(x + x^2 - \frac{1}{4}x^4 + \cdots\right)$$

and

$$y' = C_1 \left(-x - \frac{1}{2}x^2 + \frac{1}{3}x^3 + \cdots\right) + C_2 \left(1 + 2x - x^3 + \cdots\right).$$

The initial conditions imply $C_1 = 2$ and $C_2 = -1$, so

$$y = 2\left(1 - \frac{1}{2}x^2 - \frac{1}{6}x^3 + \frac{1}{12}x^4 + \cdots\right) - \left(x + x^2 - \frac{1}{4}x^4 + \cdots\right)$$

$$= 2 - x - 2x^2 - \frac{1}{3}x^3 + \frac{5}{12}x^4 + \cdots .$$

33. Substituting $y = \sum_{n=0}^{\infty} c_n x^n$ into the DE we have

$$y'' + (\sin x)y = \sum_{n=2}^{\infty} n(n-1)c_n x^{n-2} + \left(x - \frac{1}{6}x^3 + \frac{1}{120}x^5 - \cdots \right)\left(c_0 + c_1 x + c_2 x^2 + \cdots \right)$$

$$= \left[2c_2 + 6c_3 x + 12c_4 x^2 + 20c_5 x^3 + \cdots \right] + \left[c_0 x + c_1 x^2 + \left(c_2 - \frac{1}{6}c_0 \right)x^3 + \cdots \right]$$

$$= 2c_2 + (6c_3 + c_0)x + (12c_4 + c_1)x^2 + \left(20c_5 + c_2 - \frac{1}{6}c_0 \right)x^3 + \cdots = 0.$$

Thus

$$2c_2 = 0$$

$$6c_3 + c_0 = 0$$

$$12c_4 + c_1 = 0$$

$$20c_5 + c_2 - \frac{1}{6}c_0 = 0$$

and

$$c_2 = 0$$

$$c_3 = -\frac{1}{6}c_0$$

$$c_4 = -\frac{1}{12}c_1$$

$$c_5 = -\frac{1}{20}c_2 + \frac{1}{120}c_0.$$

Choosing $c_0 = 1$ and $c_1 = 0$ we find

$$c_2 = 0, \qquad c_3 = -\frac{1}{6}, \qquad c_4 = 0, \qquad c_5 = \frac{1}{120}$$

and so on. For $c_0 = 0$ and $c_1 = 1$ we obtain

$$c_2 = 0, \qquad c_3 = 0, \qquad c_4 = -\frac{1}{12}, \qquad c_5 = 0$$

and so on. Thus, two solutions are

$$y_1 = 1 - \frac{1}{6}x^3 + \frac{1}{120}x^5 + \cdots \qquad \text{and} \qquad y_2 = x - \frac{1}{12}x^4 + \cdots .$$

Section 6.2

Solutions About Singular Points

The terminology and concepts listed below provide an outline of the main ideas encountered in this section. These can be useful when preparing for a quiz or test.

Terminology and Concepts

- regular singular point of a DE

- irregular singular point of a DE

- method of Frobenius for finding a series solution about a regular singular point of a DE

- indicial equation

The basic skills listed below summarize the more mechanical types of problems encountered in the exercise set for this section.

Basic Skills

- determine the singular points of a linear DE and classify them as regular or irregular

- use the method of Frobenius to obtain two linearly independent solutions about $x = 0$ of a homogeneous linear DE with regular singular point $x = 0$ when the indicial roots

  1. do not differ by an integer, or
  2. differ by an integer.

Exercises 6.2   *Hints, Suggestions, Solutions, and Examples*

3. Writing the DE in the form

$$y'' + \frac{1}{(x-3)^2(x+3)} \, y' + \frac{1}{(x-3)^2(x+3)^2} \, y = 0$$

we see that $x = 3$ is an irregular singular point and $x = -3$ is a regular singular point.

6. Writing the DE in the form

$$y'' + \frac{4}{x(x-5)^2} \, y' + \frac{x+5}{x^2(x-5)} \, y = 0$$

we see that $x = 0$ is a regular singular point and $x = 5$ is an irregular singular point.

9. Writing the DE in the form

$$y'' + \frac{3}{x^2(x-5)(x+5)(x-2)} \, y' + \frac{7}{x^3(x-5)(x-2)^2} \, y = 0$$

we see that $x = 0$ is an irregular singular point and $x = 5$, $x = -5$, and $x = 2$ are regular singular points.

12. Writing the DE in the form

$$y'' + \frac{x+3}{x} y' + 7xy = 0$$

we see that $x_0 = 0$ is a regular singular point. Multiplying by x^2, the DE can be put in the form

$$x^2 y'' + x(x+3)y' + 7x^3 y = 0.$$

We identify $p(x) = x + 3$ and $q(x) = 7x^3$.

15. Substituting $y = \sum_{n=0}^{\infty} c_n x^{n+r}$ into the DE and collecting terms, we obtain

$$2xy'' - y' + 2y = \left(2r^2 - 3r\right) c_0 x^{r-1} + \sum_{k=1}^{\infty} [2(k+r-1)(k+r)c_k - (k+r)c_k + 2c_{k-1}]x^{k+r-1} = 0,$$

which implies

$$2r^2 - 3r = r(2r - 3) = 0$$

and

$$(k+r)(2k+2r-3)c_k + 2c_{k-1} = 0.$$

The indicial roots are $r = 0$ and $r = 3/2$. For $r = 0$ the recurrence relation is

$$c_k = -\frac{2c_{k-1}}{k(2k-3)}, \quad k = 1, 2, 3, \ldots,$$

and

$$c_1 = 2c_0, \qquad c_2 = -2c_0, \qquad c_3 = \frac{4}{9}c_0.$$

For $r = 3/2$ the recurrence relation is

$$c_k = -\frac{2c_{k-1}}{(2k+3)k}, \quad k = 1, 2, 3, \ldots,$$

and

$$c_1 = -\frac{2}{5}c_0, \qquad c_2 = \frac{2}{35}c_0, \qquad c_3 = -\frac{4}{945}c_0.$$

The general solution on $(0, \infty)$ is

$$y = C_1 \left(1 + 2x - 2x^2 + \frac{4}{9}x^3 + \cdots\right) + C_2 x^{3/2}\left(1 - \frac{2}{5}x + \frac{2}{35}x^2 - \frac{4}{945}x^3 + \cdots\right).$$

18. Substituting $y = \sum_{n=0}^{\infty} c_n x^{n+r}$ into the DE and collecting terms, we obtain

$$2x^2 y'' - xy' + \left(x^2 + 1\right)y = \left(2r^2 - 3r + 1\right)c_0 x^r + \left(2r^2 + r\right)c_1 x^{r+1}$$

$$+ \sum_{k=2}^{\infty} [2(k+r)(k+r-1)c_k - (k+r)c_k + c_k + c_{k-2}]x^{k+r}$$

$$= 0,$$

133

which implies

$$2r^2 - 3r + 1 = (2r - 1)(r - 1) = 0,$$

$$\left(2r^2 + r\right)c_1 = 0,$$

and

$$[(k + r)(2k + 2r - 3) + 1]c_k + c_{k-2} = 0.$$

The indicial roots are $r = 1/2$ and $r = 1$, so $c_1 = 0$. For $r = 1/2$ the recurrence relation is

$$c_k = -\frac{c_{k-2}}{k(2k - 1)}, \quad k = 2, 3, 4, \ldots,$$

and

$$c_2 = -\frac{1}{6}c_0, \qquad c_3 = 0, \qquad c_4 = \frac{1}{168}c_0.$$

For $r = 1$ the recurrence relation is

$$c_k = -\frac{c_{k-2}}{k(2k + 1)}, \quad k = 2, 3, 4, \ldots,$$

and

$$c_2 = -\frac{1}{10}c_0, \qquad c_3 = 0, \qquad c_4 = \frac{1}{360}c_0.$$

The general solution on $(0, \infty)$ is

$$y = C_1 x^{1/2}\left(1 - \frac{1}{6}x^2 + \frac{1}{168}x^4 + \cdots\right) + C_2 x\left(1 - \frac{1}{10}x^2 + \frac{1}{360}x^4 + \cdots\right).$$

21. Substituting $y = \sum_{n=0}^{\infty} c_n x^{n+r}$ into the DE and collecting terms, we obtain

$$2xy'' - (3 + 2x)y' + y = \left(2r^2 - 5r\right)c_0 x^{r-1} + \sum_{k=1}^{\infty}[2(k + r)(k + r - 1)c_k$$

$$- 3(k + r)c_k - 2(k + r - 1)c_{k-1} + c_{k-1}]x^{k+r-1}$$

$$= 0,$$

which implies

$$2r^2 - 5r = r(2r - 5) = 0$$

and

$$(k + r)(2k + 2r - 5)c_k - (2k + 2r - 3)c_{k-1} = 0.$$

The indicial roots are $r = 0$ and $r = 5/2$. For $r = 0$ the recurrence relation is

$$c_k = \frac{(2k - 3)c_{k-1}}{k(2k - 5)}, \quad k = 1, 2, 3, \ldots,$$

and

$$c_1 = \frac{1}{3}c_0, \qquad c_2 = -\frac{1}{6}c_0, \qquad c_3 = -\frac{1}{6}c_0.$$

For $r = 5/2$ the recurrence relation is

$$c_k = \frac{2(k+1)c_{k-1}}{k(2k+5)}, \quad k = 1, 2, 3, \ldots,$$

and

$$c_1 = \frac{4}{7}c_0, \qquad c_2 = \frac{4}{21}c_0, \qquad c_3 = \frac{32}{693}c_0.$$

The general solution on $(0, \infty)$ is

$$y = C_1\left(1 + \frac{1}{3}x - \frac{1}{6}x^2 - \frac{1}{6}x^3 + \cdots\right) + C_2 x^{5/2}\left(1 + \frac{4}{7}x + \frac{4}{21}x^2 + \frac{32}{693}x^3 + \cdots\right).$$

24. Substituting $y = \sum_{n=0}^{\infty} c_n x^{n+r}$ into the DE and collecting terms, we obtain

$$2x^2 y'' + 3xy' + (2x - 1)y = \left(2r^2 + r - 1\right)c_0 x^r$$

$$+ \sum_{k=1}^{\infty}[2(k+r)(k+r-1)c_k + 3(k+r)c_k - c_k + 2c_{k-1}]x^{k+r}$$

$$= 0,$$

which implies

$$2r^2 + r - 1 = (2r - 1)(r + 1) = 0$$

and

$$[(k+r)(2k+2r+1) - 1]c_k + 2c_{k-1} = 0.$$

The indicial roots are $r = -1$ and $r = 1/2$. For $r = -1$ the recurrence relation is

$$c_k = -\frac{2c_{k-1}}{k(2k-3)}, \quad k = 1, 2, 3, \ldots,$$

and

$$c_1 = 2c_0, \qquad c_2 = -2c_0, \qquad c_3 = \frac{4}{9}c_0.$$

For $r = 1/2$ the recurrence relation is

$$c_k = -\frac{2c_{k-1}}{k(2k+3)}, \quad k = 1, 2, 3, \ldots,$$

and

$$c_1 = -\frac{2}{5}c_0, \qquad c_2 = \frac{2}{35}c_0, \qquad c_3 = -\frac{4}{945}c_0.$$

The general solution on $(0, \infty)$ is

$$y = C_1 x^{-1}\left(1 + 2x - 2x^2 + \frac{4}{9}x^3 + \cdots\right) + C_2 x^{1/2}\left(1 - \frac{2}{5}x + \frac{2}{35}x^2 - \frac{4}{945}x^3 + \cdots\right).$$

27. Substituting $y = \sum_{n=0}^{\infty} c_n x^{n+r}$ into the DE and collecting terms, we obtain

$$xy'' - xy' + y = \left(r^2 - r\right)c_0 x^{r-1} + \sum_{k=0}^{\infty}[(k+r+1)(k+r)c_{k+1} - (k+r)c_k + c_k]x^{k+r} = 0$$

which implies

$$r^2 - r = r(r-1) = 0$$

and

$$(k+r+1)(k+r)c_{k+1} - (k+r-1)c_k = 0.$$

The indicial roots are $r_1 = 1$ and $r_2 = 0$. For $r_1 = 1$ the recurrence relation is

$$c_{k+1} = \frac{kc_k}{(k+2)(k+1)}, \quad k = 0, 1, 2, \ldots,$$

and one solution is $y_1 = c_0 x$. A second solution is

$$y_2 = x \int \frac{e^{-\int -1\,dx}}{x^2}\,dx = x \int \frac{e^x}{x^2}\,dx = x \int \frac{1}{x^2}\left(1 + x + \frac{1}{2}x^2 + \frac{1}{3!}x^3 + \cdots\right) dx$$

$$= x \int \left(\frac{1}{x^2} + \frac{1}{x} + \frac{1}{2} + \frac{1}{3!}x + \frac{1}{4!}x^2 + \cdots\right) dx = x\left[-\frac{1}{x} + \ln x + \frac{1}{2}x + \frac{1}{12}x^2 + \frac{1}{72}x^3 + \cdots\right]$$

$$= x\ln x - 1 + \frac{1}{2}x^2 + \frac{1}{12}x^3 + \frac{1}{72}x^4 + \cdots.$$

The general solution on $(0, \infty)$ is

$$y = C_1 x + C_2 y_2(x).$$

28. (*Example*) Suppose a power series $\sum_{n=0}^{\infty} c_k x^k$ is determined by the recurrence relation

$$c_k = c_{k-2}/(k+1), \quad c_0 = 1, \; c_1 = 0.$$

Find the power series representation of $1/(\sum_{k=0}^{\infty} c_k x^k)^2$.

```
Clear[c]                                    (Mathematica)
c[k_]:=c[k-2]/(k+1)
c[0]=1; c[1]=0;
ser=Sum[c[k]x^k, {k, 0, 6}]+0[x]^7
1/ser^2

with(powseries);                           (Maple)
c:='c':
powcreate(c(k)=c(k-2)/(k+1),c(0)=1,c(1)=0):
inv=inverse(multiply(c,c)):
tpsform(inv,x,7);
```

30. Substituting $y = \sum_{n=0}^{\infty} c_n x^{n+r}$ into the DE and collecting terms, we obtain

$$xy'' + y' + y = r^2 c_0 x^{r-1} + \sum_{k=1}^{\infty}[(k+r)(k+r-1)c_k + (k+r)c_k + c_{k-1}]x^{k+r-1} = 0$$

136

which implies $r^2 = 0$ and

$$(k + r)^2 c_k + c_{k-1} = 0.$$

The indicial roots are $r_1 = r_2 = 0$ and the recurrence relation is

$$c_k = -\frac{c_{k-1}}{k^2}, \quad k = 1, 2, 3, \ldots.$$

One solution is

$$y_1 = c_0 \left(1 - x + \frac{1}{2^2} x^2 - \frac{1}{(3!)^2} x^3 + \frac{1}{(4!)^2} x^4 - \cdots \right) = c_0 \sum_{n=0}^{\infty} \frac{(-1)^n}{(n!)^2} x^n.$$

A second solution is

$$y_2 = y_1 \int \frac{e^{-\int (1/x)dx}}{y_1^2} \, dx = y_1 \int \frac{dx}{x \left(1 - x + \frac{1}{4}x^2 - \frac{1}{36}x^3 + \cdots \right)^2}$$

$$= y_1 \int \frac{dx}{x \left(1 - 2x + \frac{3}{2}x^2 - \frac{5}{9}x^3 + \frac{35}{288}x^4 - \cdots \right)}$$

$$= y_1 \int \frac{1}{x} \left(1 + 2x + \frac{5}{2}x^2 + \frac{23}{9}x^3 + \frac{677}{288}x^4 + \cdots \right) dx$$

$$= y_1 \int \left(\frac{1}{x} + 2 + \frac{5}{2}x + \frac{23}{9}x^2 + \frac{677}{288}x^3 + \cdots \right) dx$$

$$= y_1 \left[\ln x + 2x + \frac{5}{4}x^2 + \frac{23}{27}x^3 + \frac{677}{1{,}152}x^4 + \cdots \right]$$

$$= y_1 \ln x + y_1 \left(2x + \frac{5}{4}x^2 + \frac{23}{27}x^3 + \frac{677}{1{,}152}x^4 + \cdots \right).$$

The general solution on $(0, \infty)$ is

$$y = C_1 y_1(x) + C_2 y_2(x).$$

33. (a) From $t = 1/x$ we have $dt/dx = -1/x^2 = -t^2$. Then

$$\frac{dy}{dx} = \frac{dy}{dt} \frac{dt}{dx} = -t^2 \frac{dy}{dt}$$

and

$$\frac{d^2y}{dx^2} = \frac{d}{dx}\left(\frac{dy}{dx} \right) = \frac{d}{dx}\left(-t^2 \frac{dy}{dt} \right) = -t^2 \frac{d^2y}{dt^2} \frac{dt}{dx} - \frac{dy}{dt}\left(2t \frac{dt}{dx} \right) = t^4 \frac{d^2y}{dt^2} + 2t^3 \frac{dy}{dt}.$$

Now

$$x^4 \frac{d^2y}{dx^2} + \lambda y = \frac{1}{t^4}\left(t^4 \frac{d^2y}{dt^2} + 2t^3 \frac{dy}{dt} \right) + \lambda y = \frac{d^2y}{dt^2} + \frac{2}{t}\frac{dy}{dt} + \lambda y = 0$$

137

becomes

$$t\frac{d^2y}{dt^2} + 2\frac{dy}{dt} + \lambda ty = 0.$$

(b) Substituting $y = \sum_{n=0}^{\infty} c_n t^{n+r}$ into the DE and collecting terms, we obtain

$$t\frac{d^2y}{dt^2} + 2\frac{dy}{dt} + \lambda ty = (r^2 + r)c_0 t^{r-1} + (r^2 + 3r + 2)c_1 t^r$$

$$+ \sum_{k=2}^{\infty}[(k+r)(k+r-1)c_k + 2(k+r)c_k + \lambda c_{k-2}]t^{k+r-1}$$

$$= 0,$$

which implies

$$r^2 + r = r(r+1) = 0,$$

$$\left(r^2 + 3r + 2\right)c_1 = 0,$$

and

$$(k+r)(k+r+1)c_k + \lambda c_{k-2} = 0.$$

The indicial roots are $r_1 = 0$ and $r_2 = -1$, so $c_1 = 0$. For $r_1 = 0$ the recurrence relation is

$$c_k = -\frac{\lambda c_{k-2}}{k(k+1)}, \quad k = 2, 3, 4, \ldots,$$

and

$$c_2 = -\frac{\lambda}{3!}c_0$$

$$c_3 = c_5 = c_7 = \cdots = 0$$

$$c_4 = \frac{\lambda^2}{5!}c_0$$

$$c_{2n} = (-1)^n\frac{\lambda^n}{(2n+1)!}c_0.$$

For $r_2 = -1$ the recurrence relation is

$$c_k = -\frac{\lambda c_{k-2}}{k(k-1)}, \quad k = 2, 3, 4, \ldots,$$

138

and

$$c_2 = -\frac{\lambda}{2!}c_0$$

$$c_3 = c_5 = c_7 = \cdots = 0$$

$$c_4 = \frac{\lambda^2}{4!}c_0$$

$$c_{2n} = (-1)^n \frac{\lambda^n}{(2n)!}c_0.$$

The general solution on $(0, \infty)$ is

$$y(t) = c_1 \sum_{n=0}^{\infty} \frac{(-1)^n}{(2n+1)!}(\sqrt{\lambda}\,t)^{2n} + c_2 t^{-1} \sum_{n=0}^{\infty} \frac{(-1)^n}{(2n)!}(\sqrt{\lambda}\,t)^{2n}$$

$$= \frac{1}{t}\left[C_1 \sum_{n=0}^{\infty} \frac{(-1)^n}{(2n+1)!}(\sqrt{\lambda}\,t)^{2n+1} + C_2 \sum_{n=0}^{\infty} \frac{(-1)^n}{(2n)!}(\sqrt{\lambda}\,t)^{2n}\right]$$

$$= \frac{1}{t}[C_1 \sin\sqrt{\lambda}\,t + C_2 \cos\sqrt{\lambda}\,t].$$

(c) Using $t = 1/x$, the solution of the original equation is

$$y(x) = C_1 x \sin\frac{\sqrt{\lambda}}{x} + C_2 x \cos\frac{\sqrt{\lambda}}{x}.$$

Section 6.3   Special Functions

The terminology and concepts listed below provide an outline of the main ideas encountered in this section. These can be useful when preparing for a quiz or test.

Terminology and Concepts

- Bessel's DE of order ν
- Bessel functions of the first kind: $J_\nu(x)$ and $J_{-\nu}(x)$
- Bessel functions of the second kind: $Y_\nu(x)$
- parametric Bessel equation of order ν
- modified Bessel equation of order ν
- modified Bessel function of the first kind: $I_\nu(x)$
- modified Bessel function of the second kind: $K_\nu(x)$

139

- aging spring

- Euler's constant: γ

- differential recurrence relation

- spherical Bessel functions

- Legendre's DE of order n

- Legendre polynomials

- Rodrigues' formula

The basic skills listed below summarize the more mechanical types of problems encountered in the exercise set for this section.

Basic Skills

- solve Bessel's DE of order ν

- solve DEs of the form $x^2 y'' + xy' + (\alpha^2 x^2 - \nu^2)y = 0$

- solve DEs of the form $y'' + [(1 - 2a)/x]y' + [b^2 c^2 x^{2c-2} + (a^2 - p^2 c^2)/x^2]y = 0,\ \ p \geq 0$

- verify various recurrence relations involving Bessel functions

- use a recurrence relation to generate Legendre polynomials

- use Rodrigues' formula to generate Legendre polynomials

Use of Computers   See Section 2.3 in this manual for the syntax used to obtain Bessel functions and Legendre polynomials in *Mathematica* and *Maple*.

Exercises 6.3   *Hints, Suggestions, Solutions, and Examples*

3. Since $\nu^2 = 25/4$ the general solution is $y = c_1 J_{5/2}(x) + c_2 J_{-5/2}(x)$.

6. Since $\nu^2 = 4$ the general solution is $y = c_1 J_2(x) + c_2 Y_2(x)$.

9. We identify $\alpha = 5$ and $\nu = \frac{2}{3}$. Then the general solution is $y = c_1 J_{2/3}(5x) + c_2 J_{-2/3}(5x)$.

12. If $y = \sqrt{x}\, v(x)$ then

$$y' = x^{1/2} v'(x) + \frac{1}{2} x^{-1/2} v(x)$$

$$y'' = x^{1/2} v''(x) + x^{-1/2} v'(x) - \frac{1}{4} x^{-3/2} v(x)$$

and

$$x^2 y'' + \left(\alpha^2 x^2 - \nu^2 + \frac{1}{4}\right) y = x^{5/2} v''(x) + x^{3/2} v'(x) - \frac{1}{4} x^{1/2} v(x) + \left(\alpha^2 x^2 - \nu^2 + \frac{1}{4}\right) x^{1/2} v(x)$$

$$= x^{5/2} v''(x) + x^{3/2} v'(x) + (\alpha^2 x^{5/2} - \nu^2 x^{1/2}) v(x) = 0.$$

Multiplying by $x^{-1/2}$ we obtain

$$x^2 v''(x) + x v'(x) + (\alpha^2 x^2 - \nu^2) v(x) = 0,$$

whose solution is $v(x) = c_1 J_\nu(\alpha x) + c_2 Y_\nu(\alpha x)$. Then $y = c_1 \sqrt{x}\, J_\nu(\alpha x) + c_2 \sqrt{x}\, Y_\nu(\alpha x)$.

15. Write the DE in the form $y'' - (1/x)y' + y = 0$. This is the form of (18) in the text with

$$1 - 2a = -1 \implies a = 1$$

$$2c - 2 = 0 \implies c = 1$$

$$b^2 c^2 = 1 \implies b = 1$$

$$a^2 - p^2 c^2 = 0 \implies p = 1.$$

Then, by (19) in the text, the general solution is

$$y = x[c_1 J_1(x) + c_2 Y_1(x)].$$

18. Write the DE in the form $y'' + (4 + 1/4x^2)y = 0$. This is the form of (18) in the text with

$$1 - 2a = 0 \implies a = \frac{1}{2}$$

$$2c - 2 = 0 \implies c = 1$$

$$b^2 c^2 = 4 \implies b = 2$$

$$a^2 - p^2 c^2 = \frac{1}{4} \implies p = 0.$$

Then, by (19) in the text, the general solution is

$$y = x^{1/2}[c_1 J_0(2x) + c_2 Y_0(2x)].$$

141

21. Using the fact that $i^2 = -1$, along with the definition of $J_\nu(x)$ in (7) in the text, we have

$$I_\nu(x) = i^{-\nu} J_\nu(ix) = i^{-\nu} \sum_{n=0}^{\infty} \frac{(-1)^n}{n!\Gamma(1+\nu+n)} \left(\frac{ix}{2}\right)^{2n+\nu}$$

$$= \sum_{n=0}^{\infty} \frac{(-1)^n}{n!\Gamma(1+\nu+n)} i^{2n+\nu-\nu} \left(\frac{x}{2}\right)^{2n+\nu}$$

$$= \sum_{n=0}^{\infty} \frac{(-1)^n}{n!\Gamma(1+\nu+n)} (i^2)^n \left(\frac{x}{2}\right)^{2n+\nu}$$

$$= \sum_{n=0}^{\infty} \frac{(-1)^{2n}}{n!\Gamma(1+\nu+n)} \left(\frac{x}{2}\right)^{2n+\nu}$$

$$= \sum_{n=0}^{\infty} \frac{1}{n!\Gamma(1+\nu+n)} \left(\frac{x}{2}\right)^{2n+\nu},$$

which is a real function.

24. Write the DE in the form $y'' + (4/x)y' + (1+2/x^2)y = 0$. This is the form of (18) in the text with

$$1 - 2a = 4 \implies a = -\frac{3}{2}$$

$$2c - 2 = 0 \implies c = 1$$

$$b^2 c^2 = 1 \implies b = 1$$

$$a^2 - p^2 c^2 = 2 \implies p = \frac{1}{2}.$$

Then, by (19), (23), and (24) in the text,

$$y = x^{-3/2} [c_1 J_{1/2}(x) + c_2 J_{-1/2}(x)] = x^{-3/2} \left[c_1 \sqrt{\frac{2}{\pi x}} \sin x + c_2 \sqrt{\frac{2}{\pi x}} \cos x \right]$$

$$= C_1 \frac{1}{x^2} \sin x + C_2 \frac{1}{x^2} \cos x.$$

27. (a) The recurrence relation follows from

$$-\nu J_\nu(x) + x J_{\nu-1}(x) = -\sum_{n=0}^{\infty} \frac{(-1)^n \nu}{n!\Gamma(1+\nu+n)} \left(\frac{x}{2}\right)^{2n+\nu} + x \sum_{n=0}^{\infty} \frac{(-1)^n}{n!\Gamma(\nu+n)} \left(\frac{x}{2}\right)^{2n+\nu-1}$$

$$= -\sum_{n=0}^{\infty} \frac{(-1)^n \nu}{n!\Gamma(1+\nu+n)} \left(\frac{x}{2}\right)^{2n+\nu} + \sum_{n=0}^{\infty} \frac{(-1)^n (\nu+n)}{n!\Gamma(1+\nu+n)} \cdot 2 \left(\frac{x}{2}\right) \left(\frac{x}{2}\right)^{2n+\nu-1}$$

$$= \sum_{n=0}^{\infty} \frac{(-1)^n (2n+\nu)}{n!\Gamma(1+\nu+n)} \left(\frac{x}{2}\right)^{2n+\nu} = x J_\nu'(x).$$

142

(b) The formula in part (a) is a linear first-order DE in $J_\nu(x)$. An integrating factor for this equation is x^ν, so

$$\frac{d}{dx}[x^\nu J_\nu(x)] = x^\nu J_{\nu-1}(x).$$

30. From (20) we obtain $J_0'(x) = -J_1(x)$, and from (21) we obtain $J_0'(x) = J_{-1}(x)$. Thus $J_0'(x) = J_{-1}(x) = -J_1(x)$.

33. Letting

$$s = \frac{2}{\alpha}\sqrt{\frac{k}{m}}\,e^{-\alpha t/2},$$

we have

$$\frac{dx}{dt} = \frac{dx}{ds}\frac{ds}{dt} = \frac{dx}{dt}\left[\frac{2}{\alpha}\sqrt{\frac{k}{m}}\left(-\frac{\alpha}{2}\right)e^{-\alpha t/2}\right] = \frac{dx}{ds}\left(-\sqrt{\frac{k}{m}}\,e^{-\alpha t/2}\right)$$

and

$$\frac{d^2x}{dt^2} = \frac{d}{dt}\left(\frac{dx}{dt}\right) = \frac{dx}{ds}\left(\frac{\alpha}{2}\sqrt{\frac{k}{m}}\,e^{-\alpha t/2}\right) + \frac{d}{dt}\left(\frac{dx}{ds}\right)\left(-\sqrt{\frac{k}{m}}\,e^{-\alpha t/2}\right)$$

$$= \frac{dx}{ds}\left(\frac{\alpha}{2}\sqrt{\frac{k}{m}}\,e^{-\alpha t/2}\right) + \frac{d^2x}{ds^2}\frac{ds}{dt}\left(-\sqrt{\frac{k}{m}}\,e^{-\alpha t/2}\right)$$

$$= \frac{dx}{ds}\left(\frac{\alpha}{2}\sqrt{\frac{k}{m}}\,e^{-\alpha t/2}\right) + \frac{d^2x}{ds^2}\left(\frac{k}{m}\,e^{-\alpha t}\right).$$

Then

$$m\frac{d^2x}{dt^2} + ke^{-\alpha t}x = ke^{-\alpha t}\frac{d^2x}{ds^2} + \frac{m\alpha}{2}\sqrt{\frac{k}{m}}\,e^{-\alpha t/2}\frac{dx}{ds} + ke^{-\alpha t}x = 0.$$

Multiplying by $2^2/\alpha^2 m$ we have

$$\frac{2^2}{\alpha^2}\frac{k}{m}e^{-\alpha t}\frac{d^2x}{ds^2} + \frac{2}{\alpha}\sqrt{\frac{k}{m}}\,e^{-\alpha t/2}\frac{dx}{ds} + \frac{2^2}{\alpha^2}\frac{k}{m}e^{-\alpha t}x = 0$$

or, since $s = (2/\alpha)\sqrt{k/m}\,e^{-\alpha t/2}$,

$$s^2\frac{d^2x}{ds^2} + s\frac{dx}{ds} + s^2x = 0.$$

36. The general solution of the DE is

$$y(x) = c_1 J_0(\alpha x) + c_2 Y_0(\alpha x).$$

In order to satisfy the conditions that $\lim_{x\to 0^+} y(x)$ and $\lim_{x\to 0^+} y'(x)$ are finite we are forced to define $c_2 = 0$. Thus, $y(x) = c_1 J_0(\alpha x)$. The second boundary condition, $y(2) = 0$, implies $c_1 = 0$ or $J_0(2\alpha) = 0$. In order to have a nontrivial solution we require that $J_0(2\alpha) = 0$. From Table 6.1, the first three positive zeros of J_0 are found to be

$$2\alpha_1 = 2.4048, \quad 2\alpha_2 = 5.5201, \quad 2\alpha_3 = 8.6537$$

143

and so $\alpha_1 = 1.2024$, $\alpha_2 = 2.7601$, $\alpha_3 = 4.3269$. The eigenfunctions corresponding to the eigenvalues $\lambda_1 = \alpha_1^2$, $\lambda_2 = \alpha_2^2$, $\lambda_3 = \alpha_3^2$ are $J_0(1.2024x)$, $J_0(2.7601x)$, and $J_0(4.3269x)$.

38. (*Suggestion*) See the table of special functions in Section 2.3 of this manual for syntax in *Mathematica* and *Maple* for the modified Bessel functions.

45. The recurrence relation can be written

$$P_{k+1}(x) = \frac{2k+1}{k+1}\, xP_k(x) - \frac{k}{k+1}\, P_{k-1}(x), \qquad k = 2,\ 3,\ 4,\ \ldots\ .$$

$k=1$: $P_2(x) = \dfrac{3}{2}x^2 - \dfrac{1}{2}$

$k=2$: $P_3(x) = \dfrac{5}{3}x\left(\dfrac{3}{2}x^2 - \dfrac{1}{2}\right) - \dfrac{2}{3}x = \dfrac{5}{2}x^3 - \dfrac{3}{2}x$

$k=3$: $P_4(x) = \dfrac{7}{4}x\left(\dfrac{5}{2}x^3 - \dfrac{3}{2}x\right) - \dfrac{3}{4}\left(\dfrac{3}{2}x^2 - \dfrac{1}{2}\right) = \dfrac{35}{8}x^4 - \dfrac{30}{8}x^2 + \dfrac{3}{8}$

$k=4$: $P_5(x) = \dfrac{9}{5}x\left(\dfrac{35}{8}x^4 - \dfrac{30}{8}x^2 + \dfrac{3}{8}\right) - \dfrac{4}{5}\left(\dfrac{5}{2}x^3 - \dfrac{3}{2}x\right) = \dfrac{63}{8}x^5 - \dfrac{35}{4}x^3 + \dfrac{15}{8}x$

$k=5$: $P_6(x) = \dfrac{11}{6}x\left(\dfrac{63}{8}x^5 - \dfrac{35}{4}x^3 + \dfrac{15}{8}x\right) - \dfrac{5}{6}\left(\dfrac{35}{8}x^4 - \dfrac{30}{8}x^2 + \dfrac{3}{8}\right) = \dfrac{231}{16}x^6 - \dfrac{315}{16}x^4 + \dfrac{105}{16}x^2 - \dfrac{5}{16}$

$k=6$: $P_7(x) = \dfrac{13}{7}x\left(\dfrac{231}{16}x^6 - \dfrac{315}{16}x^4 + \dfrac{105}{16}x^2 - \dfrac{5}{16}\right) - \dfrac{6}{7}\left(\dfrac{63}{8}x^5 - \dfrac{35}{4}x^3 + \dfrac{15}{8}x\right)$

$$= \frac{429}{16}x^7 - \frac{693}{16}x^5 + \frac{315}{16}x^3 - \frac{35}{16}x$$

48. Using a CAS we find

$$P_1(x) = \frac{1}{2}\frac{d}{dx}\left(x^2 - 1\right)^1 = x$$

$$P_2(x) = \frac{1}{2^2 2!}\frac{d^2}{dx^2}\left(x^2 - 1\right)^2 = \frac{1}{2}(3x^2 - 1)$$

$$P_3(x) = \frac{1}{2^3 3!}\frac{d^3}{dx^3}\left(x^2 - 1\right)^3 = \frac{1}{2}(5x^3 - 3x)$$

$$P_4(x) = \frac{1}{2^4 4!}\frac{d^4}{dx^4}\left(x^2 - 1\right)^4 = \frac{1}{8}(35x^4 - 30x^2 + 3)$$

$$P_5(x) = \frac{1}{2^5 5!}\frac{d^5}{dx^5}\left(x^2 - 1\right)^5 = \frac{1}{8}(63x^5 - 70x^3 + 15x)$$

$$P_6(x) = \frac{1}{2^6 6!}\frac{d^6}{dx^6}\left(x^2 - 1\right)^6 = \frac{1}{16}(231x^6 - 315x^4 + 105x^2 - 5)$$

$$P_7(x) = \frac{1}{2^7 7!}\frac{d^7}{dx^7}\left(x^2 - 1\right)^7 = \frac{1}{16}(429x^7 - 693x^5 + 315x^3 - 35x)$$

3. $x = -1$ is the nearest singular point to the ordinary point $x = 0$. Theorem 6.1 guarantees the existence of two power series solutions $y = \sum_{n=1}^{\infty} c_n x^n$ of the DE that converge at least for $-1 < x < 1$. Since $-\frac{1}{2} \le x \le \frac{1}{2}$ is properly contained in $-1 < x < 1$, both power series must converge for all points contained in $-\frac{1}{2} \le x \le \frac{1}{2}$.

6. We have

$$f(x) = \frac{\sin x}{\cos x} = \frac{x - \dfrac{x^3}{6} + \dfrac{x^5}{120} - \cdots}{1 - \dfrac{x^2}{2} + \dfrac{x^4}{24} - \cdots} = x + \frac{x^3}{3} + \frac{2x^5}{15} + \cdots.$$

9. Substituting $y = \sum_{n=0}^{\infty} c_n x^{n+r}$ into the DE we obtain

$$2xy'' + y' + y = \left(2r^2 - r\right)c_0 x^{r-1} + \sum_{k=1}^{\infty}[2(k+r)(k+r-1)c_k + (k+r)c_k + c_{k-1}]x^{k+r-1} = 0$$

which implies

$$2r^2 - r = r(2r - 1) = 0$$

and

$$(k+r)(2k + 2r - 1)c_k + c_{k-1} = 0.$$

The indicial roots are $r = 0$ and $r = 1/2$. For $r = 0$ the recurrence relation is

$$c_k = -\frac{c_{k-1}}{k(2k-1)}, \quad k = 1, 2, 3, \ldots,$$

so

$$c_1 = -c_0, \qquad c_2 = \frac{1}{6}c_0, \qquad c_3 = -\frac{1}{90}c_0.$$

For $r = 1/2$ the recurrence relation is

$$c_k = -\frac{c_{k-1}}{k(2k+1)}, \quad k = 1, 2, 3, \ldots,$$

so

$$c_1 = -\frac{1}{3}c_0, \qquad c_2 = \frac{1}{30}c_0, \qquad c_3 = -\frac{1}{630}c_0.$$

Two linearly independent solutions are

$$y_1 = 1 - x + \frac{1}{6}x^2 - \frac{1}{90}x^3 + \cdots$$

and

$$y_2 = x^{1/2}\left(1 - \frac{1}{3}x + \frac{1}{30}x^2 - \frac{1}{630}x^3 + \cdots\right).$$

Chapter 6 in Review

12. Substituting $y = \sum_{n=0}^{\infty} c_n x^n$ into the DE we obtain

$$y'' - x^2 y' + xy = 2c_2 + (6c_3 + c_0)x + \sum_{k=1}^{\infty} [(k+3)(k+2)c_{k+3} - (k-1)c_k]x^{k+1} = 0$$

which implies $c_2 = 0$, $c_3 = -c_0/6$, and

$$c_{k+3} = \frac{k-1}{(k+3)(k+2)}c_k, \quad k = 1, 2, 3, \dots .$$

Choosing $c_0 = 1$ and $c_1 = 0$ we find

$$c_3 = -\frac{1}{6}$$

$$c_4 = c_7 = c_{10} = \dots = 0$$

$$c_5 = c_8 = c_{11} = \dots = 0$$

$$c_6 = -\frac{1}{90}$$

and so on. For $c_0 = 0$ and $c_1 = 1$ we obtain

$$c_3 = c_6 = c_9 = \dots = 0$$

$$c_4 = c_7 = c_{10} = \dots = 0$$

$$c_5 = c_8 = c_{11} = \dots = 0$$

and so on. Thus, two solutions are

$$y_1 = 1 - \frac{1}{6}x^3 - \frac{1}{90}x^6 - \cdots \quad \text{and} \quad y_2 = x.$$

15. Substituting $y = \sum_{n=0}^{\infty} c_n x^n$ into the DE we have

$$y'' + xy' + 2y = \underbrace{\sum_{n=2}^{\infty} n(n-1)c_n x^{n-2}}_{k=n-2} + \underbrace{\sum_{n=1}^{\infty} nc_n x^n}_{k=n} + 2\underbrace{\sum_{n=0}^{\infty} c_n x^n}_{k=n}$$

$$= \sum_{k=0}^{\infty} (k+2)(k+1)c_{k+2}x^k + \sum_{k=1}^{\infty} kc_k x^k + 2\sum_{k=0}^{\infty} c_k x^k$$

$$= 2c_2 + 2c_0 + \sum_{k=1}^{\infty} [(k+2)(k+1)c_{k+2} + (k+2)c_k]x^k = 0.$$

Thus

$$2c_2 + 2c_0 = 0$$

$$(k+2)(k+1)c_{k+2} + (k+2)c_k = 0$$

and

$$c_2 = -c_0$$

$$c_{k+2} = -\frac{1}{k+1}c_k, \quad k = 1, 2, 3, \ldots.$$

Choosing $c_0 = 1$ and $c_1 = 0$ we find

$$c_2 = -1$$

$$c_3 = c_5 = c_7 = \cdots = 0$$

$$c_4 = \frac{1}{3}$$

$$c_6 = -\frac{1}{15}$$

and so on. For $c_0 = 0$ and $c_1 = 1$ we obtain

$$c_2 = c_4 = c_6 = \cdots = 0$$

$$c_3 = -\frac{1}{2}$$

$$c_5 = \frac{1}{8}$$

$$c_7 = -\frac{1}{48}$$

and so on. Thus, the general solution is

$$y = C_0\left(1 - x^2 + \frac{1}{3}x^4 - \frac{1}{15}x^6 + \cdots\right) + C_1\left(x - \frac{1}{2}x^3 + \frac{1}{8}x^5 - \frac{1}{48}x^7 + \cdots\right)$$

and

$$y' = C_0\left(-2x + \frac{4}{3}x^3 - \frac{2}{5}x^5 + \cdots\right) + C_1\left(1 - \frac{3}{2}x^2 + \frac{5}{8}x^4 - \frac{7}{48}x^6 + \cdots\right).$$

Setting $y(0) = 3$ and $y'(0) = -2$ we find $c_0 = 3$ and $c_1 = -2$. Therefore, the solution of the IVP is

$$y = 3 - 2x - 3x^2 + x^3 + x^4 - \frac{1}{4}x^5 - \frac{1}{5}x^6 + \frac{1}{24}x^7 + \cdots.$$

18. While we can find two solutions of the form

$$y_1 = c_0[1 + \cdots] \quad \text{and} \quad y_2 = c_1[x + \cdots],$$

the initial conditions at $x = 1$ give solutions for c_0 and c_1 in terms of infinite series. Letting $t = x - 1$ the IVP becomes

$$\frac{d^2y}{dt^2} + (t+1)\frac{dy}{dt} + y = 0, \qquad y(0) = -6, \ y'(0) = 3.$$

Chapter 6 in Review

Substituting $y = \sum_{n=0}^{\infty} c_n t^n$ into the DE we have

$$\frac{d^2y}{dt^2} + (t+1)\frac{dy}{dt} + y = \sum_{n=2}^{\infty} n(n-1)c_n t^{n-2} + \underbrace{\sum_{n=1}^{\infty} nc_n t^n}_{k=n} + \underbrace{\sum_{n=1}^{\infty} nc_n t^{n-1}}_{k=n-1} + \underbrace{\sum_{n=0}^{\infty} c_n t^n}_{k=n}$$

$$\underbrace{\phantom{\sum_{n=2}^{\infty} n(n-1)c_n t^{n-2}}}_{k=n-2}$$

$$= \sum_{k=0}^{\infty} (k+2)(k+1)c_{k+2} t^k + \sum_{k=1}^{\infty} kc_k t^k + \sum_{k=0}^{\infty} (k+1)c_{k+1} t^k + \sum_{k=0}^{\infty} c_k t^k$$

$$= 2c_2 + c_1 + c_0 + \sum_{k=1}^{\infty} [(k+2)(k+1)c_{k+2} + (k+1)c_{k+1} + (k+1)c_k]t^k = 0.$$

Thus
$$2c_2 + c_1 + c_0 = 0$$

$$(k+2)(k+1)c_{k+2} + (k+1)c_{k+1} + (k+1)c_k = 0$$

and
$$c_2 = -\frac{c_1 + c_0}{2}$$

$$c_{k+2} = -\frac{c_{k+1} + c_k}{k+2}, \quad k = 1, 2, 3, \ldots.$$

Choosing $c_0 = 1$ and $c_1 = 0$ we find

$$c_2 = -\frac{1}{2}, \quad c_3 = \frac{1}{6}, \quad c_4 = \frac{1}{12},$$

and so on. For $c_0 = 0$ and $c_1 = 1$ we find

$$c_2 = -\frac{1}{2}, \quad c_3 = -\frac{1}{6}, \quad c_4 = \frac{1}{6},$$

and so on. Thus, the general solution is

$$y = c_0 \left[1 - \frac{1}{2}t^2 + \frac{1}{6}t^3 + \frac{1}{12}t^4 + \cdots \right] + c_1 \left[t - \frac{1}{2}t^2 - \frac{1}{6}t^3 + \frac{1}{6}t^4 + \cdots \right].$$

The initial conditions then imply $c_0 = -6$ and $c_1 = 3$. Thus the solution of the IVP is

$$y = -6 \left[1 - \frac{1}{2}(x-1)^2 + \frac{1}{6}(x-1)^3 + \frac{1}{12}(x-1)^4 + \cdots \right]$$

$$+ 3 \left[(x-1) - \frac{1}{2}(x-1)^2 - \frac{1}{6}(x-1)^3 + \frac{1}{6}(x-1)^4 + \cdots \right].$$

21. Substituting $y = \sum_{n=0}^{\infty} c_n x^n$ into the DE we have

$$y'' + x^2 y' + 2xy = \sum_{n=2}^{\infty} n(n-1)c_n x^{n-2} + \underbrace{\sum_{n=1}^{\infty} nc_n x^{n+1}}_{k=n+1} + 2\underbrace{\sum_{n=0}^{\infty} c_n x^{n+1}}_{k=n+1}$$

$$\underbrace{\phantom{\sum_{n=2}^{\infty} n(n-1)c_n x^{n-2}}}_{k=n-2}$$

$$= \sum_{k=0}^{\infty} (k+2)(k+1)c_{k+2} x^k + \sum_{k=2}^{\infty} (k-1)c_{k-1} x^k + 2\sum_{k=1}^{\infty} c_{k-1} x^k$$

$$= 2c_2 + (6c_3 + 2c_0)x + \sum_{k=2}^{\infty} [(k+2)(k+1)c_{k+2} + (k+1)c_{k-1}]x^k = 5 - 2x + 10x^3.$$

Thus, equating coefficients of like powers of x gives

$$2c_2 = 5$$

$$6c_3 + 2c_0 = -2$$

$$12c_4 + 3c_1 = 0$$

$$20c_5 + 4c_2 = 10$$

$$(k+2)(k+1)c_{k+2} + (k+1)c_{k-1} = 0, \quad k = 4, 5, 6, \ldots,$$

and

$$c_2 = \frac{5}{2}$$

$$c_3 = -\frac{1}{3}c_0 - \frac{1}{3}$$

$$c_4 = -\frac{1}{4}c_1$$

$$c_5 = \frac{1}{2} - \frac{1}{5}c_2 = \frac{1}{2} - \frac{1}{5}\left(\frac{5}{2}\right) = 0$$

$$c_{k+2} = -\frac{1}{k+2}c_{k-1}.$$

Using the recurrence relation, we find

$$c_6 = -\frac{1}{6}c_3 = \frac{1}{3 \cdot 6}(c_0 + 1) = \frac{1}{3^2 \cdot 2!}c_0 + \frac{1}{3^2 \cdot 2!}$$

$$c_7 = -\frac{1}{7}c_4 = \frac{1}{4 \cdot 7}c_1$$

$$c_8 = c_{11} = c_{14} = \cdots = 0$$

$$c_9 = -\frac{1}{9}c_6 = -\frac{1}{3^3 \cdot 3!}c_0 - \frac{1}{3^3 \cdot 3!}$$

$$c_{10} = -\frac{1}{10}c_7 = -\frac{1}{4 \cdot 7 \cdot 10}c_1$$

$$c_{12} = -\frac{1}{12}c_9 = \frac{1}{3^4 \cdot 4!}c_0 + \frac{1}{3^4 \cdot 4!}$$

$$c_{13} = -\frac{1}{13}c_0 = \frac{1}{4 \cdot 7 \cdot 10 \cdot 13}c_1$$

and so on. Thus

$$y = c_0 \left[1 - \frac{1}{3}x^3 + \frac{1}{3^2 \cdot 2!}x^6 - \frac{1}{3^3 \cdot 3!}x^9 + \frac{1}{3^4 \cdot 4!}x^{12} - \cdots \right]$$

$$+ c_1 \left[x - \frac{1}{4}x^4 + \frac{1}{4 \cdot 7}x^7 - \frac{1}{4 \cdot 7 \cdot 10}x^{10} + \frac{1}{4 \cdot 7 \cdot 10 \cdot 13}x^{13} - \cdots \right]$$

$$+ \left[\frac{5}{2}x^2 - \frac{1}{3}x^3 + \frac{1}{3^2 \cdot 2!}x^6 - \frac{1}{3^3 \cdot 3!}x^9 + \frac{1}{3^4 \cdot 4!}x^{12} - \cdots \right].$$

24. **(a)** Using formula (5) of Section 4.2 in the text, we find that a second solution of $(1-x^2)y'' - 2xy' = 0$ is

$$y_2(x) = 1 \cdot \int \frac{e^{\int 2x\,dx/(1-x^2)}}{1^2}\,dx = \int e^{-\ln(1-x^2)}\,dx$$

$$= \int \frac{dx}{1-x^2} = \frac{1}{2}\ln\left(\frac{1+x}{1-x}\right),$$

where partial fractions was used to obtain the last integral. (The formula for this integral is also given in Section 2.2 of this manual.)

(b) Using formula (5) of Section 4.2 in the text, we find that a second solution of $(1-x^2)y'' - 2xy' + 2y = 0$ is

$$y_2(x) = x \cdot \int \frac{e^{\int 2x\,dx/(1-x^2)}}{x^2}\,dx = x \int \frac{e^{-\ln(1-x^2)}}{x^2}\,dx$$

$$= x \int \frac{dx}{x^2(1-x^2)}\,dx = x \left[\frac{1}{2}\ln\left(\frac{1+x}{1-x}\right) - \frac{1}{x} \right]$$

$$= \frac{x}{2}\ln\left(\frac{1+x}{1-x}\right) - 1,$$

where partial fractions was used to obtain the last integral.

(c)

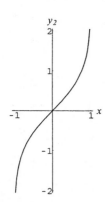

$$y_2(x) = \frac{1}{2}\ln\left(\frac{1+x}{1-x}\right) \qquad\qquad y_2 = \frac{x}{2}\ln\left(\frac{1+x}{1-x}\right) - 1$$

7

Laplace Transform

Definition of the Laplace Transform

The terminology and concepts listed below provide an outline of the main ideas encountered in this section. These can be useful when preparing for a quiz or test.

Terminology and Concepts

- integral transform

- definition of the Laplace transform

- linearity of the Laplace transform

- exponential order

- piecewise-continuous function

- transform of a piecewise-continuous function

- behavior of the Laplace transform $F(s)$ as $s \to \infty$

The basic skills listed below summarize the more mechanical types of problems encountered in the exercise set for this section.

Basic Skills

- find the Laplace transform of a function using the definition

- find the Laplace transform of a function using linearity and known transforms of some basic functions

Piecewise Continuity   A function f that is piecewise continuous on $[0, \infty)$ has only a finite number of *finite* discontinuities on any finite interval of $[0, \infty)$. Another name for a **finite discontinuity** is a **jump discontinuity**; this means that at any point of discontinuity t_1, the left- and right-hand limits, $\lim_{t \to t_1^-} f(t)$ and $\lim_{t \to t_1^+} f(t)$, respectively, exist but are different. Put another way, f cannot have any vertical asymptotes on the interval $[0, \infty)$.

151

Section 7.1 Definition of the Laplace Transform

Use of Computers   The Laplace transform function is built into version 5 of *Mathematica*. In *Maple* it is contained in the integral transforms package `inttrans`. The routines below define the Laplace transform of $f(t) = t^2 \cos t$ as a function $F(s)$.

$$\text{Clear[f]} \hspace{4cm} (\textit{Mathematica})$$

```
Clear[f]                                    (Mathematica)
F[s_]:=Evaluate[LaplaceTransform[t^2 Cos[t], t, s]]
F[s]
```

```
with(inttrans);                             (Maple)
F:=s->laplace(t^2*cos(t),t,s);
F(s);
```

Exercises 7.1   *Hints, Suggestions, Solutions, and Examples*

3. $\mathcal{L}\{f(t)\} = \displaystyle\int_0^1 te^{-st}dt + \int_1^\infty e^{-st}dt = \left(-\frac{1}{s}te^{-st} - \frac{1}{s^2}e^{-st}\right)\Big|_0^1 - \frac{1}{s}e^{-st}\Big|_1^\infty$

$$= \left(-\frac{1}{s}e^{-s} - \frac{1}{s^2}e^{-s}\right) - \left(0 - \frac{1}{s^2}\right) - \frac{1}{s}(0 - e^{-s}) = \frac{1}{s^2}(1 - e^{-s}), \quad s > 0$$

6. $\mathcal{L}\{f(t)\} = \displaystyle\int_{\pi/2}^\infty (\cos t)e^{-st}dt = \left(-\frac{s}{s^2+1}e^{-st}\cos t + \frac{1}{s^2+1}e^{-st}\sin t\right)\Big|_{\pi/2}^\infty$

$$= 0 - \left(0 + \frac{1}{s^2+1}e^{-\pi s/2}\right) = -\frac{1}{s^2+1}e^{-\pi s/2}, \quad s > 0$$

9. The function is $f(t) = \begin{cases} 1-t, & 0 < t < 1 \\ 0, & t > 0 \end{cases}$ so

$$\mathcal{L}\{f(t)\} = \int_0^1 (1-t)e^{-st}\,dt + \int_1^\infty 0e^{-st}\,dt = \int_0^1 (1-t)e^{-st}\,dt = \left(-\frac{1}{s}(1-t)e^{-st} + \frac{1}{s^2}e^{-st}\right)\Big|_0^1$$

$$= \frac{1}{s^2}e^{-s} + \frac{1}{s} - \frac{1}{s^2}, \quad s > 0$$

12. $\mathcal{L}\{f(t)\} = \displaystyle\int_0^\infty e^{-2t-5}e^{-st}dt = e^{-5}\int_0^\infty e^{-(s+2)t}dt = -\frac{e^{-5}}{s+2}e^{-(s+2)t}\Big|_0^\infty = \frac{e^{-5}}{s+2}, \quad s > -2$

15. $\mathcal{L}\{f(t)\} = \displaystyle\int_0^\infty e^{-t}(\sin t)e^{-st}dt = \int_0^\infty (\sin t)e^{-(s+1)t}dt$

$$= \left(\frac{-(s+1)}{(s+1)^2+1}e^{-(s+1)t}\sin t - \frac{1}{(s+1)^2+1}e^{-(s+1)t}\cos t\right)\Big|_0^\infty$$

$$= \frac{1}{(s+1)^2+1} = \frac{1}{s^2+2s+2}, \quad s > -1$$

18. $\mathcal{L}\{f(t)\} = \displaystyle\int_0^\infty t(\sin t)e^{-st}\,dt$

$$= \left[\left(-\frac{t}{s^2+1} - \frac{2s}{(s^2+1)^2}\right)(\cos t)e^{-st} - \left(\frac{st}{s^2+1} + \frac{s^2-1}{(s^2+1)^2}\right)(\sin t)e^{-st}\right]_0^\infty$$

$$= \frac{2s}{(s^2+1)^2}, \quad s > 0$$

21. $\mathcal{L}\{4t - 10\} = \dfrac{4}{s^2} - \dfrac{10}{s}$

24. $\mathcal{L}\{-4t^2 + 16t + 9\} = -4\dfrac{2}{s^3} + \dfrac{16}{s^2} + \dfrac{9}{s}$

27. $\mathcal{L}\{1 + e^{4t}\} = \dfrac{1}{s} + \dfrac{1}{s-4}$

30. $\mathcal{L}\{e^{2t} - 2 + e^{-2t}\} = \dfrac{1}{s-2} - \dfrac{2}{s} + \dfrac{1}{s+2}$

33. $\mathcal{L}\{\sinh kt\} = \dfrac{1}{2}\mathcal{L}\{e^{kt} - e^{-kt}\} = \dfrac{1}{2}\left[\dfrac{1}{s-k} - \dfrac{1}{s+k}\right] = \dfrac{k}{s^2 - k^2}$

36. $\mathcal{L}\{e^{-t}\cosh t\} = \mathcal{L}\left\{e^{-t}\dfrac{e^t + e^{-t}}{2}\right\} = \mathcal{L}\left\{\dfrac{1}{2} + \dfrac{1}{2}e^{-2t}\right\} = \dfrac{1}{2s} + \dfrac{1}{2(s+2)}$

39. From the addition formula for the sine function, $\sin(4t + 5) = \sin 4t \cos 5 + \cos 4t \sin 5$ so

$$\mathcal{L}\{\sin(4t + 5)\} = \cos 5 \,\mathcal{L}\{\sin 4t\} + \sin 5 \,\mathcal{L}\{\cos 4t\}$$

$$= \cos 5 \,\frac{4}{s^2 + 16} + \sin 5 \,\frac{s}{s^2 + 16}$$

$$= \frac{4\cos 5 + (\sin 5)s}{s^2 + 16}.$$

42. (a) $\mathcal{L}\{t^{-1/2}\} = \dfrac{\Gamma(1/2)}{s^{1/2}} = \sqrt{\dfrac{\pi}{s}}$

(b) $\mathcal{L}\{t^{1/2}\} = \dfrac{\Gamma(3/2)}{s^{3/2}} = \dfrac{\sqrt{\pi}}{2s^{3/2}}$

(c) $\mathcal{L}\{t^{3/2}\} = \dfrac{\Gamma(5/2)}{s^{5/2}} = \dfrac{3\sqrt{\pi}}{4s^{5/2}}$

43. (*Hint*) Think of a differentiable function that is of exponential order but whose derivative is not.

48. (*Hint*) Use integration by parts.

Section 7.2

Inverse Transform and Transforms of Derivatives

The terminology and concepts listed below provide an outline of the main ideas encountered in this section. These can be useful when preparing for a quiz or test.

Terminology and Concepts

- definition of the inverse Laplace transform
- linearity of the inverse Laplace transform
- Laplace transform of the derivative of a function
- use of the Laplace transform to solve an IVP

The basic skills listed below summarize the more mechanical types of problems encountered in the exercise set for this section.

Basic Skills

- find the inverse Laplace transform of a function (frequently using partial fractions)
- solve a linear IVP using the Laplace transform

Partial Fractions Again   The key to success in working with the Laplace transform is the ability to do algebra quickly and accurately. Partial fraction decomposition is used almost every time you have to solve some sort of equation. In this manual, check out the review of partial fractions given in Section 2.2.

Use of Computers   As can be seen in this section, partial fractions are used extensively in the computation of inverse Laplace transforms. This process can be very computationally intensive, and as such, is ideally suited to a CAS. The examples below show how to find the partial fraction decomposition of $(s^2 + 1)/(s^3 + 4s)$.

> Apart[(s^2+1)/(s^3+4s)]   (*Mathematica*)
>
> convert((s^2+1)/(s^3+4*s),parfrac,s);   (*Maple*)

The routines below define the inverse Laplace transform of $F(s) = (2s^3 - 6s)/(1 + s^2)^3$ as a function $f(t)$.

> Clear[f]   (*Mathematica*)
> f[t_]:=Evaluate[InverseLaplaceTransform[(2s^3 - 6s)/(s^2 + 1)^3, s, t]]
> f[t]

```
with(inttrans);                                    (Maple)
f:=t->invlaplace((2*s^3-6*s)/(s^2+1)^3,s,t);
f(t);
```

Exercises 7.2 *Hints, Suggestions, Solutions, and Examples*

3. $\mathscr{L}^{-1}\left\{\dfrac{1}{s^2} - \dfrac{48}{s^5}\right\} = \mathscr{L}^{-1}\left\{\dfrac{1}{s^2} - \dfrac{48}{24} \cdot \dfrac{4!}{s^5}\right\} = t - 2t^4$

6. $\mathscr{L}^{-1}\left\{\dfrac{(s+2)^2}{s^3}\right\} = \mathscr{L}^{-1}\left\{\dfrac{1}{s} + 4 \cdot \dfrac{1}{s^2} + 2 \cdot \dfrac{2}{s^3}\right\} = 1 + 4t + 2t^2$

9. $\mathscr{L}^{-1}\left\{\dfrac{1}{4s+1}\right\} = \dfrac{1}{4}\mathscr{L}^{-1}\left\{\dfrac{1}{s+1/4}\right\} = \dfrac{1}{4}e^{-t/4}$

12. $\mathscr{L}^{-1}\left\{\dfrac{10s}{s^2+16}\right\} = 10\cos 4t$

15. $\mathscr{L}^{-1}\left\{\dfrac{2s-6}{s^2+9}\right\} = \mathscr{L}^{-1}\left\{2 \cdot \dfrac{s}{s^2+9} - 2 \cdot \dfrac{3}{s^2+9}\right\} = 2\cos 3t - 2\sin 3t$

18. $\mathscr{L}^{-1}\left\{\dfrac{s+1}{s^2-4s}\right\} = \mathscr{L}^{-1}\left\{-\dfrac{1}{4} \cdot \dfrac{1}{s} + \dfrac{5}{4} \cdot \dfrac{1}{s-4}\right\} = -\dfrac{1}{4} + \dfrac{5}{4}e^{4t}$

21. $\mathscr{L}^{-1}\left\{\dfrac{0.9s}{(s-0.1)(s+0.2)}\right\} = \mathscr{L}^{-1}\left\{(0.3) \cdot \dfrac{1}{s-0.1} + (0.6) \cdot \dfrac{1}{s+0.2}\right\} = 0.3e^{0.1t} + 0.6e^{-0.2t}$

24. $\mathscr{L}^{-1}\left\{\dfrac{s^2+1}{s(s-1)(s+1)(s-2)}\right\} = \mathscr{L}^{-1}\left\{\dfrac{1}{2} \cdot \dfrac{1}{s} - \dfrac{1}{s-1} - \dfrac{1}{3} \cdot \dfrac{1}{s+1} + \dfrac{5}{6} \cdot \dfrac{1}{s-2}\right\}$

$$= \dfrac{1}{2} - e^t - \dfrac{1}{3}e^{-t} + \dfrac{5}{6}e^{2t}$$

27. $\mathscr{L}^{-1}\left\{\dfrac{2s-4}{(s^2+s)(s^2+1)}\right\} = \mathscr{L}^{-1}\left\{\dfrac{2s-4}{s(s^2+1)^2}\right\} = \mathscr{L}^{-1}\left\{-\dfrac{4}{s} + \dfrac{3}{s+1} + \dfrac{s}{s^2+1} + \dfrac{3}{s^2+1}\right\}$

$$= -4 + 3e^{-t} + \cos t + 3\sin t$$

30. $\mathscr{L}^{-1}\left\{\dfrac{6s+3}{(s^2+1)(s^2+4)}\right\} = \mathscr{L}^{-1}\left\{2 \cdot \dfrac{s}{s^2+1} + \dfrac{1}{s^2+1} - 2 \cdot \dfrac{s}{s^2+4} - \dfrac{1}{2} \cdot \dfrac{2}{s^2+4}\right\}$

$$= 2\cos t + \sin t - 2\cos 2t - \dfrac{1}{2}\sin 2t$$

33. The Laplace transform of the IVP is

$$s\mathscr{L}\{y\} - y(0) + 6\mathscr{L}\{y\} = \dfrac{1}{s-4}.$$

155

Solving for $\mathscr{L}\{y\}$ we obtain

$$\mathscr{L}\{y\} = \frac{1}{(s-4)(s+6)} + \frac{2}{s+6} = \frac{1}{10} \cdot \frac{1}{s-4} + \frac{19}{10} \cdot \frac{1}{s+6}.$$

Thus

$$y = \frac{1}{10}e^{4t} + \frac{19}{10}e^{-6t}.$$

36. The Laplace transform of the IVP is

$$s^2 \mathscr{L}\{y\} - sy(0) - y'(0) - 4\left[s\mathscr{L}\{y\} - y(0)\right] = \frac{6}{s-3} - \frac{3}{s+1}.$$

Solving for $\mathscr{L}\{y\}$ we obtain

$$\mathscr{L}\{y\} = \frac{6}{(s-3)(s^2-4s)} - \frac{3}{(s+1)(s^2-4s)} + \frac{s-5}{s^2-4s}$$

$$= \frac{5}{2} \cdot \frac{1}{s} - \frac{2}{s-3} - \frac{3}{5} \cdot \frac{1}{s+1} + \frac{11}{10} \cdot \frac{1}{s-4}.$$

Thus

$$y = \frac{5}{2} - 2e^{3t} - \frac{3}{5}e^{-t} + \frac{11}{10}e^{4t}.$$

39. The Laplace transform of the IVP is

$$2\left[s^3 \mathscr{L}\{y\} - s^2(0) - sy'(0) - y''(0)\right] + 3[s^2 \mathscr{L}\{y\} - sy(0) - y'(0)] - 3[s\mathscr{L}\{y\} - y(0)] - 2\mathscr{L}\{y\} = \frac{1}{s+1}.$$

Solving for $\mathscr{L}\{y\}$ we obtain

$$\mathscr{L}\{y\} = \frac{2s+3}{(s+1)(s-1)(2s+1)(s+2)} = \frac{1}{2}\frac{1}{s+1} + \frac{5}{18}\frac{1}{s-1} - \frac{8}{9}\frac{1}{s+1/2} + \frac{1}{9}\frac{1}{s+2}.$$

Thus

$$y = \frac{1}{2}e^{-t} + \frac{5}{18}e^{t} - \frac{8}{9}e^{-t/2} + \frac{1}{9}e^{-2t}.$$

42. The Laplace transform of the IVP is

$$[s^2 \mathscr{L}\{y\} - s\cdot 1 - 3] - 2[s\mathscr{L}\{y\} - 1] + 5\mathscr{L}\{y\} = (s^2 - 2s + 5)\mathscr{L}\{y\} - s - 1 = 0.$$

Solving for $\mathscr{L}\{y\}$ we obtain

$$\mathscr{L}\{y\} = \frac{s+1}{s^2 - 2s + 5} = \frac{s-1+2}{(s-1)^2 + 2^2} = \frac{s-1}{(s-1)^2 + 2^2} + \frac{2}{(s-1)^2 + 2^2}.$$

Thus

$$y = e^{t}\cos 2t + e^{t}\sin 2t.$$

43. (*Hint*) In part (b) compute $f''(t)$.

Operational Properties I

The terminology and concepts listed below provide an outline of the main ideas encountered in this section. These can be useful when preparing for a quiz or test.

Terminology and Concepts

- first translation (or shifting) theorem
- unit step function (Heaviside function)
- second translation (or shifting) theorem

The basic skills listed below summarize the more mechanical types of problems encountered in the exercise set for this section.

Basic Skills

- use the first translation theorem to find the Laplace transform of a function that is a multiple of e^{at}.
- use the first translation theorem to find the inverse Laplace transform of the shifted function $F(s-a)$
- use the Laplace transform in conjunction with the first translation theorem to solve an IVP
- express a piecewise-defined function in terms of unit step functions
- find the Laplace transform of a piecewise-defined function
- find the inverse Laplace transform of a function that is a multiple of e^{-as}
- use the Laplace transform to solve an IVP involving piecewise-defined functions

Use of Computers
See Section 2.3 in this manual for *Mathematica* and *Maple* syntax for the unit step function.

Exercises 7.3   *Hints, Suggestions, Solutions, and Examples*

3. $\mathscr{L}\left\{t^3 e^{-2t}\right\} = \dfrac{3!}{(s+2)^4}$

6. $\mathscr{L}\left\{e^{2t}(t-1)^2\right\} = \mathscr{L}\left\{t^2 e^{2t} - 2t e^{2t} + e^{2t}\right\} = \dfrac{2}{(s-2)^3} - \dfrac{2}{(s-2)^2} + \dfrac{1}{s-2}$

9. $\mathscr{L}\{(1 - e^t + 3e^{-4t})\cos 5t\} = \mathscr{L}\{\cos 5t - e^t \cos 5t + 3e^{-4t}\cos 5t\}$

$$= \frac{s}{s^2 + 25} - \frac{s - 1}{(s - 1)^2 + 25} + \frac{3(s + 4)}{(s + 4)^2 + 25}$$

12. $\mathscr{L}^{-1}\left\{\dfrac{1}{(s-1)^4}\right\} = \dfrac{1}{6}\mathscr{L}^{-1}\left\{\dfrac{3!}{(s-1)^4}\right\} = \dfrac{1}{6}t^3 e^t$

15. $\mathscr{L}^{-1}\left\{\dfrac{s}{s^2 + 4s + 5}\right\} = \mathscr{L}^{-1}\left\{\dfrac{s + 2}{(s+2)^2 + 1^2} - 2\dfrac{1}{(s+2)^2 + 1^2}\right\} = e^{-2t}\cos t - 2e^{-2t}\sin t$

18. $\mathscr{L}^{-1}\left\{\dfrac{5s}{(s-2)^2}\right\} = \mathscr{L}^{-1}\left\{\dfrac{5(s-2) + 10}{(s-2)^2}\right\} = \mathscr{L}^{-1}\left\{\dfrac{5}{s - 2} + \dfrac{10}{(s-2)^2}\right\} = 5e^{2t} + 10te^{2t}$

21. The Laplace transform of the DE is

$$s\mathscr{L}\{y\} - y(0) + 4\mathscr{L}\{y\} = \frac{1}{s + 4}.$$

Solving for $\mathscr{L}\{y\}$ we obtain

$$\mathscr{L}\{y\} = \frac{1}{(s+4)^2} + \frac{2}{s + 4}.$$

Thus

$$y = te^{-4t} + 2e^{-4t}.$$

24. The Laplace transform of the DE is

$$s^2\mathscr{L}\{y\} - sy(0) - y'(0) - 4\left[s\mathscr{L}\{y\} - y(0)\right] + 4\mathscr{L}\{y\} = \frac{6}{(s-2)^4}.$$

Solving for $\mathscr{L}\{y\}$ we obtain $\mathscr{L}\{y\} = \dfrac{1}{20}\dfrac{5!}{(s-2)^6}$. Thus, $y = \dfrac{1}{20}t^5 e^{2t}$.

27. The Laplace transform of the DE is

$$s^2\mathscr{L}\{y\} - sy(0) - y'(0) - 6\left[s\mathscr{L}\{y\} - y(0)\right] + 13\mathscr{L}\{y\} = 0.$$

Solving for $\mathscr{L}\{y\}$ we obtain

$$\mathscr{L}\{y\} = -\frac{3}{s^2 - 6s + 13} = -\frac{3}{2}\frac{2}{(s-3)^2 + 2^2}.$$

Thus

$$y = -\frac{3}{2}e^{3t}\sin 2t.$$

30. The Laplace transform of the DE is

$$s^2\mathscr{L}\{y\} - sy(0) - y'(0) - 2\left[s\mathscr{L}\{y\} - y(0)\right] + 5\mathscr{L}\{y\} = \frac{1}{s} + \frac{1}{s^2}.$$

Solving for $\mathscr{L}\{y\}$ we obtain

$$\mathscr{L}\{y\} = \frac{4s^2 + s + 1}{s^2(s^2 - 2s + 5)} = \frac{7}{25}\frac{1}{s} + \frac{1}{5}\frac{1}{s^2} + \frac{-7s/25 + 109/25}{s^2 - 2s + 5}$$

$$= \frac{7}{25}\frac{1}{s} + \frac{1}{5}\frac{1}{s^2} - \frac{7}{25}\frac{s-1}{(s-1)^2 + 2^2} + \frac{51}{25}\frac{2}{(s-1)^2 + 2^2}.$$

Thus

$$y = \frac{7}{25} + \frac{1}{5}t - \frac{7}{25}e^t\cos 2t + \frac{51}{25}e^t\sin 2t.$$

33. Recall from Section 5.1 that $mx'' = -kx - \beta x'$. Now $m = W/g = 4/32 = \frac{1}{8}$ slug, and $4 = 2k$ so that $k = 2$ lb/ft. Thus, the DE is $x'' + 7x' + 16x = 0$. The initial conditions are $x(0) = -3/2$ and $x'(0) = 0$. The Laplace transform of the DE is

$$s^2\mathscr{L}\{x\} + \frac{3}{2}s + 7s\mathscr{L}\{x\} + \frac{21}{2} + 16\mathscr{L}\{x\} = 0.$$

Solving for $\mathscr{L}\{x\}$ we obtain

$$\mathscr{L}\{x\} = \frac{-3s/2 - 21/2}{s^2 + 7s + 16} = -\frac{3}{2}\frac{s + 7/2}{(s+7/2)^2 + (\sqrt{15}/2)^2} - \frac{7\sqrt{15}}{10}\frac{\sqrt{15}/2}{(s+7/2)^2 + (\sqrt{15}/2)^2}.$$

Thus

$$x = -\frac{3}{2}e^{-7t/2}\cos\frac{\sqrt{15}}{2}t - \frac{7\sqrt{15}}{10}e^{-7t/2}\sin\frac{\sqrt{15}}{2}t.$$

36. The DE is

$$R\frac{dq}{dt} + \frac{1}{C}q = E_0 e^{-kt}, \quad q(0) = 0.$$

The Laplace transform of this equation is

$$Rs\mathscr{L}\{q\} + \frac{1}{C}\mathscr{L}\{q\} = E_0\frac{1}{s+k}.$$

Solving for $\mathscr{L}\{q\}$ we obtain

$$\mathscr{L}\{q\} = \frac{E_0 C}{(s+k)(RCs+1)} = \frac{E_0/R}{(s+k)(s+1/RC)}.$$

When $1/RC \neq k$ we have by partial fractions

$$\mathscr{L}\{q\} = \frac{E_0}{R}\left(\frac{1/(1/RC - k)}{s+k} - \frac{1/(1/RC - k)}{s + 1/RC}\right) = \frac{E_0}{R}\frac{1}{1/RC - k}\left(\frac{1}{s+k} - \frac{1}{s+1/RC}\right).$$

Thus

$$q(t) = \frac{E_0 C}{1 - kRC}\left(e^{-kt} - e^{-t/RC}\right).$$

When $1/RC = k$ we have

$$\mathscr{L}\{q\} = \frac{E_0}{R}\frac{1}{(s+k)^2}.$$

159

Section 7.3   Operational Properties I

Thus

$$q(t) = \frac{E_0}{R} te^{-kt} = \frac{E_0}{R} te^{-t/RC}.$$

39. $\mathscr{L}\{t\,\mathscr{U}(t-2)\} = \mathscr{L}\{(t-2)\,\mathscr{U}(t-2) + 2\,\mathscr{U}(t-2)\} = \dfrac{e^{-2s}}{s^2} + \dfrac{2e^{-2s}}{s}$

Alternatively, (16) of this section could be used:

$$\mathscr{L}\{t\,\mathscr{U}(t-2)\} = e^{-2s}\mathscr{L}\{t+2\} = e^{-2s}\left(\frac{1}{s^2} + \frac{2}{s}\right).$$

42. $\mathscr{L}\left\{\sin t\,\mathscr{U}\left(t-\dfrac{\pi}{2}\right)\right\} = \mathscr{L}\left\{\cos\left(t-\dfrac{\pi}{2}\right)\mathscr{U}\left(t-\dfrac{\pi}{2}\right)\right\} = \dfrac{se^{-\pi s}}{s^2+1}$

45. $\mathscr{L}^{-1}\left\{\dfrac{e^{-\pi s}}{s^2+1}\right\} = \sin(t-\pi)\,\mathscr{U}(t-\pi)$

48. $\mathscr{L}^{-1}\left\{\dfrac{e^{-2s}}{s^2(s-1)}\right\} = \mathscr{L}^{-1}\left\{-\dfrac{e^{-2s}}{s} - \dfrac{e^{-2s}}{s^2} + \dfrac{e^{-2s}}{s-1}\right\} = -\mathscr{U}(t-2) - (t-2)\,\mathscr{U}(t-2) + e^{t-2}\,\mathscr{U}(t-2)$

51. The graph in Figure 7.22 is the portion of the graph in Figure 7.19 from $t=0$ to $t=b-a$ shifted a units to the right, with the rest of the graph zeroed out. This is the function in (f).

54. The graph in Figure 7.25 is the portion of the graph in Figure 7.19 from $t=0$ to $t=b$ with the rest of the graph zeroed out. This is the function in (d).

57. $\mathscr{L}\{t^2\,\mathscr{U}(t-1)\} = \mathscr{L}\{[(t-1)^2 + 2t - 1]\,\mathscr{U}(t-1)\} = \mathscr{L}\{[(t-1)^2 + 2(t-1) - 1]\,\mathscr{U}(t-1)\}$

$$= \left(\frac{2}{s^3} + \frac{2}{s^2} + \frac{1}{s}\right)e^{-s}$$

Alternatively, by (16),

$$\mathscr{L}\{t^2\,\mathscr{U}(t-1)\} = e^{-s}\mathscr{L}\{t^2 + 2t + 1\} = e^{-s}\left(\frac{2}{s^3} + \frac{2}{s^2} + \frac{1}{s}\right).$$

60. $\mathscr{L}\{\sin t - \sin t\,\mathscr{U}(t-2\pi)\} = \mathscr{L}\{\sin t - \sin(t-2\pi)\,\mathscr{U}(t-2\pi)\} = \dfrac{1}{s^2+1} - \dfrac{e^{-2\pi s}}{s^2+1}$

63. The Laplace transform of the DE is

$$s\mathscr{L}\{y\} - y(0) + \mathscr{L}\{y\} = \frac{5}{s}e^{-s}.$$

Solving for $\mathscr{L}\{y\}$ we obtain

$$\mathscr{L}\{y\} = \frac{5e^{-s}}{s(s+1)} = 5e^{-s}\left[\frac{1}{s} - \frac{1}{s+1}\right].$$

Thus

$$y = 5\,\mathscr{U}(t-1) - 5e^{-(t-1)}\,\mathscr{U}(t-1).$$

66. The Laplace transform of the DE is

$$s^2 \mathscr{L}\{y\} - sy(0) - y'(0) + 4\,\mathscr{L}\{y\} = \frac{1}{s} - \frac{e^{-s}}{s}.$$

Solving for $\mathscr{L}\{y\}$ we obtain

$$\mathscr{L}\{y\} = \frac{1-s}{s(s^2+4)} - e^{-s}\frac{1}{s(s^2+4)} = \frac{1}{4}\frac{1}{s} - \frac{1}{4}\frac{s}{s^2+4} - \frac{1}{2}\frac{2}{s^2+4} - e^{-s}\left[\frac{1}{4}\frac{1}{s} - \frac{1}{4}\frac{s}{s^2+4}\right].$$

Thus

$$y = \frac{1}{4} - \frac{1}{4}\cos 2t - \frac{1}{2}\sin 2t - \left[\frac{1}{4} - \frac{1}{4}\cos 2(t-1)\right]\mathscr{U}(t-1).$$

69. The Laplace transform of the DE is

$$s^2 \mathscr{L}\{y\} - sy(0) - y'(0) + \mathscr{L}\{y\} = \frac{e^{-\pi s}}{s} - \frac{e^{-2\pi s}}{s}.$$

Solving for $\mathscr{L}\{y\}$ we obtain

$$\mathscr{L}\{y\} = e^{-\pi s}\left[\frac{1}{s} - \frac{s}{s^2+1}\right] - e^{-2\pi s}\left[\frac{1}{s} - \frac{s}{s^2+1}\right] + \frac{1}{s^2+1}.$$

Thus

$$y = [1 - \cos(t-\pi)]\,\mathscr{U}(t-\pi) - [1 - \cos(t-2\pi)]\,\mathscr{U}(t-2\pi) + \sin t.$$

72. Recall from Section 5.1 that $mx'' = -kx + f(t)$. Now $m = W/g = 32/32 = 1$ slug, and $32 = 2k$ so that $k = 16$ lb/ft. Thus, the differential equation is $x'' + 16x = f(t)$. The initial conditions are $x(0) = 0$, $x'(0) = 0$. Also, since

$$f(t) = \begin{cases} \sin t, & 0 \le t < 2\pi \\ 0, & t \ge 2\pi \end{cases}$$

and $\sin t = \sin(t - 2\pi)$ we can write

$$f(t) = \sin t - \sin(t - 2\pi)\mathscr{U}(t - 2\pi).$$

The Laplace transform of the DE is

$$s^2\mathscr{L}\{x\} + 16\,\mathscr{L}\{x\} = \frac{1}{s^2+1} - \frac{1}{s^2+1}e^{-2\pi s}.$$

Solving for $\mathscr{L}\{x\}$ we obtain

$$\mathscr{L}\{x\} = \frac{1}{(s^2+16)(s^2+1)} - \frac{1}{(s^2+16)(s^2+1)}e^{-2\pi s}$$

$$= \frac{-1/15}{s^2+16} + \frac{1/15}{s^2+1} - \left[\frac{-1/15}{s^2+16} + \frac{1/15}{s^2+1}\right]e^{-2\pi s}.$$

Thus

$$x(t) = -\frac{1}{60}\sin 4t + \frac{1}{15}\sin t + \frac{1}{60}\sin 4(t-2\pi)\mathscr{U}(t-2\pi) - \frac{1}{15}\sin(t-2\pi)\mathscr{U}(t-2\pi)$$

$$= \begin{cases} -\frac{1}{60}\sin 4t + \frac{1}{15}\sin t, & 0 \le t < 2\pi \\ 0, & t \ge 2\pi. \end{cases}$$

75. (a) The DE is

$$\frac{di}{dt} + 10i = \sin t + \cos\left(t - \frac{3\pi}{2}\right)\mathscr{U}\left(t - \frac{3\pi}{2}\right), \quad i(0) = 0.$$

The Laplace transform of this equation is

$$s\mathscr{L}\{i\} + 10\mathscr{L}\{i\} = \frac{1}{s^2 + 1} + \frac{se^{-3\pi s/2}}{s^2 + 1}.$$

Solving for $\mathscr{L}\{i\}$ we obtain

$$\mathscr{L}\{i\} = \frac{1}{(s^2 + 1)(s + 10)} + \frac{s}{(s^2 + 1)(s + 10)}e^{-3\pi s/2}$$

$$= \frac{1}{101}\left(\frac{1}{s + 10} - \frac{s}{s^2 + 1} + \frac{10}{s^2 + 1}\right) + \frac{1}{101}\left(\frac{-10}{s + 10} + \frac{10s}{s^2 + 1} + \frac{1}{s^2 + 1}\right)e^{-3\pi s/2}.$$

Thus

$$i(t) = \frac{1}{101}\left(e^{-10t} - \cos t + 10\sin t\right)$$

$$+ \frac{1}{101}\left[-10e^{-10(t - 3\pi/2)} + 10\cos\left(t - \frac{3\pi}{2}\right) + \sin\left(t - \frac{3\pi}{2}\right)\right]\mathscr{U}\left(t - \frac{3\pi}{2}\right).$$

(b)

The maximum value of $i(t)$ is approximately 0.1 at $t = 1.7$, the minimum is approximately -0.1 at 4.7.

78. The DE is

$$EI\frac{d^4 y}{dx^4} = w_0[\mathscr{U}(x - L/3) - \mathscr{U}(x - 2L/3)].$$

Taking the Laplace transform of both sides and using $y(0) = y'(0) = 0$ we obtain

$$s^4\mathscr{L}\{y\} - sy''(0) - y'''(0) = \frac{w_0}{EI}\frac{1}{s}\left(e^{-Ls/3} - e^{-2Ls/3}\right).$$

Letting $y''(0) = c_1$ and $y'''(0) = c_2$ we have

$$\mathscr{L}\{y\} = \frac{c_1}{s^3} + \frac{c_2}{s^4} + \frac{w_0}{EI}\frac{1}{s^5}\left(e^{-Ls/3} - e^{-2Ls/3}\right)$$

so that

$$y(x) = \frac{1}{2}c_1 x^2 + \frac{1}{6}c_2 x^3 + \frac{1}{24}\frac{w_0}{EI}\left[\left(x - \frac{L}{3}\right)^4\mathscr{U}\left(x - \frac{L}{3}\right) - \left(x - \frac{2L}{3}\right)^4\mathscr{U}\left(x - \frac{2L}{3}\right)\right].$$

To find c_1 and c_2 we compute

$$y''(x) = c_1 + c_2 x + \frac{1}{2}\frac{w_0}{EI}\left[\left(x-\frac{L}{3}\right)^2 \mathcal{U}\left(x-\frac{L}{3}\right) - \left(x-\frac{2L}{3}\right)^2 \mathcal{U}\left(x-\frac{2L}{3}\right)\right]$$

and

$$y'''(x) = c_2 + \frac{w_0}{EI}\left[\left(x-\frac{L}{3}\right)\mathcal{U}\left(x-\frac{L}{3}\right) - \left(x-\frac{2L}{3}\right)\mathcal{U}\left(x-\frac{2L}{3}\right)\right].$$

Then $y''(L) = y'''(L) = 0$ yields the system

$$c_1 + c_2 L + \frac{1}{2}\frac{w_0}{EI}\left[\left(\frac{2L}{3}\right)^2 - \left(\frac{L}{3}\right)^2\right] = c_1 + c_2 L + \frac{1}{6}\frac{w_0 L^2}{EI} = 0$$

$$c_2 + \frac{w_0}{EI}\left[\frac{2L}{3} - \frac{L}{3}\right] = c_2 + \frac{1}{3}\frac{w_0 L}{EI} = 0.$$

Solving for c_1 and c_2 we obtain $c_1 = \frac{1}{6}w_0 L^2/EI$ and $c_2 = -\frac{1}{3}w_0 L/EI$. Thus

$$y(x) = \frac{w_0}{EI}\left(\frac{1}{12}L^2 x^2 - \frac{1}{18}Lx^3 + \frac{1}{24}\left[\left(x-\frac{L}{3}\right)^4\mathcal{U}\left(x-\frac{L}{3}\right) - \left(x-\frac{2L}{3}\right)^4\mathcal{U}\left(x-\frac{2L}{3}\right)\right]\right).$$

Section 7.4 — Operational Properties II

The terminology and concepts listed below provide an outline of the main ideas encountered in this section. These can be useful when preparing for a quiz or test.

Terminology and Concepts

- the relationship between $t^n f(t)$ and the derivative of the Laplace transform of $f(t)$
- convolution of two functions
- the Laplace transform of the convolution f*g of two functions
- the Laplace transform of the integral of a function
- Volterra integral equation
- integrodifferential differential equation
- the Laplace transform of a periodic function

The basic skills listed below summarize the more mechanical types of problems encountered in the exercise set for this section.

Basic Skills

- find the Laplace transform of the product $t^n f(t)$ in terms of the Laplace transform of $f(t)$
- find the Laplace transform of the convolution of two functions
- find the Laplace transform of the integral of a function
- solve an integral equation
- solve an integrodifferential equation
- find the Laplace transform of a periodic function

DEs With Variable Coefficients The Laplace transform is not well-suited to solving linear differential equations with variable coefficients. Nevertheless, it can be used in *some* instances. If the DE has coefficients that are at most linear polynomials in t, then its transform is a linear first-order DE in the transformed function $Y(s)$. For example, let's find a solution of $ty'' + 2ty' + 2y = t$ subject to $y(0) = 0$. From Theorem 7.8 observe

$$\mathscr{L}\{ty''\} = -\frac{d}{ds}\mathscr{L}\{y''\} = -\frac{d}{ds}[s^2Y(s) - sy(0) - y'(0)] = -s^2Y'(s) - 2sY(s) + y(0) \qquad (1)$$

$$\mathscr{L}\{ty'\} = -\frac{d}{ds}\mathscr{L}\{y'\} = -\frac{d}{ds}[sY(s) - y(0)] = -sY'(s) - Y(s). \qquad (2)$$

Note here that we have used the product rule to differentiate $s^2Y(s)$ and $sY(s)$. Hence the transform of the second-order DE is the first-order DE,

$$-s^2Y'(s) - 2sY(s) - 2sY'(s) - 2Y(s) + 2Y(s) = \frac{1}{s^2}$$

or

$$\frac{dY}{ds} + \frac{2}{s+2}Y = -\frac{1}{s^3(s+2)},$$

where we have used $y(0) = 0$. Since this last equation is recognized as a linear first-order DE in Y, we find from Section 2.3 in the text that the integrating factor is $(s+2)^2$. Multiplying the last equation by this factor gives

$$\frac{d}{ds}[(s+2)^2Y] = -\frac{s+2}{s^3} = -\frac{1}{s^2} - \frac{2}{s^3}.$$

Integrating, we obtain

$$(s+2)^2Y = \frac{1}{s} + \frac{1}{s^2} + c = \frac{s+1}{s^2} + c.$$

Now solve for Y and expand into partial fractions:

$$Y = \frac{s+1}{s^2(s+2)^2} + \frac{c}{(s+2)^2} = \frac{1}{4}\frac{1}{s^2} - \frac{1}{4}\frac{1}{(s+2)^2} + \frac{c}{(s+2)^2}.$$

The inverse of the preceding expression is then

$$y(t) = \frac{1}{4}t - \frac{1}{4}te^{-2t} + cte^{-2t} \qquad \text{or} \qquad y(t) = \frac{1}{4}t + c_1 te^{-2t},$$

where we have replaced the term $c - \frac{1}{4}$ by c_1.

Exercises 7.4 *Hints, Suggestions, Solutions, and Examples*

3. $\mathcal{L}\{t\cos 2t\} = -\dfrac{d}{ds}\left(\dfrac{s}{s^2+4}\right) = \dfrac{s^2-4}{(s^2+4)^2}$

6. $\mathcal{L}\{t^2\cos t\} = \dfrac{d^2}{ds^2}\left(\dfrac{s}{s^2+1}\right) = \dfrac{d}{ds}\left(\dfrac{1-s^2}{(s^2+1)^2}\right) = \dfrac{2s\left(s^2-3\right)}{(s^2+1)^3}$

9. The Laplace transform of the DE is

$$s\,\mathcal{L}\{y\} + \mathcal{L}\{y\} = \frac{2s}{(s^2+1)^2}.$$

Solving for $\mathcal{L}\{y\}$ we obtain

$$\mathcal{L}\{y\} = \frac{2s}{(s+1)(s^2+1)^2} = -\frac{1}{2}\frac{1}{s+1} - \frac{1}{2}\frac{1}{s^2+1} + \frac{1}{2}\frac{s}{s^2+1} + \frac{1}{(s^2+1)^2} + \frac{s}{(s^2+1)^2}.$$

Thus

$$y(t) = -\frac{1}{2}e^{-t} - \frac{1}{2}\sin t + \frac{1}{2}\cos t + \frac{1}{2}(\sin t - t\cos t) + \frac{1}{2}t\sin t$$

$$= -\frac{1}{2}e^{-t} + \frac{1}{2}\cos t - \frac{1}{2}t\cos t + \frac{1}{2}t\sin t.$$

12. The Laplace transform of the DE is

$$s^2\mathcal{L}\{y\} - sy(0) - y'(0) + \mathcal{L}\{y\} = \frac{1}{s^2+1}.$$

Solving for $\mathcal{L}\{y\}$ we obtain

$$\mathcal{L}\{y\} = \frac{s^3 - s^2 + s}{(s^2+1)^2} = \frac{s}{s^2+1} - \frac{1}{s^2+1} + \frac{1}{(s^2+1)^2}.$$

Thus

$$y = \cos t - \sin t + \left(\frac{1}{2}\sin t - \frac{1}{2}t\cos t\right) = \cos t - \frac{1}{2}\sin t - \frac{1}{2}t\cos t.$$

15.

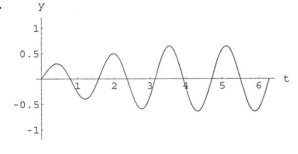

17. (*Hint*) See the discussion **DEs With Variable Coefficients** in this section for an example of how to solve this type of problem.

18. From the discussion **DEs With Variable Coefficients** in this section

$$\mathscr{L}\{ty'\} = -\frac{d}{ds}\mathscr{L}\{y'\} = -\frac{d}{ds}[sY(s) - y(0)] = -s\frac{dY}{ds} - Y$$

so that the transform of the given second-order DE is the linear first-order DE in $Y(s)$:

$$Y' + \left(\frac{3}{s} - 2s\right)Y = -\frac{10}{s}.$$

Using the integrating factor $s^3 e^{-s^2}$, the last equation yields

$$Y(s) = \frac{5}{s^3} + \frac{c}{s^3}e^{s^2}.$$

But if $Y(s)$ is the Laplace transform of a piecewise-continuous function of exponential order, we must have, in view of Theorem 7.2, $\lim_{s\to\infty} Y(s) = 0$. In order to obtain this condition we require $c = 0$. Hence

$$y(t) = \mathscr{L}^{-1}\left\{\frac{5}{s^3}\right\} = \frac{5}{2}t^2.$$

21. $\mathscr{L}\left\{e^{-t} * e^t \cos t\right\} = \dfrac{s-1}{(s+1)\left[(s-1)^2 + 1\right]}$

24. $\mathscr{L}\left\{\displaystyle\int_0^t \cos\tau\, d\tau\right\} = \dfrac{1}{s}\mathscr{L}\{\cos t\} = \dfrac{s}{s(s^2+1)} = \dfrac{1}{s^2+1}$

27. $\mathscr{L}\left\{\displaystyle\int_0^t \tau e^{t-\tau}\, d\tau\right\} = \mathscr{L}\{t\}\,\mathscr{L}\{e^t\} = \dfrac{1}{s^2(s-1)}$

30. $\mathscr{L}\left\{t\displaystyle\int_0^t \tau e^{-\tau}\, d\tau\right\} = -\dfrac{d}{ds}\mathscr{L}\left\{\displaystyle\int_0^t \tau e^{-\tau}\, d\tau\right\} = -\dfrac{d}{ds}\left(\dfrac{1}{s}\dfrac{1}{(s+1)^2}\right) = \dfrac{3s+1}{s^2(s+1)^3}$

33. $\mathscr{L}^{-1}\left\{\dfrac{1}{s^3(s-1)}\right\} = \mathscr{L}^{-1}\left\{\dfrac{1/s^2(s-1)}{s}\right\} = \displaystyle\int_0^t (e^\tau - \tau - 1)d\tau = e^t - \dfrac{1}{2}t^2 - t - 1$

36. The Laplace transform of the DE is

$$s^2\mathscr{L}\{y\} + \mathscr{L}\{y\} = \frac{1}{(s^2+1)} + \frac{2s}{(s^2+1)^2}.$$

Thus

$$\mathscr{L}\{y\} = \frac{1}{(s^2+1)^2} + \frac{2s}{(s^2+1)^3}$$

and, using Problem 35,

$$y = \frac{1}{2}(\sin t - t\cos t) + \frac{1}{4}(t\sin t - t^2 \cos t).$$

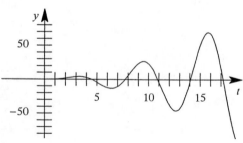

166

39. The Laplace transform of the given equation is

$$\mathscr{L}\{f\} = \mathscr{L}\{te^t\} + \mathscr{L}\{t\}\mathscr{L}\{f\}.$$

Solving for $\mathscr{L}\{f\}$ we obtain

$$\mathscr{L}\{f\} = \frac{s^2}{(s-1)^3(s+1)} = \frac{1}{8}\frac{1}{s-1} + \frac{3}{4}\frac{1}{(s-1)^2} + \frac{1}{4}\frac{2}{(s-1)^3} - \frac{1}{8}\frac{1}{s+1}.$$

Thus

$$f(t) = \frac{1}{8}e^t + \frac{3}{4}te^t + \frac{1}{4}t^2e^t - \frac{1}{8}e^{-t}$$

42. The Laplace transform of the given equation is

$$\mathscr{L}\{f\} = \mathscr{L}\{\cos t\} + \mathscr{L}\{e^{-t}\}\mathscr{L}\{f\}.$$

Solving for $\mathscr{L}\{f\}$ we obtain

$$\mathscr{L}\{f\} = \frac{s}{s^2+1} + \frac{1}{s^2+1}.$$

Thus

$$f(t) = \cos t + \sin t.$$

45. The Laplace transform of the given equation is

$$s\mathscr{L}\{y\} - y(0) = \mathscr{L}\{1\} - \mathscr{L}\{\sin t\} - \mathscr{L}\{1\}\mathscr{L}\{y\}.$$

Solving for $\mathscr{L}\{f\}$ we obtain

$$\mathscr{L}\{y\} = \frac{s^2 - s + 1}{(s^2+1)^2} = \frac{1}{s^2+1} - \frac{1}{2}\frac{2s}{(s^2+1)^2}.$$

Thus

$$y = \sin t - \frac{1}{2}t\sin t.$$

167

48. The DE is

$$0.005\frac{di}{dt} + i + \frac{1}{0.02}\int_0^t i(\tau)d\tau = 100[t - (t-1)\,\mathcal{U}(t-1)]$$

or

$$\frac{di}{dt} + 200i + 10,000\int_0^t i(\tau)d\tau = 20,000[t - (t-1)\,\mathcal{U}(t-1)],$$

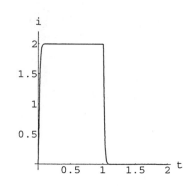

where $i(0) = 0$. The Laplace transform of the DE is

$$s\,\mathcal{L}\{i\} + 200\,\mathcal{L}\{i\} + \frac{10,000}{s}\,\mathcal{L}\{i\} = 20,000\left(\frac{1}{s^2} - \frac{1}{s^2}e^{-s}\right).$$

Solving for $\mathcal{L}\{i\}$ we obtain

$$\mathcal{L}\{i\} = \frac{20,000}{s(s+100)^2}(1 - e^{-s}) = \left[\frac{2}{s} - \frac{2}{s+100} - \frac{200}{(s+100)^2}\right](1 - e^{-s}).$$

Thus

$$i(t) = 2 - 2e^{-100t} - 200te^{-100t} - 2\,\mathcal{U}(t-1) + 2e^{-100(t-1)}\,\mathcal{U}(t-1) + 200(t-1)e^{-100(t-1)}\,\mathcal{U}(t-1).$$

51. Using integration by parts,

$$\mathcal{L}\{f(t)\} = \frac{1}{1 - e^{-bs}}\int_0^b \frac{a}{b}te^{-st}dt = \frac{a}{s}\left(\frac{1}{bs} - \frac{1}{e^{bs}-1}\right).$$

54. $\mathcal{L}\{f(t)\} = \dfrac{1}{1 - e^{-2\pi s}}\displaystyle\int_0^\pi e^{-st}\sin t\, dt = \dfrac{1}{s^2+1}\cdot\dfrac{1}{1 - e^{-\pi s}}$

57. The DE is $x'' + 2x' + 10x = 20f(t)$, where $f(t)$ is the meander function in Problem 49 with $a = \pi$. Using the initial conditions $x(0) = x'(0) = 0$ and taking the Laplace transform we obtain

$$(s^2 + 2s + 10)\,\mathcal{L}\{x(t)\} = \frac{20}{s}(1 - e^{-\pi s})\frac{1}{1 + e^{-\pi s}}$$

$$= \frac{20}{s}(1 - e^{-\pi s})(1 - e^{-\pi s} + e^{-2\pi s} - e^{-3\pi s} + \cdots)$$

$$= \frac{20}{s}(1 - 2e^{-\pi s} + 2e^{-2\pi s} - 2e^{-3\pi s} + \cdots)$$

$$= \frac{20}{s} + \frac{40}{s}\sum_{n=1}^{\infty}(-1)^n e^{-n\pi s}.$$

Then

$$\mathcal{L}\{x(t)\} = \frac{20}{s(s^2 + 2s + 10)} + \frac{40}{s(s^2 + 2s + 10)} \sum_{n=1}^{\infty} (-1)^n e^{-n\pi s}$$

$$= \frac{2}{s} - \frac{2s+4}{s^2+2s+10} + \sum_{n=1}^{\infty} (-1)^n \left[\frac{4}{s} - \frac{4s+8}{s^2+2s+10} \right] e^{-n\pi s}$$

$$= \frac{2}{s} - \frac{2(s+1)+2}{(s+1)^2+9} + 4 \sum_{n=1}^{\infty} (-1)^n \left[\frac{1}{s} - \frac{(s+1)+1}{(s+1)^2+9} \right] e^{-n\pi s}$$

and

$$x(t) = 2 \left(1 - e^{-t} \cos 3t - \frac{1}{3} e^{-t} \sin 3t \right) + 4 \sum_{n=1}^{\infty} (-1)^n \left[1 - e^{-(t-n\pi)} \cos 3(t-n\pi) \right.$$

$$\left. - \frac{1}{3} e^{-(t-n\pi)} \sin 3(t-n\pi) \right] \mathcal{U}(t-n\pi).$$

The graph of $x(t)$ on the interval $[0, 2\pi)$ is shown below.

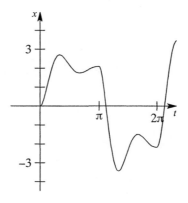

Section 7.5
The Dirac Delta Function

The terminology and concepts listed below provide an outline of the main ideas encountered in this section. These can be useful when preparing for a quiz or test.

Terminology and Concepts

- unit impulse function
- Dirac delta function
- Laplace transform of the Dirac delta function

Section 7.5 The Dirac Delta Function

The basic skills listed below summarize the more mechanical types of problems encountered in the exercise set for this section.

Basic Skills

- solve a linear second-order DE whose input function involves the Dirac delta function

---

Exercises 7.5   *Hints, Suggestions, Solutions, and Examples*

In Problems 3, 6, and 9 we use the fact that $\sin t$ is 2π-periodic, that is, $\sin(t - 2\pi) = \sin t$.

3. The Laplace transform of the DE yields

$$\mathscr{L}\{y\} = \frac{1}{s^2 + 1}\left(1 + e^{-2\pi s}\right)$$

so that

$$y = \sin t + \sin t \,\mathscr{U}(t - 2\pi).$$

6. The Laplace transform of the DE yields

$$\mathscr{L}\{y\} = \frac{s}{s^2 + 1} + \frac{1}{s^2 + 1}(e^{-2\pi s} + e^{-4\pi s})$$

so that

$$y = \cos t + \sin t\left[\mathscr{U}(t - 2\pi) + \mathscr{U}(t - 4\pi)\right].$$

9. The Laplace transform of the DE yields

$$\mathscr{L}\{y\} = \frac{1}{(s + 2)^2 + 1}e^{-2\pi s}$$

so that

$$y = e^{-2(t - 2\pi)}\sin t \,\mathscr{U}(t - 2\pi).$$

12. The Laplace transform of the DE yields

$$\mathscr{L}\{y\} = \frac{1}{(s - 1)^2(s - 6)} + \frac{e^{-2s} + e^{-4s}}{(s - 1)(s - 6)}$$

$$= -\frac{1}{25}\frac{1}{s - 1} - \frac{1}{5}\frac{1}{(s - 1)^2} + \frac{1}{25}\frac{1}{s - 6} + \left[-\frac{1}{5}\frac{1}{s - 1} + \frac{1}{5}\frac{1}{s - 6}\right]\left(e^{-2s} + e^{-4s}\right)$$

so that

$$y = -\frac{1}{25}e^t - \frac{1}{5}te^t + \frac{1}{25}e^{6t} + \left[-\frac{1}{5}e^{t-2} + \frac{1}{5}e^{6(t-2)}\right]\mathscr{U}(t - 2)$$

$$+ \left[-\frac{1}{5}e^{t-4} + \frac{1}{5}e^{6(t-4)}\right]\mathscr{U}(t - 4).$$

15. (*Hint*) Consider the initial condition $y'(0) = 0$.

Systems of Linear Differential Equations

The terminology and concepts listed below provide an outline of the main ideas encountered in this section. These can be useful when preparing for a quiz or test.

Terminology and Concepts

  • coupled spring systems

  • doubled pendulum

The basic skills listed below summarize the more mechanical types of problems encountered in the exercise set for this section.

Basic Skills

  • use Laplace transform methods to solve a linear system of differential equations

  • solve algebraic systems of equations by elimination

Exercises 7.6   *Hints, Suggestions, Solutions, and Examples*

3. Taking the Laplace transform of the system gives

$$s\mathcal{L}\{x\} + 1 = \mathcal{L}\{x\} - 2\mathcal{L}\{y\}$$

$$s\mathcal{L}\{y\} - 2 = 5\mathcal{L}\{x\} - \mathcal{L}\{y\}$$

so that

$$\mathcal{L}\{x\} = \frac{-s-5}{s^2+9} = -\frac{s}{s^2+9} - \frac{5}{3}\frac{3}{s^2+9}$$

and

$$x = -\cos 3t - \frac{5}{3}\sin 3t.$$

Then

$$y = \frac{1}{2}x - \frac{1}{2}x' = 2\cos 3t - \frac{7}{3}\sin 3t.$$

6. Taking the Laplace transform of the system gives

$$(s+1)\,\mathcal{L}\{x\} - (s-1)\mathcal{L}\{y\} = -1$$

$$s\mathcal{L}\{x\} + (s+2)\,\mathcal{L}\{y\} = 1$$

171

so that

$$\mathcal{L}\{y\} = \frac{s + 1/2}{s^2 + s + 1} = \frac{s + 1/2}{(s + 1/2)^2 + (\sqrt{3}/2)^2}$$

and

$$\mathcal{L}\{x\} = \frac{-3/2}{s^2 + s + 1} = -\sqrt{3}\,\frac{\sqrt{3}/2}{(s + 1/2)^2 + (\sqrt{3}/2)^2}.$$

Then

$$y = e^{-t/2}\cos\frac{\sqrt{3}}{2}t \qquad \text{and} \qquad x = -\sqrt{3}\,e^{-t/2}\sin\frac{\sqrt{3}}{2}t.$$

9. Adding the equations and then subtracting them gives

$$\frac{d^2x}{dt^2} = \frac{1}{2}t^2 + 2t$$

$$\frac{d^2y}{dt^2} = \frac{1}{2}t^2 - 2t.$$

Taking the Laplace transform of the system gives

$$\mathcal{L}\{x\} = 8\frac{1}{s} + \frac{1}{24}\frac{4!}{s^5} + \frac{1}{3}\frac{3!}{s^4}$$

and

$$\mathcal{L}\{y\} = \frac{1}{24}\frac{4!}{s^5} - \frac{1}{3}\frac{3!}{s^4}$$

so that

$$x = 8 + \frac{1}{24}t^4 + \frac{1}{3}t^3 \qquad \text{and} \qquad y = \frac{1}{24}t^4 - \frac{1}{3}t^3.$$

12. Taking the Laplace transform of the system gives

$$(s - 4)\,\mathcal{L}\{x\} + 2\mathcal{L}\{y\} = \frac{2e^{-s}}{s}$$

$$-3\,\mathcal{L}\{x\} + (s + 1)\,\mathcal{L}\{y\} = \frac{1}{2} + \frac{e^{-s}}{s}$$

so that

$$\mathcal{L}\{x\} = \frac{-1/2}{(s - 1)(s - 2)} + e^{-s}\frac{1}{(s - 1)(s - 2)}$$

$$= \frac{1}{2}\frac{1}{s - 1} - \frac{1}{2}\frac{1}{s - 2} + e^{-s}\left[-\frac{1}{s - 1} + \frac{1}{s - 2}\right]$$

and

$$\mathcal{L}\{y\} = \frac{e^{-s}}{s} + \frac{s/4 - 1}{(s - 1)(s - 2)} + e^{-s}\frac{-s/2 + 2}{(s - 1)(s - 2)}$$

$$= \frac{3}{4}\frac{1}{s - 1} - \frac{1}{2}\frac{1}{s - 2} + e^{-s}\left[\frac{1}{s} - \frac{3}{2}\frac{1}{s - 1} + \frac{1}{s - 2}\right].$$

Then

$$x = \frac{1}{2}e^t - \frac{1}{2}e^{2t} + \left[-e^{t-1} + e^{2(t-1)} \right] \mathcal{U}(t-1)$$

$$y = \frac{3}{4}e^t - \frac{1}{2}e^{2t} + \left[1 - \frac{3}{2}e^{t-1} + e^{2(t-1)} \right] \mathcal{U}(t-1).$$

15. (a) By Kirchhoff's first law we have $i_1 = i_2 + i_3$. By Kirchhoff's second law, on each loop we have
$E(t) = Ri_1 + L_1 i_2'$ and $E(t) = Ri_1 + L_2 i_3'$ or $L_1 i_2' + Ri_2 + Ri_3 = E(t)$ and $L_2 i_3' + Ri_2 + Ri_3 = E(t)$.

(b) Taking the Laplace transform of the system

$$0.01 i_2' + 5 i_2 + 5 i_3 = 100$$

$$0.0125 i_3' + 5 i_2 + 5 i_3 = 100$$

gives

$$(s + 500) \mathcal{L}\{i_2\} + 500 \mathcal{L}\{i_3\} = \frac{10{,}000}{s}$$

$$400 \mathcal{L}\{i_2\} + (s + 400) \mathcal{L}\{i_3\} = \frac{8{,}000}{s}$$

so that

$$\mathcal{L}\{i_3\} = \frac{8{,}000}{s^2 + 900s} = \frac{80}{9}\frac{1}{s} - \frac{80}{9}\frac{1}{s + 900}.$$

Then

$$i_3 = \frac{80}{9} - \frac{80}{9}e^{-900t} \qquad \text{and} \qquad i_2 = 20 - 0.0025 i_3' - i_3 = \frac{100}{9} - \frac{100}{9}e^{-900t}.$$

(c) $i_1 = i_2 + i_3 = 20 - 20e^{-900t}$

18. Taking the Laplace transform of the system

$$0.5 i_1' + 50 i_2 = 60$$

$$0.005 i_2' + i_2 - i_1 = 0$$

gives

$$s \mathcal{L}\{i_1\} + 100 \mathcal{L}\{i_2\} = \frac{120}{s}$$

$$-200 \mathcal{L}\{i_1\} + (s + 200) \mathcal{L}\{i_2\} = 0$$

so that

$$\mathcal{L}\{i_2\} = \frac{24{,}000}{s(s^2 + 200s + 20{,}000)} = \frac{6}{5}\frac{1}{s} - \frac{6}{5}\frac{s + 100}{(s + 100)^2 + 100^2} - \frac{6}{5}\frac{100}{(s + 100)^2 + 100^2}.$$

Then

$$i_2 = \frac{6}{5} - \frac{6}{5}e^{-100t}\cos 100t - \frac{6}{5}e^{-100t}\sin 100t$$

$$i_1 = 0.005 i_2' + i_2 = \frac{6}{5} - \frac{6}{5}e^{-100t}\cos 100t.$$

173

3. False; consider $f(t) = t^{-1/2}$.

6. False; consider $f(t) = 1$ and $g(t) = 1$.

9. $\mathscr{L}\{\sin 2t\} = \dfrac{2}{s^2 + 4}$

12. $\mathscr{L}\{\sin 2t \,\mathscr{U}(t - \pi)\} = \mathscr{L}\{\sin 2(t - \pi)\mathscr{U}(t - \pi)\} = \dfrac{2}{s^2 + 4}e^{-\pi s}$

15. $\mathscr{L}^{-1}\left\{\dfrac{1}{(s - 5)^3}\right\} = \dfrac{1}{2}\mathscr{L}^{-1}\left\{\dfrac{2}{(s - 5)^3}\right\} = \dfrac{1}{2}t^2 e^{5t}$

18. $\mathscr{L}^{-1}\left\{\dfrac{1}{s^2}e^{-5s}\right\} = (t - 5)\mathscr{U}(t - 5)$

21. $\mathscr{L}\{e^{-5t}\}$ exists for $s > -5$.

24. $\mathscr{L}\left\{\displaystyle\int_0^t e^{a\tau}f(\tau)\,d\tau\right\} = \dfrac{1}{s}\mathscr{L}\{e^{at}f(t)\} = \dfrac{F(s - a)}{s}$, whereas

$$\mathscr{L}\left\{e^{at}\int_0^t f(\tau)\,d\tau\right\} = \mathscr{L}\left\{\int_0^t f(\tau)\,d\tau\right\}\bigg|_{s \to s-a} = \dfrac{F(s)}{s}\bigg|_{s \to s-a} = \dfrac{F(s - a)}{s - a}.$$

27. $f(t - t_0)\mathscr{U}(t - t_0)$

30. $f(t) = \sin t\,\mathscr{U}(t - \pi) - \sin t\,\mathscr{U}(t - 3\pi) = -\sin(t - \pi)\mathscr{U}(t - \pi) + \sin(t - 3\pi)\mathscr{U}(t - 3\pi)$

$$\mathscr{L}\{f(t)\} = -\dfrac{1}{s^2 + 1}e^{-\pi s} + \dfrac{1}{s^2 + 1}e^{-3\pi s}$$

$$\mathscr{L}\{e^t f(t)\} = -\dfrac{1}{(s - 1)^2 + 1}e^{-\pi(s-1)} + \dfrac{1}{(s - 1)^2 + 1}e^{-3\pi(s-1)}$$

33. Taking the Laplace transform of the DE we obtain

$$\mathscr{L}\{y\} = \dfrac{5}{(s - 1)^2} + \dfrac{1}{2}\dfrac{2}{(s - 1)^3}$$

so that

$$y = 5te^t + \dfrac{1}{2}t^2 e^t.$$

36. Taking the Laplace transform of the DE we obtain

$$\mathscr{L}\{y\} = \frac{s^3+2}{s^3(s-5)} - \frac{2+2s+s^2}{s^3(s-5)}e^{-s}$$

$$= -\frac{2}{125}\frac{1}{s} - \frac{2}{25}\frac{1}{s^2} - \frac{1}{5}\frac{2}{s^3} + \frac{127}{125}\frac{1}{s-5} - \left[-\frac{37}{125}\frac{1}{s} - \frac{12}{25}\frac{1}{s^2} - \frac{1}{5}\frac{2}{s^3} + \frac{37}{125}\frac{1}{s-5}\right]e^{-s}$$

so that

$$y = -\frac{2}{125} - \frac{2}{25}t - \frac{1}{5}t^2 + \frac{127}{125}e^{5t} - \left[-\frac{37}{125} - \frac{12}{25}(t-1) - \frac{1}{5}(t-1)^2 + \frac{37}{125}e^{5(t-1)}\right]\mathscr{U}(t-1).$$

39. Taking the Laplace transform of the system gives

$$s\mathscr{L}\{x\} + \mathscr{L}\{y\} = \frac{1}{s^2} + 1$$

$$4\mathscr{L}\{x\} + s\mathscr{L}\{y\} = 2$$

so that

$$\mathscr{L}\{x\} = \frac{s^2-2s+1}{s(s-2)(s+2)} = -\frac{1}{4}\frac{1}{s} + \frac{1}{8}\frac{1}{s-2} + \frac{9}{8}\frac{1}{s+2}.$$

Then

$$x = -\frac{1}{4} + \frac{1}{8}e^{2t} + \frac{9}{8}e^{-2t} \quad\text{and}\quad y = -x'+t = \frac{9}{4}e^{-2t} - \frac{1}{4}e^{2t} + t.$$

42. The DE is

$$\frac{1}{2}\frac{d^2q}{dt^2} + 10\frac{dq}{dt} + 100q = 10 - 10\mathscr{U}(t-5).$$

Taking the Laplace transform we obtain

$$\mathscr{L}\{q\} = \frac{20}{s(s^2+20s+200)}\left(1-e^{-5s}\right)$$

$$= \left[\frac{1}{10}\frac{1}{s} - \frac{1}{10}\frac{s+10}{(s+10)^2+10^2} - \frac{1}{10}\frac{10}{(s+10)^2+10^2}\right]\left(1-e^{-5s}\right)$$

so that

$$q(t) = \frac{1}{10} - \frac{1}{10}e^{-10t}\cos 10t - \frac{1}{10}e^{-10t}\sin 10t$$

$$- \left[\frac{1}{10} - \frac{1}{10}e^{-10(t-5)}\cos 10(t-5) - \frac{1}{10}e^{-10(t-5)}\sin 10(t-5)\right]\mathscr{U}(t-5).$$

45. (a) With $\omega^2 = g/l$ and $K = k/m$ the system of DEs is

$$\theta_1'' + \omega^2\theta_1 = -K(\theta_1-\theta_2)$$

$$\theta_2'' + \omega^2\theta_2 = K(\theta_1-\theta_2).$$

Denoting the Laplace transform of $\theta(t)$ by $\Theta(s)$ we have that the Laplace transform of the system is

$$(s^2 + \omega^2)\Theta_1(s) = -K\Theta_1(s) + K\Theta_2(s) + s\theta_0$$

$$(s^2 + \omega^2)\Theta_2(s) = K\Theta_1(s) - K\Theta_2(s) + s\psi_0.$$

If we add the two equations, we get

$$\Theta_1(s) + \Theta_2(s) = (\theta_0 + \psi_0)\frac{s}{s^2 + \omega^2}$$

which implies

$$\theta_1(t) + \theta_2(t) = (\theta_0 + \psi_0)\cos\omega t.$$

This enables us to solve for first, say, $\theta_1(t)$ and then find $\theta_2(t)$ from

$$\theta_2(t) = -\theta_1(t) + (\theta_0 + \psi_0)\cos\omega t.$$

Now solving

$$(s^2 + \omega^2 + K)\Theta_1(s) - K\Theta_2(s) = s\theta_0$$

$$-k\Theta_1(s) + (s^2 + \omega^2 + K)\Theta_2(s) = s\psi_0$$

gives

$$[(s^2 + \omega^2 + K)^2 - K^2]\Theta_1(s) = s(s^2 + \omega^2 + K)\theta_0 + Ks\psi_0.$$

Factoring the difference of two squares and using partial fractions we get

$$\Theta_1(s) = \frac{s(s^2 + \omega^2 + K)\theta_0 + Ks\psi_0}{(s^2 + \omega^2)(s^2 + \omega^2 + 2K)} = \frac{\theta_0 + \psi_0}{2}\frac{s}{s^2 + \omega^2} + \frac{\theta_0 - \psi_0}{2}\frac{s}{s^2 + \omega^2 + 2K},$$

so

$$\theta_1(t) = \frac{\theta_0 + \psi_0}{2}\cos\omega t + \frac{\theta_0 - \psi_0}{2}\cos\sqrt{\omega^2 + 2K}\,t.$$

Then from $\theta_2(t) = -\theta_1(t) + (\theta_0 + \psi_0)\cos\omega t$ we get

$$\theta_2(t) = \frac{\theta_0 + \psi_0}{2}\cos\omega t - \frac{\theta_0 - \psi_0}{2}\cos\sqrt{\omega^2 + 2K}\,t.$$

(b) With the initial conditions $\theta_1(0) = \theta_0$, $\theta_1'(0) = 0$, $\theta_2(0) = \theta_0$, $\theta_2'(0) = 0$ we have

$$\theta_1(t) = \theta_0\cos\omega t, \qquad \theta_2(t) = \theta_0\cos\omega t.$$

Physically this means that both pendulums swing in the same direction as if they were free since the spring exerts no influence on the motion ($\theta_1(t)$ and $\theta_2(t)$ are free of K).

With the initial conditions $\theta_1(0) = \theta_0$, $\theta_1'(0) = 0$, $\theta_2(0) = -\theta_0$, $\theta_2'(0) = 0$ we have

$$\theta_1(t) = \theta_0\cos\sqrt{\omega^2 + 2K}\,t, \qquad \theta_2(t) = -\theta_0\cos\sqrt{\omega^2 + 2K}\,t.$$

Physically this means that both pendulums swing in the opposite directions, stretching and compressing the spring. The amplitude of both displacements is $|\theta_0|$. Moreover, $\theta_1(t) = \theta_0$ and $\theta_2(t) = -\theta_0$ at precisely the same times. At these times the spring is stretched to its maximum.

8 Systems of Linear First-Order Differential Equations

Preliminary Theory

The terminology and concepts listed below provide an outline of the main ideas encountered in this section. These can be useful when preparing for a quiz or test.

Terminology and Concepts

- linear system of first-order DEs

- homogeneous linear system of DEs

- matrix form of a linear system of DEs

- superposition principle

- linear independence of a set of solutions

- Wronskian

- fundamental set of solutions

- general solution of a linear system of DEs

- particular solution of a nonhomogeneous linear system of DEs

- complementary function of a homogeneous linear system of DEs

The basic skills listed below summarize the more mechanical types of problems encountered in the exercise set for this section.

Basic Skills

- express a linear first-order system in both normal form and matrix form

- determine whether a set of vectors forms a fundamental set

177

3. Let $\mathbf{X} = \begin{pmatrix} x \\ y \\ z \end{pmatrix}$. Then $\mathbf{X}' = \begin{pmatrix} -3 & 4 & -9 \\ 6 & -1 & 0 \\ 10 & 4 & 3 \end{pmatrix} \mathbf{X}$.

6. Let $\mathbf{X} = \begin{pmatrix} x \\ y \\ z \end{pmatrix}$. Then $\mathbf{X}' = \begin{pmatrix} -3 & 4 & 0 \\ 5 & 9 & 0 \\ 0 & 1 & 6 \end{pmatrix} \mathbf{X} + \begin{pmatrix} e^{-t}\sin 2t \\ 4e^{-t}\cos 2t \\ -e^{-t} \end{pmatrix}$.

9. $\dfrac{dx}{dt} = x - y + 2z + e^{-t} - 3t$

$\dfrac{dy}{dt} = 3x - 4y + z + 2e^{-t} + t$

$\dfrac{dz}{dt} = -2x + 5y + 6z + 2e^{-t} - t$

12. Since

$$\mathbf{X}' = \begin{pmatrix} 5\cos t - 5\sin t \\ 2\cos t - 4\sin t \end{pmatrix} e^t \quad \text{and} \quad \begin{pmatrix} -2 & 5 \\ -2 & 4 \end{pmatrix} \mathbf{X} = \begin{pmatrix} 5\cos t - 5\sin t \\ 2\cos t - 4\sin t \end{pmatrix} e^t$$

we see that

$$\mathbf{X}' = \begin{pmatrix} -2 & 5 \\ -2 & 4 \end{pmatrix} \mathbf{X}.$$

15. Since

$$\mathbf{X}' = \begin{pmatrix} 0 \\ 0 \\ 0 \end{pmatrix} \quad \text{and} \quad \begin{pmatrix} 1 & 2 & 1 \\ 6 & -1 & 0 \\ -1 & -2 & -1 \end{pmatrix} \mathbf{X} = \begin{pmatrix} 0 \\ 0 \\ 0 \end{pmatrix}$$

we see that

$$\mathbf{X}' = \begin{pmatrix} 1 & 2 & 1 \\ 6 & -1 & 0 \\ -1 & -2 & -1 \end{pmatrix} \mathbf{X}.$$

18. Yes, since $W(\mathbf{X}_1, \mathbf{X}_2) = 8e^{2t} \neq 0$ the set $\mathbf{X}_1, \mathbf{X}_2$ is linearly independent on $-\infty < t < \infty$.

21. Since

$$\mathbf{X}'_p = \begin{pmatrix} 2 \\ -1 \end{pmatrix} \quad \text{and} \quad \begin{pmatrix} 1 & 4 \\ 3 & 2 \end{pmatrix} \mathbf{X}_p + \begin{pmatrix} 2 \\ -4 \end{pmatrix} t + \begin{pmatrix} -7 \\ -18 \end{pmatrix} = \begin{pmatrix} 2 \\ -1 \end{pmatrix}$$

we see that

$$\mathbf{X}'_p = \begin{pmatrix} 1 & 4 \\ 3 & 2 \end{pmatrix} \mathbf{X}_p + \begin{pmatrix} 2 \\ -4 \end{pmatrix} t + \begin{pmatrix} -7 \\ -18 \end{pmatrix}.$$

24. Since

$$\mathbf{X}'_p = \begin{pmatrix} 3\cos 3t \\ 0 \\ -3\sin 3t \end{pmatrix} \quad \text{and} \quad \begin{pmatrix} 1 & 2 & 3 \\ -4 & 2 & 0 \\ -6 & 1 & 0 \end{pmatrix} \mathbf{X}_p + \begin{pmatrix} -1 \\ 4 \\ 3 \end{pmatrix} \sin 3t = \begin{pmatrix} 3\cos 3t \\ 0 \\ -3\sin 3t \end{pmatrix}$$

we see that

$$\mathbf{X}'_p = \begin{pmatrix} 1 & 2 & 3 \\ -4 & 2 & 0 \\ -6 & 1 & 0 \end{pmatrix} \mathbf{X}_p + \begin{pmatrix} -1 \\ 4 \\ 3 \end{pmatrix} \sin 3t.$$

Section 8.2

Homogeneous Linear Systems

The terminology and concepts listed below provide an outline of the main ideas encountered in this section. These can be useful when preparing for a quiz or test.

Terminology and Concepts

- characteristic equation of a matrix
- eigenvalues of a matrix
- eigenvector of a matrix corresponding to an eigenvalue
- general solution of a homogeneous linear system of DEs with constant coefficients
- phase plane
- trajectory
- phase portrait of a linear system of DEs
- repeller
- attractor
- multiplicity of an eigenvalue

The basic skills listed below summarize the more mechanical types of problems encountered in the exercise set for this section.

Basic Skills

- find the general solution of a homogeneous linear system of DEs with constant coefficients when the eigenvalues are real and distinct, real and repeated, or complex

Use of Computers To find eigenvalues and eigenvectors of a matrix using *Mathematica* or *Maple*, we can input the matrix by rows. For example,

$$m = \begin{pmatrix} -4 & 1 & 1 \\ 1 & 5 & -1 \\ 0 & 1 & -3 \end{pmatrix}$$

is the matrix of the coefficients of the linear system (6) in Example 2 in the text. Using *Mathematica*, we can write **m** as

$$m=\{\{-4, 1, 1\}, \{1, 5, -1\}, \{0, 1, -3\}\}$$

The commands **Eigenvalues[m]** and **Eigenvectors[m]** given in sequence yield

$$[-4, -3, 5] \quad \text{and} \quad \{\{-10, -1, 1\}, \{1, 0, 1\}, \{1, 8, 1\}\}.$$

Translated this means $\lambda = -4$ is an eigenvalue with corresponding eigenvector

$$\mathbf{K} = \begin{pmatrix} -10 \\ -1 \\ 1 \end{pmatrix},$$

and so on. In *Mathematica* eigenvalues and eigenvectors can be obtained at the same time using the single command **Eigensystem[m]**. In *Maple*, the matrix **m** is written

```
m:=matrix(3,3,[-4,1,1,1,5,-1,0,1,-3])
```

The commands for eigenvalues and eigenvectors are, respectively, `eigenvals(m)` and `eigenvects(m)`.

Exercises 8.2 *Hints, Suggestions, Solutions, and Examples*

3. The system is

$$\mathbf{X}' = \begin{pmatrix} -4 & 2 \\ -5/2 & 2 \end{pmatrix} \mathbf{X}$$

and $\det(\mathbf{A} - \lambda\mathbf{I}) = (\lambda - 1)(\lambda + 3) = 0$. For $\lambda_1 = 1$ we obtain

$$\begin{pmatrix} -5 & 2 & | & 0 \\ -5/2 & 1 & | & 0 \end{pmatrix} \implies \begin{pmatrix} -5 & 2 & | & 0 \\ 0 & 0 & | & 0 \end{pmatrix} \quad \text{so that} \quad \mathbf{K}_1 = \begin{pmatrix} 2 \\ 5 \end{pmatrix}.$$

For $\lambda_2 = -3$ we obtain

$$\begin{pmatrix} -1 & 2 & | & 0 \\ -5/2 & 5 & | & 0 \end{pmatrix} \implies \begin{pmatrix} -1 & 2 & | & 0 \\ 0 & 0 & | & 0 \end{pmatrix} \quad \text{so that} \quad \mathbf{K}_2 = \begin{pmatrix} 2 \\ 1 \end{pmatrix}.$$

Then

$$\mathbf{X} = c_1 \begin{pmatrix} 2 \\ 5 \end{pmatrix} e^t + c_2 \begin{pmatrix} 2 \\ 1 \end{pmatrix} e^{-3t}.$$

6. The system is

$$\mathbf{X}' = \begin{pmatrix} -6 & 2 \\ -3 & 1 \end{pmatrix} \mathbf{X}$$

and $\det(\mathbf{A} - \lambda \mathbf{I}) = \lambda(\lambda + 5) = 0$. For $\lambda_1 = 0$ we obtain

$$\begin{pmatrix} -6 & 2 & | & 0 \\ -3 & 1 & | & 0 \end{pmatrix} \implies \begin{pmatrix} 1 & -1/3 & | & 0 \\ 0 & 0 & | & 0 \end{pmatrix} \quad \text{so that} \quad \mathbf{K}_1 = \begin{pmatrix} 1 \\ 3 \end{pmatrix}.$$

For $\lambda_2 = -5$ we obtain

$$\begin{pmatrix} -1 & 2 & | & 0 \\ -3 & 6 & | & 0 \end{pmatrix} \implies \begin{pmatrix} 1 & -2 & | & 0 \\ 0 & 0 & | & 0 \end{pmatrix} \quad \text{so that} \quad \mathbf{K}_2 = \begin{pmatrix} 2 \\ 1 \end{pmatrix}.$$

Then

$$\mathbf{X} = c_1 \begin{pmatrix} 1 \\ 3 \end{pmatrix} + c_2 \begin{pmatrix} 2 \\ 1 \end{pmatrix} e^{-5t}.$$

9. We have $\det(\mathbf{A} - \lambda \mathbf{I}) = -(\lambda + 1)(\lambda - 3)(\lambda + 2) = 0$. For $\lambda_1 = -1$, $\lambda_2 = 3$, and $\lambda_3 = -2$ we obtain

$$\mathbf{K}_1 = \begin{pmatrix} -1 \\ 0 \\ 1 \end{pmatrix}, \quad \mathbf{K}_2 = \begin{pmatrix} 1 \\ 4 \\ 3 \end{pmatrix}, \quad \text{and} \quad \mathbf{K}_3 = \begin{pmatrix} 1 \\ -1 \\ 3 \end{pmatrix},$$

so that

$$\mathbf{X} = c_1 \begin{pmatrix} -1 \\ 0 \\ 1 \end{pmatrix} e^{-t} + c_2 \begin{pmatrix} 1 \\ 4 \\ 3 \end{pmatrix} e^{3t} + c_3 \begin{pmatrix} 1 \\ -1 \\ 3 \end{pmatrix} e^{-2t}.$$

12. We have $\det(\mathbf{A} - \lambda \mathbf{I}) = (\lambda - 3)(\lambda + 5)(6 - \lambda) = 0$. For $\lambda_1 = 3$, $\lambda_2 = -5$, and $\lambda_3 = 6$ we obtain

$$\mathbf{K}_1 = \begin{pmatrix} 1 \\ 1 \\ 0 \end{pmatrix}, \quad \mathbf{K}_2 = \begin{pmatrix} 1 \\ -1 \\ 0 \end{pmatrix}, \quad \text{and} \quad \mathbf{K}_3 = \begin{pmatrix} 2 \\ -2 \\ 11 \end{pmatrix},$$

so that

$$\mathbf{X} = c_1 \begin{pmatrix} 1 \\ 1 \\ 0 \end{pmatrix} e^{3t} + c_2 \begin{pmatrix} 1 \\ -1 \\ 0 \end{pmatrix} e^{-5t} + c_3 \begin{pmatrix} 2 \\ -2 \\ 11 \end{pmatrix} e^{6t}.$$

21. We have $\det(\mathbf{A} - \lambda \mathbf{I}) = (\lambda - 2)^2 = 0$. For $\lambda_1 = 2$ we obtain

$$\mathbf{K} = \begin{pmatrix} 1 \\ 1 \end{pmatrix}.$$

A solution of $(\mathbf{A} - \lambda_1\mathbf{I})\mathbf{P} = \mathbf{K}$ is

$$\mathbf{P} = \begin{pmatrix} -1/3 \\ 0 \end{pmatrix}$$

so that

$$\mathbf{X} = c_1 \begin{pmatrix} 1 \\ 1 \end{pmatrix} e^{2t} + c_2 \left[\begin{pmatrix} 1 \\ 1 \end{pmatrix} te^{2t} + \begin{pmatrix} -1/3 \\ 0 \end{pmatrix} e^{2t} \right].$$

24. We have $\det(\mathbf{A} - \lambda\mathbf{I}) = (\lambda - 8)(\lambda + 1)^2 = 0$. For $\lambda_1 = 8$ we obtain

$$\mathbf{K}_1 = \begin{pmatrix} 2 \\ 1 \\ 2 \end{pmatrix}.$$

For $\lambda_2 = -1$ we obtain

$$\mathbf{K}_2 = \begin{pmatrix} 0 \\ -2 \\ 1 \end{pmatrix} \quad \text{and} \quad \mathbf{K}_3 = \begin{pmatrix} 1 \\ -2 \\ 0 \end{pmatrix}.$$

Then

$$\mathbf{X} = c_1 \begin{pmatrix} 2 \\ 1 \\ 2 \end{pmatrix} e^{8t} + c_2 \begin{pmatrix} 0 \\ -2 \\ 1 \end{pmatrix} e^{-t} + c_3 \begin{pmatrix} 1 \\ -2 \\ 0 \end{pmatrix} e^{-t}.$$

27. We have $\det(\mathbf{A} - \lambda\mathbf{I}) = -(\lambda - 1)^3 = 0$. For $\lambda_1 = 1$ we obtain

$$\mathbf{K} = \begin{pmatrix} 0 \\ 1 \\ 1 \end{pmatrix}.$$

Solutions of $(\mathbf{A} - \lambda_1\mathbf{I})\mathbf{P} = \mathbf{K}$ and $(\mathbf{A} - \lambda_1\mathbf{I})\mathbf{Q} = \mathbf{P}$ are

$$\mathbf{P} = \begin{pmatrix} 0 \\ 1 \\ 0 \end{pmatrix} \quad \text{and} \quad \mathbf{Q} = \begin{pmatrix} 1/2 \\ 0 \\ 0 \end{pmatrix}$$

so that

$$\mathbf{X} = c_1 \begin{pmatrix} 0 \\ 1 \\ 1 \end{pmatrix} e^t + c_2 \left[\begin{pmatrix} 0 \\ 1 \\ 1 \end{pmatrix} te^t + \begin{pmatrix} 0 \\ 1 \\ 0 \end{pmatrix} e^t \right] + c_3 \left[\begin{pmatrix} 0 \\ 1 \\ 1 \end{pmatrix} \frac{t^2}{2} e^t + \begin{pmatrix} 0 \\ 1 \\ 0 \end{pmatrix} te^t + \begin{pmatrix} 1/2 \\ 0 \\ 0 \end{pmatrix} e^t \right].$$

30. We have $\det(\mathbf{A} - \lambda\mathbf{I}) = -(\lambda + 1)(\lambda - 1)^2 = 0$. For $\lambda_1 = -1$ we obtain

$$\mathbf{K}_1 = \begin{pmatrix} -1 \\ 0 \\ 1 \end{pmatrix}.$$

For $\lambda_2 = 1$ we obtain

$$\mathbf{K}_2 = \begin{pmatrix} 1 \\ 0 \\ 1 \end{pmatrix} \quad \text{and} \quad \mathbf{K}_3 = \begin{pmatrix} 0 \\ 1 \\ 0 \end{pmatrix}$$

so that

$$\mathbf{X} = c_1 \begin{pmatrix} -1 \\ 0 \\ 1 \end{pmatrix} e^{-t} + c_2 \begin{pmatrix} 1 \\ 0 \\ 1 \end{pmatrix} e^t + c_3 \begin{pmatrix} 0 \\ 1 \\ 0 \end{pmatrix} e^t.$$

If

$$\mathbf{X}(0) = \begin{pmatrix} 1 \\ 2 \\ 5 \end{pmatrix}$$

then $c_1 = 2$, $c_2 = 3$, and $c_3 = 2$.

In Problems 33-46 the form of the answer will vary according to the choice of eigenvector. For example, in Problem 33, if $\mathbf{K}_1$ is chosen to be $\begin{pmatrix} 1 \\ 2 - i \end{pmatrix}$ the solution has the form

$$\mathbf{X} = c_1 \begin{pmatrix} \cos t \\ 2\cos t + \sin t \end{pmatrix} e^{4t} + c_2 \begin{pmatrix} \sin t \\ 2\sin t - \cos t \end{pmatrix} e^{4t}.$$

33. We have $\det(\mathbf{A} - \lambda\mathbf{I}) = \lambda^2 - 8\lambda + 17 = 0$. For $\lambda_1 = 4 + i$ we obtain

$$\mathbf{K}_1 = \begin{pmatrix} 2 + i \\ 5 \end{pmatrix}$$

so that

$$\mathbf{X}_1 = \begin{pmatrix} 2 + i \\ 5 \end{pmatrix} e^{(4+i)t} = \begin{pmatrix} 2\cos t - \sin t \\ 5\cos t \end{pmatrix} e^{4t} + i \begin{pmatrix} \cos t + 2\sin t \\ 5\sin t \end{pmatrix} e^{4t}.$$

Then

$$\mathbf{X} = c_1 \begin{pmatrix} 2\cos t - \sin t \\ 5\cos t \end{pmatrix} e^{4t} + c_2 \begin{pmatrix} \cos t + 2\sin t \\ 5\sin t \end{pmatrix} e^{4t}.$$

36. We have $\det(\mathbf{A} - \lambda\mathbf{I}) = \lambda^2 - 10\lambda + 34 = 0$. For $\lambda_1 = 5 + 3i$ we obtain

$$\mathbf{K}_1 = \begin{pmatrix} 1 - 3i \\ 2 \end{pmatrix}$$

so that

$$\mathbf{X}_1 = \begin{pmatrix} 1 - 3i \\ 2 \end{pmatrix} e^{(5+3i)t} = \begin{pmatrix} \cos 3t + 3\sin 3t \\ 2\cos 3t \end{pmatrix} e^{5t} + i \begin{pmatrix} \sin 3t - 3\cos 3t \\ 2\sin 3t \end{pmatrix} e^{5t}.$$

Then

$$\mathbf{X} = c_1 \begin{pmatrix} \cos 3t + 3\sin 3t \\ 2\cos 3t \end{pmatrix} e^{5t} + c_2 \begin{pmatrix} \sin 3t - 3\cos 3t \\ 2\sin 3t \end{pmatrix} e^{5t}.$$

39. We have $\det(\mathbf{A} - \lambda\mathbf{I}) = -\lambda\left(\lambda^2 + 1\right) = 0$. For $\lambda_1 = 0$ we obtain

$$\mathbf{K}_1 = \begin{pmatrix} 1 \\ 0 \\ 0 \end{pmatrix}.$$

For $\lambda_2 = i$ we obtain

$$\mathbf{K}_2 = \begin{pmatrix} -i \\ i \\ 1 \end{pmatrix}$$

so that

$$\mathbf{X}_2 = \begin{pmatrix} -i \\ i \\ 1 \end{pmatrix} e^{it} = \begin{pmatrix} \sin t \\ -\sin t \\ \cos t \end{pmatrix} + i \begin{pmatrix} -\cos t \\ \cos t \\ \sin t \end{pmatrix}.$$

Then

$$\mathbf{X} = c_1 \begin{pmatrix} 1 \\ 0 \\ 0 \end{pmatrix} + c_2 \begin{pmatrix} \sin t \\ -\sin t \\ \cos t \end{pmatrix} + c_3 \begin{pmatrix} -\cos t \\ \cos t \\ \sin t \end{pmatrix}.$$

42. We have $\det(\mathbf{A} - \lambda\mathbf{I}) = -(\lambda - 6)(\lambda^2 - 8\lambda + 20) = 0$. For $\lambda_1 = 6$ we obtain

$$\mathbf{K}_1 = \begin{pmatrix} 0 \\ 1 \\ 0 \end{pmatrix}.$$

For $\lambda_2 = 4 + 2i$ we obtain

$$\mathbf{K}_2 = \begin{pmatrix} -i \\ 0 \\ 2 \end{pmatrix}$$

so that

$$\mathbf{X}_2 = \begin{pmatrix} -i \\ 0 \\ 2 \end{pmatrix} e^{(4+2i)t} = \begin{pmatrix} \sin 2t \\ 0 \\ 2\cos 2t \end{pmatrix} e^{4t} + i \begin{pmatrix} -\cos 2t \\ 0 \\ 2\sin 2t \end{pmatrix} e^{4t}.$$

Then

$$\mathbf{X} = c_1 \begin{pmatrix} 0 \\ 1 \\ 0 \end{pmatrix} e^{6t} + c_2 \begin{pmatrix} \sin 2t \\ 0 \\ 2\cos 2t \end{pmatrix} e^{4t} + c_3 \begin{pmatrix} -\cos 2t \\ 0 \\ 2\sin 2t \end{pmatrix} e^{4t}.$$

45. We have $\det(\mathbf{A} - \lambda\mathbf{I}) = (1 - \lambda)(\lambda^2 + 25) = 0$. For $\lambda_1 = 1$ we obtain

$$\mathbf{K}_1 = \begin{pmatrix} 25 \\ -7 \\ 6 \end{pmatrix}.$$

For $\lambda_2 = 5i$ we obtain

$$\mathbf{K}_2 = \begin{pmatrix} 1 + 5i \\ 1 \\ 1 \end{pmatrix}$$

so that

$$\mathbf{X}_2 = \begin{pmatrix} 1 + 5i \\ 1 \\ 1 \end{pmatrix} e^{5it} = \begin{pmatrix} \cos 5t - 5\sin 5t \\ \cos 5t \\ \cos 5t \end{pmatrix} + i \begin{pmatrix} \sin 5t + 5\cos 5t \\ \sin 5t \\ \sin 5t \end{pmatrix}.$$

Then

$$\mathbf{X} = c_1 \begin{pmatrix} 25 \\ -7 \\ 6 \end{pmatrix} e^t + c_2 \begin{pmatrix} \cos 5t - 5\sin 5t \\ \cos 5t \\ \cos 5t \end{pmatrix} + c_3 \begin{pmatrix} \sin 5t + 5\cos 5t \\ \sin 5t \\ \sin 5t \end{pmatrix}.$$

If

$$\mathbf{X}(0) = \begin{pmatrix} 4 \\ 6 \\ -7 \end{pmatrix}$$

then $c_1 = c_2 = -1$ and $c_3 = 6$.

Section 8.3 Nonhomogeneous Linear Systems

The terminology and concepts listed below provide an outline of the main ideas encountered in this section. These can be useful when preparing for a quiz or test.

Terminology and Concepts

- method of undetermined coefficients for finding a particular solution of a nonhomogeneous linear system of DEs

- fundamental matrix of a linear system of DEs

- method of variation of parameters for finding a particular solution of a nonhomogeneous linear system of DEs

The basic skills listed below summarize the more mechanical types of problems encountered in the exercise set for this section.

Basic Skills

- use undetermined coefficients to solve a nonhomogeneous linear system of DEs
- use variation of parameters to solve a nonhomogeneous linear system of DEs

Exercises 8.3   *Hints, Suggestions, Solutions, and Examples*

3. Solving

$$\begin{vmatrix} 1-\lambda & 3 \\ 3 & 1-\lambda \end{vmatrix} = \lambda^2 - 2\lambda - 8 = (\lambda - 4)(\lambda + 2) = 0$$

we obtain eigenvalues $\lambda_1 = -2$ and $\lambda_2 = 4$. Corresponding eigenvectors are

$$\mathbf{K}_1 = \begin{pmatrix} 1 \\ -1 \end{pmatrix} \quad \text{and} \quad \mathbf{K}_2 = \begin{pmatrix} 1 \\ 1 \end{pmatrix}.$$

Thus

$$\mathbf{X}_c = c_1 \begin{pmatrix} 1 \\ -1 \end{pmatrix} e^{-2t} + c_2 \begin{pmatrix} 1 \\ 1 \end{pmatrix} e^{4t}.$$

Substituting

$$\mathbf{X}_p = \begin{pmatrix} a_3 \\ b_3 \end{pmatrix} t^2 + \begin{pmatrix} a_2 \\ b_2 \end{pmatrix} t + \begin{pmatrix} a_1 \\ b_1 \end{pmatrix}$$

into the system yields

$$a_3 + 3b_3 = 2 \qquad a_2 + 3b_2 = 2a_3 \qquad a_1 + 3b_1 = a_2$$

$$3a_3 + b_3 = 0 \qquad 3a_2 + b_2 + 1 = 2b_3 \qquad 3a_1 + b_1 + 5 = b_2$$

from which we obtain $a_3 = -1/4$, $b_3 = 3/4$, $a_2 = 1/4$, $b_2 = -1/4$, $a_1 = -2$, and $b_1 = 3/4$. Then

$$\mathbf{X}(t) = c_1 \begin{pmatrix} 1 \\ -1 \end{pmatrix} e^{-2t} + c_2 \begin{pmatrix} 1 \\ 1 \end{pmatrix} e^{4t} + \begin{pmatrix} -1/4 \\ 3/4 \end{pmatrix} t^2 + \begin{pmatrix} 1/4 \\ -1/4 \end{pmatrix} t + \begin{pmatrix} -2 \\ 3/4 \end{pmatrix}.$$

6. Solving

$$\begin{vmatrix} -1-\lambda & 5 \\ -1 & 1-\lambda \end{vmatrix} = \lambda^2 + 4 = 0$$

we obtain the eigenvalues $\lambda_1 = 2i$ and $\lambda_2 = -2i$. Corresponding eigenvectors are

$$\mathbf{K}_1 = \begin{pmatrix} 5 \\ 1 + 2i \end{pmatrix} \quad \text{and} \quad \mathbf{K}_2 = \begin{pmatrix} 5 \\ 1 - 2i \end{pmatrix}.$$

Thus

$$\mathbf{X}_c = c_1 \begin{pmatrix} 5\cos 2t \\ \cos 2t - 2\sin 2t \end{pmatrix} + c_2 \begin{pmatrix} 5\sin 2t \\ 2\cos 2t + \sin 2t \end{pmatrix}.$$

Substituting

$$\mathbf{X}_p = \begin{pmatrix} a_2 \\ b_2 \end{pmatrix} \cos t + \begin{pmatrix} a_1 \\ b_1 \end{pmatrix} \sin t$$

into the system yields

$$-a_2 + 5b_2 - a_1 = 0$$

$$-a_2 + b_2 - b_1 - 2 = 0$$

$$-a_1 + 5b_1 + a_2 + 1 = 0$$

$$-a_1 + b_1 + b_2 = 0$$

from which we obtain $a_2 = -3$, $b_2 = -2/3$, $a_1 = -1/3$, and $b_1 = 1/3$. Then

$$\mathbf{X}(t) = c_1 \begin{pmatrix} 5\cos 2t \\ \cos 2t - 2\sin 2t \end{pmatrix} + c_2 \begin{pmatrix} 5\sin 2t \\ 2\cos 2t + \sin 2t \end{pmatrix} + \begin{pmatrix} -3 \\ -2/3 \end{pmatrix} \cos t + \begin{pmatrix} -1/3 \\ 1/3 \end{pmatrix} \sin t.$$

9. Solving

$$\begin{vmatrix} -1-\lambda & -2 \\ 3 & 4-\lambda \end{vmatrix} = \lambda^2 - 3\lambda + 2 = (\lambda - 1)(\lambda - 2) = 0$$

we obtain the eigenvalues $\lambda_1 = 1$ and $\lambda_2 = 2$. Corresponding eigenvectors are

$$\mathbf{K}_1 = \begin{pmatrix} 1 \\ -1 \end{pmatrix} \quad \text{and} \quad \mathbf{K}_2 = \begin{pmatrix} -4 \\ 6 \end{pmatrix}.$$

Thus

$$\mathbf{X}_c = c_1 \begin{pmatrix} 1 \\ -1 \end{pmatrix} e^t + c_2 \begin{pmatrix} -4 \\ 6 \end{pmatrix} e^{2t}.$$

Substituting

$$\mathbf{X}_p = \begin{pmatrix} a_1 \\ b_1 \end{pmatrix}$$

into the system yields

$$-a_1 - 2b_1 = -3$$

$$3a_1 + 4b_1 = -3$$

from which we obtain $a_1 = -9$ and $b_1 = 6$. Then

$$\mathbf{X}(t) = c_1 \begin{pmatrix} 1 \\ -1 \end{pmatrix} e^t + c_2 \begin{pmatrix} -4 \\ 6 \end{pmatrix} e^{2t} + \begin{pmatrix} -9 \\ 6 \end{pmatrix}.$$

Setting

$$\mathbf{X}(0) = \begin{pmatrix} -4 \\ 5 \end{pmatrix}$$

187

we obtain

$$c_1 - 4c_2 - 9 = -4$$

$$-c_1 + 6c_2 + 6 = 5.$$

Then $c_1 = 13$ and $c_2 = 2$ so

$$\mathbf{X}(t) = 13 \begin{pmatrix} 1 \\ -1 \end{pmatrix} e^t + 2 \begin{pmatrix} -4 \\ 6 \end{pmatrix} e^{2t} + \begin{pmatrix} -9 \\ 6 \end{pmatrix}.$$

12. From

$$\mathbf{X}' = \begin{pmatrix} 2 & -1 \\ 3 & -2 \end{pmatrix} \mathbf{X} + \begin{pmatrix} 0 \\ 4 \end{pmatrix} t$$

we obtain

$$\mathbf{X}_c = c_1 \begin{pmatrix} 1 \\ 1 \end{pmatrix} e^t + c_2 \begin{pmatrix} 1 \\ 3 \end{pmatrix} e^{-t}.$$

Then

$$\boldsymbol{\Phi} = \begin{pmatrix} e^t & e^{-t} \\ e^t & 3e^{-t} \end{pmatrix} \quad \text{and} \quad \boldsymbol{\Phi}^{-1} = \begin{pmatrix} \frac{3}{2}e^{-t} & -\frac{1}{2}e^{-t} \\ -\frac{1}{2}e^t & \frac{1}{2}e^t \end{pmatrix}$$

so that

$$\mathbf{U} = \int \boldsymbol{\Phi}^{-1} \mathbf{F}\, dt = \int \begin{pmatrix} -2te^{-t} \\ 2te^t \end{pmatrix} dt = \begin{pmatrix} 2te^{-t} + 2e^{-t} \\ 2te^t - 2e^t \end{pmatrix}$$

and

$$\mathbf{X}_p = \boldsymbol{\Phi}\mathbf{U} = \begin{pmatrix} 4 \\ 8 \end{pmatrix} t + \begin{pmatrix} 0 \\ -4 \end{pmatrix}.$$

15. From

$$\mathbf{X}' = \begin{pmatrix} 0 & 2 \\ -1 & 3 \end{pmatrix} \mathbf{X} + \begin{pmatrix} 1 \\ -1 \end{pmatrix} e^t$$

we obtain

$$\mathbf{X}_c = c_1 \begin{pmatrix} 2 \\ 1 \end{pmatrix} e^t + c_2 \begin{pmatrix} 1 \\ 1 \end{pmatrix} e^{2t}.$$

Then

$$\boldsymbol{\Phi} = \begin{pmatrix} 2e^t & e^{2t} \\ e^t & e^{2t} \end{pmatrix} \quad \text{and} \quad \boldsymbol{\Phi}^{-1} = \begin{pmatrix} e^{-t} & -e^{-t} \\ -e^{-2t} & 2e^{-2t} \end{pmatrix}$$

so that

$$\mathbf{U} = \int \boldsymbol{\Phi}^{-1} \mathbf{F}\, dt = \int \begin{pmatrix} 2 \\ -3e^{-t} \end{pmatrix} dt = \begin{pmatrix} 2t \\ 3e^{-t} \end{pmatrix}$$

and

$$\mathbf{X}_p = \boldsymbol{\Phi}\mathbf{U} = \begin{pmatrix} 4 \\ 2 \end{pmatrix} te^t + \begin{pmatrix} 3 \\ 3 \end{pmatrix} e^t.$$

18. From

$$\mathbf{X}' = \begin{pmatrix} 1 & 8 \\ 1 & -1 \end{pmatrix} \mathbf{X} + \begin{pmatrix} e^{-t} \\ te^t \end{pmatrix}$$

188

we obtain

$$\mathbf{X}_c = c_1 \begin{pmatrix} 4 \\ 1 \end{pmatrix} e^{3t} + c_2 \begin{pmatrix} -2 \\ 1 \end{pmatrix} e^{-3t}.$$

Then

$$\Phi = \begin{pmatrix} 4e^{3t} & -2e^{3t} \\ e^{3t} & e^{-3t} \end{pmatrix} \quad \text{and} \quad \Phi^{-1} = \begin{pmatrix} \frac{1}{6}e^{-3t} & \frac{1}{3}e^{-3t} \\ -\frac{1}{6}e^{3t} & \frac{2}{3}e^{3t} \end{pmatrix}$$

so that

$$\mathbf{U} = \int \Phi^{-1}\mathbf{F}\, dt = \int \begin{pmatrix} \frac{1}{6}e^{-4t} + \frac{1}{3}te^{-2t} \\ -\frac{1}{6}e^{2t} + \frac{2}{3}te^{4t} \end{pmatrix} dt = \begin{pmatrix} -\frac{1}{24}e^{-4t} - \frac{1}{6}te^{-2t} - \frac{1}{12}e^{-2t} \\ -\frac{1}{12}e^{2t} + \frac{1}{6}te^{4t} - \frac{1}{24}e^{4t} \end{pmatrix}$$

and

$$\mathbf{X}_p = \Phi\mathbf{U} = \begin{pmatrix} -te^{t} - \frac{1}{4}e^{t} \\ -\frac{1}{8}e^{-t} - \frac{1}{8}e^{t} \end{pmatrix}.$$

21. From

$$\mathbf{X}' = \begin{pmatrix} 0 & -1 \\ 1 & 0 \end{pmatrix} \mathbf{X} + \begin{pmatrix} \sec t \\ 0 \end{pmatrix}$$

we obtain

$$\mathbf{X}_c = c_1 \begin{pmatrix} \cos t \\ \sin t \end{pmatrix} + c_2 \begin{pmatrix} \sin t \\ -\cos t \end{pmatrix}.$$

Then

$$\Phi = \begin{pmatrix} \cos t & \sin t \\ \sin t & -\cos t \end{pmatrix} \quad \text{and} \quad \Phi^{-1} = \begin{pmatrix} \cos t & \sin t \\ \sin t & -\cos t \end{pmatrix}$$

so that

$$\mathbf{U} = \int \Phi^{-1}\mathbf{F}\, dt = \int \begin{pmatrix} 1 \\ \tan t \end{pmatrix} dt = \begin{pmatrix} t \\ -\ln|\cos t| \end{pmatrix}$$

and

$$\mathbf{X}_p = \Phi\mathbf{U} = \begin{pmatrix} t\cos t - \sin t \ln|\cos t| \\ t\sin t + \cos t \ln|\cos t| \end{pmatrix}.$$

24. From

$$\mathbf{X}' = \begin{pmatrix} 2 & -2 \\ 8 & -6 \end{pmatrix} \mathbf{X} + \begin{pmatrix} 1 \\ 3 \end{pmatrix} \frac{1}{t}e^{-2t}$$

we obtain

$$\mathbf{X}_c = c_1 \begin{pmatrix} 1 \\ 2 \end{pmatrix} e^{-2t} + c_2 \left[\begin{pmatrix} 1 \\ 2 \end{pmatrix} te^{-2t} + \begin{pmatrix} 1/2 \\ 1/2 \end{pmatrix} e^{-2t} \right].$$

Then

$$\Phi = \begin{pmatrix} 1 & t + \frac{1}{2} \\ 2 & 2t + \frac{1}{2} \end{pmatrix} e^{-2t} \quad \text{and} \quad \Phi^{-1} = \begin{pmatrix} -4t - 1 & 2t + 1 \\ 4 & -2 \end{pmatrix} e^{2t}$$

so that

$$\mathbf{U} = \int \Phi^{-1}\mathbf{F}\, dt = \int \begin{pmatrix} 2 + 2/t \\ -2/t \end{pmatrix} dt = \begin{pmatrix} 2t + 2\ln t \\ -2\ln t \end{pmatrix}$$

and

$$\mathbf{X}_p = \mathbf{\Phi U} = \begin{pmatrix} 2t + \ln t - 2t \ln t \\ 4t + 3 \ln t - 4t \ln t \end{pmatrix} e^{-2t}.$$

27. From

$$\mathbf{X}' = \begin{pmatrix} 1 & 2 \\ -1/2 & 1 \end{pmatrix} \mathbf{X} + \begin{pmatrix} \csc t \\ \sec t \end{pmatrix} e^t$$

we obtain

$$\mathbf{X}_c = c_1 \begin{pmatrix} 2 \sin t \\ \cos t \end{pmatrix} e^t + c_2 \begin{pmatrix} 2 \cos t \\ -\sin t \end{pmatrix} e^t.$$

Then

$$\mathbf{\Phi} = \begin{pmatrix} 2 \sin t & 2 \cos t \\ \cos t & -\sin t \end{pmatrix} e^t \quad \text{and} \quad \mathbf{\Phi}^{-1} = \begin{pmatrix} \frac{1}{2} \sin t & \cos t \\ \frac{1}{2} \cos t & -\sin t \end{pmatrix} e^{-t}$$

so that

$$\mathbf{U} = \int \mathbf{\Phi}^{-1} \mathbf{F} \, dt = \int \begin{pmatrix} \frac{3}{2} \\ \frac{1}{2} \cot t - \tan t \end{pmatrix} dt = \begin{pmatrix} \frac{3}{2} t \\ \frac{1}{2} \ln|\sin t| + \ln|\cos t| \end{pmatrix}$$

and

$$\mathbf{X}_p = \mathbf{\Phi U} = \begin{pmatrix} 3 \sin t \\ \frac{3}{2} \cos t \end{pmatrix} te^t + \begin{pmatrix} \cos t \\ -\frac{1}{2} \sin t \end{pmatrix} e^t \ln|\sin t| + \begin{pmatrix} 2 \cos t \\ -\sin t \end{pmatrix} e^t \ln|\cos t|.$$

30. From

$$\mathbf{X}' = \begin{pmatrix} 3 & -1 & -1 \\ 1 & 1 & -1 \\ 1 & -1 & 1 \end{pmatrix} \mathbf{X} + \begin{pmatrix} 0 \\ t \\ 2e^t \end{pmatrix}$$

we obtain

$$\mathbf{X}_c = c_1 \begin{pmatrix} 1 \\ 1 \\ 1 \end{pmatrix} e^t + c_2 \begin{pmatrix} 1 \\ 1 \\ 0 \end{pmatrix} e^{2t} + c_3 \begin{pmatrix} 1 \\ 0 \\ 1 \end{pmatrix} e^{2t}.$$

Then

$$\mathbf{\Phi} = \begin{pmatrix} e^t & e^{2t} & e^{2t} \\ e^t & e^{2t} & 0 \\ e^t & 0 & e^{2t} \end{pmatrix} \quad \text{and} \quad \mathbf{\Phi}^{-1} = \begin{pmatrix} -e^{-t} & e^{-t} & e^{-t} \\ e^{-2t} & 0 & -e^{-2t} \\ e^{-2t} & -e^{-2t} & 0 \end{pmatrix}$$

so that

$$\mathbf{U} = \int \mathbf{\Phi}^{-1} \mathbf{F} \, dt = \int \begin{pmatrix} te^{-t} + 2 \\ -2e^{-t} \\ -te^{-2t} \end{pmatrix} dt = \begin{pmatrix} -te^{-t} - e^{-t} + 2t \\ 2e^{-t} \\ \frac{1}{2}te^{-2t} + \frac{1}{4}e^{-2t} \end{pmatrix}$$

and

$$\mathbf{X}_p = \mathbf{\Phi U} = \begin{pmatrix} -1/2 \\ -1 \\ -1/2 \end{pmatrix} t + \begin{pmatrix} -3/4 \\ -1 \\ -3/4 \end{pmatrix} + \begin{pmatrix} 2 \\ 2 \\ 0 \end{pmatrix} e^t + \begin{pmatrix} 2 \\ 2 \\ 2 \end{pmatrix} te^t.$$

33. Let $\mathbf{I} = \begin{pmatrix} i_1 \\ i_2 \end{pmatrix}$ so that

$$\mathbf{I}' = \begin{pmatrix} -11 & 3 \\ 3 & -3 \end{pmatrix} \mathbf{I} + \begin{pmatrix} 100 \sin t \\ 0 \end{pmatrix}$$

and

$$\mathbf{X}_c = c_1 \begin{pmatrix} 1 \\ 3 \end{pmatrix} e^{-2t} + c_2 \begin{pmatrix} 3 \\ -1 \end{pmatrix} e^{-12t}.$$

Then

$$\mathbf{\Phi} = \begin{pmatrix} e^{-2t} & 3e^{-12t} \\ 3e^{-2t} & -e^{-12t} \end{pmatrix}, \quad \mathbf{\Phi}^{-1} = \begin{pmatrix} \frac{1}{10}e^{2t} & \frac{3}{10}e^{2t} \\ \frac{3}{10}e^{12t} & -\frac{1}{10}e^{12t} \end{pmatrix},$$

$$\mathbf{U} = \int \mathbf{\Phi}^{-1}\mathbf{F}\,dt = \int \begin{pmatrix} 10e^{2t}\sin t \\ 30e^{12t}\sin t \end{pmatrix} dt = \begin{pmatrix} 2e^{2t}(2\sin t - \cos t) \\ \frac{6}{29}e^{12t}(12\sin t - \cos t) \end{pmatrix},$$

and

$$\mathbf{I}_p = \mathbf{\Phi}\mathbf{U} = \begin{pmatrix} \frac{332}{29}\sin t - \frac{76}{29}\cos t \\ \frac{276}{29}\sin t - \frac{168}{29}\cos t \end{pmatrix}$$

so that

$$\mathbf{I} = c_1 \begin{pmatrix} 1 \\ 3 \end{pmatrix} e^{-2t} + c_2 \begin{pmatrix} 3 \\ -1 \end{pmatrix} e^{-12t} + \mathbf{I}_p.$$

If $\mathbf{I}(0) = \begin{pmatrix} 0 \\ 0 \end{pmatrix}$ then $c_1 = 2$ and $c_2 = \frac{6}{29}$.

Section 8.4 Matrix Exponential

The terminology and concepts listed below provide an outline of the main ideas encountered in this section. These can be useful when preparing for a quiz or test.

Terminology and Concepts

- infinite series representation of the matrix exponential $e^{\mathbf{A}t}$
- the matrix exponential is a fundamental matrix

The basic skills listed below summarize the more mechanical types of problems encountered in the exercise set for this section.

Basic Skills

- compute $e^{\mathbf{A}t}$ for a given matrix $\mathbf{A}$
- use the matrix exponential to solve a nonhomogeneous linear system

Section 8.4 Matrix Exponential

Use of Computers To compute the matrix exponential for a square matrix $\mathbf{A}t$ use

MatrixExp[At]	(*Mathematica*)
with(linalg):	(*Maple*)
exponential(A,t);	
expm(At)	(MATLAB)

Exercises 8.4   *Hints, Suggestions, Solutions, and Examples*

3. For

$$\mathbf{A} = \begin{pmatrix} 1 & 1 & 1 \\ 1 & 1 & 1 \\ -2 & -2 & -2 \end{pmatrix}$$

we have

$$\mathbf{A}^2 = \begin{pmatrix} 1 & 1 & 1 \\ 1 & 1 & 1 \\ -2 & -2 & -2 \end{pmatrix} \begin{pmatrix} 1 & 1 & 1 \\ 1 & 1 & 1 \\ -2 & -2 & -2 \end{pmatrix} = \begin{pmatrix} 0 & 0 & 0 \\ 0 & 0 & 0 \\ 0 & 0 & 0 \end{pmatrix}.$$

Thus, $\mathbf{A}^3 = \mathbf{A}^4 = \mathbf{A}^5 = \cdots = \mathbf{0}$ and

$$e^{\mathbf{A}t} = \mathbf{I} + \mathbf{A}t = \begin{pmatrix} 1 & 0 & 0 \\ 0 & 1 & 0 \\ 0 & 0 & 1 \end{pmatrix} + \begin{pmatrix} t & t & t \\ t & t & t \\ -2t & -2t & -2t \end{pmatrix} = \begin{pmatrix} t+1 & t & t \\ t & t+1 & t \\ -2t & -2t & -2t+1 \end{pmatrix}.$$

6. In Problem 2 it is shown that for

$$\mathbf{A} = \begin{pmatrix} 0 & 1 \\ 1 & 0 \end{pmatrix}$$

we have

$$e^{\mathbf{A}t} = \begin{pmatrix} \cosh t & \sinh t \\ \sinh t & \cosh t \end{pmatrix}.$$

Thus, the solution of the given system is

$$\mathbf{X} = \begin{pmatrix} \cosh t & \sinh t \\ \sinh t & \cosh t \end{pmatrix} \begin{pmatrix} c_1 \\ c_2 \end{pmatrix} = c_1 \begin{pmatrix} \cosh t \\ \sinh t \end{pmatrix} + c_2 \begin{pmatrix} \sinh t \\ \cosh t \end{pmatrix}.$$

9. To solve

$$\mathbf{X}' = \begin{pmatrix} 1 & 0 \\ 0 & 2 \end{pmatrix} \mathbf{X} + \begin{pmatrix} 3 \\ -1 \end{pmatrix}$$

we identify $t_0 = 0$, $\mathbf{F}(t) = \begin{pmatrix} 3 \\ -1 \end{pmatrix}$, and use the results of Problem 1 and equation (5) in the text.

$$\mathbf{X}(t) = e^{\mathbf{A}t}\mathbf{C} + e^{\mathbf{A}t}\int_{t_0}^{t} e^{-\mathbf{A}s}\mathbf{F}(s)\,ds$$

$$= \begin{pmatrix} e^t & 0 \\ 0 & e^{2t} \end{pmatrix}\begin{pmatrix} c_1 \\ c_2 \end{pmatrix} + \begin{pmatrix} e^t & 0 \\ 0 & e^{2t} \end{pmatrix}\int_0^t \begin{pmatrix} e^{-s} & 0 \\ 0 & e^{-2s} \end{pmatrix}\begin{pmatrix} 3 \\ -1 \end{pmatrix}\,ds$$

$$= \begin{pmatrix} c_1 e^t \\ c_2 e^{2t} \end{pmatrix} + \begin{pmatrix} e^t & 0 \\ 0 & e^{2t} \end{pmatrix}\int_0^t \begin{pmatrix} 3e^{-s} \\ -e^{-2s} \end{pmatrix}\,ds$$

$$= \begin{pmatrix} c_1 e^t \\ c_2 e^{2t} \end{pmatrix} + \begin{pmatrix} e^t & 0 \\ 0 & e^{2t} \end{pmatrix}\begin{pmatrix} -3e^{-s} \\ \frac{1}{2}e^{-2s} \end{pmatrix}\Big|_0^t$$

$$= \begin{pmatrix} c_1 e^t \\ c_2 e^{2t} \end{pmatrix} + \begin{pmatrix} e^t & 0 \\ 0 & e^{2t} \end{pmatrix}\begin{pmatrix} -3e^{-t}+3 \\ \frac{1}{2}e^{-2t}-\frac{1}{2} \end{pmatrix}$$

$$= \begin{pmatrix} c_1 e^t \\ c_2 e^{2t} \end{pmatrix} + \begin{pmatrix} -3+3e^t \\ \frac{1}{2}-\frac{1}{2}e^{2t} \end{pmatrix} = c_3\begin{pmatrix} 1 \\ 0 \end{pmatrix}e^t + c_4\begin{pmatrix} 0 \\ 1 \end{pmatrix}e^{2t} + \begin{pmatrix} -3 \\ \frac{1}{2} \end{pmatrix}.$$

12. See the solution of Problem 6 above for the result of Problem 2. To solve

$$\mathbf{X}' = \begin{pmatrix} 0 & 1 \\ 1 & 0 \end{pmatrix}\mathbf{X} + \begin{pmatrix} \cosh t \\ \sinh t \end{pmatrix}$$

we identify $t_0 = 0$, $\mathbf{F}(t) = \begin{pmatrix} \cosh t \\ \sinh t \end{pmatrix}$, and use the results of Problem 2 and equation (5) in the text.

$$\mathbf{X}(t) = e^{\mathbf{A}t}\mathbf{C} + e^{\mathbf{A}t}\int_{t_0}^{t} e^{-\mathbf{A}s}\mathbf{F}(s)\,ds$$

$$= \begin{pmatrix} \cosh t & \sinh t \\ \sinh t & \cosh t \end{pmatrix}\begin{pmatrix} c_1 \\ c_2 \end{pmatrix} + \begin{pmatrix} \cosh t & \sinh t \\ \sinh t & \cosh t \end{pmatrix}\int_0^t \begin{pmatrix} \cosh s & -\sinh s \\ -\sinh s & \cosh s \end{pmatrix}\begin{pmatrix} \cosh s \\ \sinh s \end{pmatrix}\,ds$$

$$= \begin{pmatrix} c_1\cosh t + c_2\sinh t \\ c_1\sinh t + c_2\cosh t \end{pmatrix} + \begin{pmatrix} \cosh t & \sinh t \\ \sinh t & \cosh t \end{pmatrix}\int_0^t \begin{pmatrix} 1 \\ 0 \end{pmatrix}\,ds$$

$$= \begin{pmatrix} c_1\cosh t + c_2\sinh t \\ c_1\sinh t + c_2\cosh t \end{pmatrix} + \begin{pmatrix} \cosh t & \sinh t \\ \sinh t & \cosh t \end{pmatrix}\begin{pmatrix} s \\ 0 \end{pmatrix}\Big|_0^t$$

$$= \begin{pmatrix} c_1\cosh t + c_2\sinh t \\ c_1\sinh t + c_2\cosh t \end{pmatrix} + \begin{pmatrix} \cosh t & \sinh t \\ \sinh t & \cosh t \end{pmatrix}\begin{pmatrix} t \\ 0 \end{pmatrix}$$

$$= \begin{pmatrix} c_1\cosh t + c_2\sinh t \\ c_1\sinh t + c_2\cosh t \end{pmatrix} + \begin{pmatrix} t\cosh t \\ t\sinh t \end{pmatrix} = c_1\begin{pmatrix} \cosh t \\ \sinh t \end{pmatrix} + c_2\begin{pmatrix} \sinh t \\ \cosh t \end{pmatrix} + t\begin{pmatrix} \cosh t \\ \sinh t \end{pmatrix}.$$

193

15. From $s\mathbf{I} - \mathbf{A} = \begin{pmatrix} s-4 & -3 \\ 4 & s+4 \end{pmatrix}$ we find

$$(s\mathbf{I} - \mathbf{A})^{-1} = \begin{pmatrix} \dfrac{3/2}{s-2} - \dfrac{1/2}{s+2} & \dfrac{3/4}{s-2} - \dfrac{3/4}{s+2} \\ \dfrac{-1}{s-2} + \dfrac{1}{s+2} & \dfrac{-1/2}{s-2} + \dfrac{3/2}{s+2} \end{pmatrix}$$

and

$$e^{\mathbf{A}t} = \begin{pmatrix} \frac{3}{2}e^{2t} - \frac{1}{2}e^{-2t} & \frac{3}{4}e^{2t} - \frac{3}{4}e^{-2t} \\ -e^{2t} + e^{-2t} & -\frac{1}{2}e^{2t} + \frac{3}{2}e^{-2t} \end{pmatrix}.$$

The general solution of the system is then

$$\mathbf{X} = e^{\mathbf{A}t}\mathbf{C} = \begin{pmatrix} \frac{3}{2}e^{2t} - \frac{1}{2}e^{-2t} & \frac{3}{4}e^{2t} - \frac{3}{4}e^{-2t} \\ -e^{2t} + e^{-2t} & -\frac{1}{2}e^{2t} + \frac{3}{2}e^{-2t} \end{pmatrix}\begin{pmatrix} c_1 \\ c_2 \end{pmatrix}$$

$$= c_1\begin{pmatrix} 3/2 \\ -1 \end{pmatrix}e^{2t} + c_1\begin{pmatrix} -1/2 \\ 1 \end{pmatrix}e^{-2t} + c_2\begin{pmatrix} 3/4 \\ -1/2 \end{pmatrix}e^{2t} + c_2\begin{pmatrix} -3/4 \\ 3/2 \end{pmatrix}e^{-2t}$$

$$= \left(\frac{1}{2}c_1 + \frac{1}{4}c_2\right)\begin{pmatrix} 3 \\ -2 \end{pmatrix}e^{2t} + \left(-\frac{1}{2}c_1 - \frac{3}{4}c_2\right)\begin{pmatrix} 1 \\ -2 \end{pmatrix}e^{-2t}$$

$$= c_3\begin{pmatrix} 3 \\ -2 \end{pmatrix}e^{2t} + c_4\begin{pmatrix} 1 \\ -2 \end{pmatrix}e^{-2t}.$$

18. From $s\mathbf{I} - \mathbf{A} = \begin{pmatrix} s & -1 \\ 2 & s+2 \end{pmatrix}$ we find

$$(s\mathbf{I} - \mathbf{A})^{-1} = \begin{pmatrix} \dfrac{s+1+1}{(s+1)^2+1} & \dfrac{1}{(s+1)^2+1} \\ \dfrac{-2}{(s+1)^2+1} & \dfrac{s+1-1}{(s+1)^2+1} \end{pmatrix}$$

and

$$e^{\mathbf{A}t} = \begin{pmatrix} e^{-t}\cos t + e^{-t}\sin t & e^{-t}\sin t \\ -2e^{-t}\sin t & e^{-t}\cos t - e^{-t}\sin t \end{pmatrix}.$$

The general solution of the system is then

$$X = e^{At}C = \begin{pmatrix} e^{-t}\cos t + e^{-t}\sin t & e^{-t}\sin t \\ -2e^{-t}\sin t & e^{-t}\cos t - e^{-t}\sin t \end{pmatrix} \begin{pmatrix} c_1 \\ c_2 \end{pmatrix}$$

$$= c_1 \begin{pmatrix} 1 \\ 0 \end{pmatrix} e^{-t}\cos t + c_1 \begin{pmatrix} 1 \\ -2 \end{pmatrix} e^{-t}\sin t + c_2 \begin{pmatrix} 0 \\ 1 \end{pmatrix} e^{-t}\cos t + c_2 \begin{pmatrix} 1 \\ -1 \end{pmatrix} e^{-t}\sin t$$

$$= c_1 \begin{pmatrix} \cos t + \sin t \\ -2\sin t \end{pmatrix} e^{-t} + c_2 \begin{pmatrix} \sin t \\ \cos t - \sin t \end{pmatrix} e^{-t}.$$

21. From equation (3) in the text

$$e^{tA} = e^{tPDP^{-1}} = I + t(PDP^{-1}) + \frac{1}{2!}t^2(PDP^{-1})^2 + \frac{1}{3!}t^3(PDP^{-1})^3 + \cdots$$

$$= P\left[I + tD + \frac{1}{2!}(tD)^2 + \frac{1}{3!}(tD)^3 + \cdots\right]P^{-1} = Pe^{tD}P^{-1}.$$

24. In Problem 20 it is shown that when

$$A = \begin{pmatrix} 2 & 1 \\ 1 & 2 \end{pmatrix}$$

then

$$PDP^{-1} = \begin{pmatrix} 2 & 1 \\ 1 & 2 \end{pmatrix}.$$

From Problems 20-22 and equation (1) in the text

$$X = e^{tA}C = Pe^{tD}P^{-1}C$$

$$= \begin{pmatrix} -e^t & e^{3t} \\ e^t & e^{3t} \end{pmatrix} \begin{pmatrix} e^t & 0 \\ 0 & e^{3t} \end{pmatrix} \begin{pmatrix} -\frac{1}{2}e^{-t} & \frac{1}{2}e^{-t} \\ \frac{1}{2}e^{3t} & \frac{1}{2}e^{-3t} \end{pmatrix} \begin{pmatrix} c_1 \\ c_2 \end{pmatrix}$$

$$= \begin{pmatrix} \frac{1}{2}e^t + \frac{1}{2}e^{9t} & -\frac{1}{2}e^t + \frac{1}{2}e^{3t} \\ -\frac{1}{2}e^t + \frac{1}{2}e^{9t} & \frac{1}{2}e^t + \frac{1}{2}e^{3t} \end{pmatrix} \begin{pmatrix} c_1 \\ c_2 \end{pmatrix}.$$

Chapter 8 in Review   *Hints, Suggestions, Solutions, and Examples*

3. Since

$$\begin{pmatrix} 4 & 6 & 6 \\ 1 & 3 & 2 \\ -1 & -4 & -3 \end{pmatrix} \begin{pmatrix} 3 \\ 1 \\ -1 \end{pmatrix} = \begin{pmatrix} 12 \\ 4 \\ -4 \end{pmatrix} = 4 \begin{pmatrix} 3 \\ 1 \\ -1 \end{pmatrix},$$

Chapter 8 in Review

we see that $\lambda = 4$ is an eigenvalue with eigenvector $\mathbf{K}_3$. The corresponding solution is $\mathbf{X}_3 = \mathbf{K}_3 e^{4t}$.

6. We have $\det(\mathbf{A} - \lambda\mathbf{I}) = (\lambda + 6)(\lambda + 2) = 0$ so that

$$\mathbf{X} = c_1 \begin{pmatrix} 1 \\ -1 \end{pmatrix} e^{-6t} + c_2 \begin{pmatrix} 1 \\ 1 \end{pmatrix} e^{-2t}.$$

9. We have $\det(\mathbf{A} - \lambda\mathbf{I}) = -(\lambda - 2)(\lambda - 4)(\lambda + 3) = 0$ so that

$$\mathbf{X} = c_1 \begin{pmatrix} -2 \\ 3 \\ 1 \end{pmatrix} e^{2t} + c_2 \begin{pmatrix} 0 \\ 1 \\ 1 \end{pmatrix} e^{4t} + c_3 \begin{pmatrix} 7 \\ 12 \\ -16 \end{pmatrix} e^{-3t}.$$

12. We have

$$\mathbf{X}_c = c_1 \begin{pmatrix} 2\cos t \\ -\sin t \end{pmatrix} e^t + c_2 \begin{pmatrix} 2\sin t \\ \cos t \end{pmatrix} e^t.$$

Then

$$\mathbf{\Phi} = \begin{pmatrix} 2\cos t & 2\sin t \\ -\sin t & \cos t \end{pmatrix} e^t, \quad \mathbf{\Phi}^{-1} = \begin{pmatrix} \frac{1}{2}\cos t & -\sin t \\ \frac{1}{2}\sin t & \cos t \end{pmatrix} e^{-t},$$

and

$$\mathbf{U} = \int \mathbf{\Phi}^{-1}\mathbf{F}\, dt = \int \begin{pmatrix} \cos t - \sec t \\ \sin t \end{pmatrix} dt = \begin{pmatrix} \sin t - \ln|\sec t + \tan t| \\ -\cos t \end{pmatrix},$$

so that

$$\mathbf{X}_p = \mathbf{\Phi}\mathbf{U} = \begin{pmatrix} -2\cos t \ln|\sec t + \tan t| \\ -1 + \sin t \ln|\sec t + \tan t| \end{pmatrix} e^t.$$

15. (a) Letting

$$\mathbf{K} = \begin{pmatrix} k_1 \\ k_2 \\ k_3 \end{pmatrix}$$

we note that $(\mathbf{A} - 2\mathbf{I})\mathbf{K} = \mathbf{0}$ implies that $3k_1 + 3k_2 + 3k_3 = 0$, so $k_1 = -(k_2 + k_3)$. Choosing $k_2 = 0$, $k_3 = 1$ and then $k_2 = 1$, $k_3 = 0$ we get

$$\mathbf{K}_1 = \begin{pmatrix} -1 \\ 0 \\ 1 \end{pmatrix} \quad \text{and} \quad \mathbf{K}_2 = \begin{pmatrix} -1 \\ 1 \\ 0 \end{pmatrix},$$

respectively. Thus,

$$\mathbf{X}_1 = \begin{pmatrix} -1 \\ 0 \\ 1 \end{pmatrix} e^{2t} \quad \text{and} \quad \mathbf{X}_2 = \begin{pmatrix} -1 \\ 1 \\ 0 \end{pmatrix} e^{2t}$$

are two solutions.

(b) From $\det(\mathbf{A} - \lambda\mathbf{I}) = \lambda^2(3 - \lambda) = 0$ we see that $\lambda_1 = 3$, and 0 is an eigenvalue of multiplicity two. Letting

$$\mathbf{K} = \begin{pmatrix} k_1 \\ k_2 \\ k_3 \end{pmatrix},$$

as in part (a), we note that $(\mathbf{A} - 0\mathbf{I})\mathbf{K} = \mathbf{A}\mathbf{K} = \mathbf{0}$ implies that $k_1 + k_2 + k_3 = 0$, so $k_1 = -(k_2 + k_3)$. Choosing $k_2 = 0$, $k_3 = 1$, and then $k_2 = 1$, $k_3 = 0$ we get

$$\mathbf{K}_2 = \begin{pmatrix} -1 \\ 0 \\ 1 \end{pmatrix} \quad \text{and} \quad \mathbf{K}_3 = \begin{pmatrix} -1 \\ 1 \\ 0 \end{pmatrix},$$

respectively. Since the eigenvector corresponding to $\lambda_1 = 3$ is

$$\mathbf{K}_1 = \begin{pmatrix} 1 \\ 1 \\ 1 \end{pmatrix},$$

the general solution of the system is

$$\mathbf{X} = c_1 \begin{pmatrix} 1 \\ 1 \\ 1 \end{pmatrix} e^{3t} + c_2 \begin{pmatrix} -1 \\ 0 \\ 1 \end{pmatrix} + c_3 \begin{pmatrix} -1 \\ 1 \\ 0 \end{pmatrix}.$$

9 Numerical Solutions of Ordinary Differential Equations

Section 9.1
Euler Methods and Error Analysis

The terminology and concepts listed below provide an outline of the main ideas encountered in this section. These can be useful when preparing for a quiz or test.

Terminology and Concepts

- error analysis for numerical solutions of ordinary DEs

- round-off error

- local and global truncation errors

- bound for local truncation error

- Euler's method for approximating the solution of a first-order IVP

- improved Euler's method for approximating the solution of a first-order IVP

- a predictor-corrector method

The basic skills listed below summarize the more mechanical types of problems encountered in the exercise set for this section.

Basic Skills

- use the improved Euler's method to approximate the solution of a first-order IVP

- find a bound on the local truncation error when Euler's method or the improved Euler's method is used to approximate the solution of an IVP

198

Exercises 9.1   *Hints, Suggestions, Solutions, and Examples*

3. $h=0.1$ $h=0.05$ **6.** $h=0.1$ $h=0.05$

x_n	y_n
0.00	0.0000
0.10	0.1005
0.20	0.2030
0.30	0.3098
0.40	0.4234
0.50	0.5470

x_n	y_n
0.00	0.0000
0.05	0.0501
0.10	0.1004
0.15	0.1512
0.20	0.2028
0.25	0.2554
0.30	0.3095
0.35	0.3652
0.40	0.4230
0.45	0.4832
0.50	0.5465

x_n	y_n
0.00	0.0000
0.10	0.0050
0.20	0.0200
0.30	0.0451
0.40	0.0805
0.50	0.1266

x_n	y_n
0.00	0.0000
0.05	0.0013
0.10	0.0050
0.15	0.0113
0.20	0.0200
0.25	0.0313
0.30	0.0451
0.35	0.0615
0.40	0.0805
0.45	0.1022
0.50	0.1266

9. $h=0.1$ $h=0.05$

x_n	y_n
1.00	1.0000
1.10	1.0095
1.20	1.0404
1.30	1.0967
1.40	1.1866
1.50	1.3260

x_n	y_n
1.00	1.0000
1.05	1.0024
1.10	1.0100
1.15	1.0228
1.20	1.0414
1.25	1.0663
1.30	1.0984
1.35	1.1389
1.40	1.1895
1.45	1.2526
1.50	1.3315

11. (*Hint*) Use the substitution $u = x + y - 1$ to find the analytic solution of the DE.

12. (a)

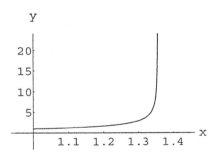

(b)

x_n	Euler	Imp. Euler
1.00	1.0000	1.0000
1.10	1.2000	1.2469
1.20	1.4938	1.6430
1.30	1.9711	2.4042
1.40	2.9060	4.5085

15. (a) Using Euler's method we obtain $y(0.1) \approx y_1 = 0.8$.

(b) Using $y'' = 5e^{-2x}$ we see that the local truncation error is

$$5e^{-2c}\frac{(0.1)^2}{2} = 0.025e^{-2c}.$$

199

Since e^{-2x} is a decreasing function, $e^{-2c} \leq e^0 = 1$ for $0 \leq c \leq 0.1$. Thus an upper bound for the local truncation error is $0.025(1) = 0.025$.

(c) Since $y(0.1) = 0.8234$, the actual error is $y(0.1) - y_1 = 0.0234$, which is less than 0.025.

(d) Using Euler's method with $h = 0.05$ we obtain $y(0.1) \approx y_2 = 0.8125$.

(e) The error in (d) is $0.8234 - 0.8125 = 0.0109$. With global truncation error $O(h)$, when the step size is halved we expect the error for $h = 0.05$ to be one-half the error when $h = 0.1$. Comparing 0.0109 with 0.0234 we see that this is the case.

18. (a) Using $y''' = -114e^{-3(x-1)}$ we see that the local truncation error is

$$\left| y'''(c) \frac{h^3}{6} \right| = 114e^{-3(x-1)} \frac{h^3}{6} = 19h^3 e^{-3(c-1)}.$$

(b) Since $e^{-3(x-1)}$ is a decreasing function for $1 \leq x \leq 1.5$, $e^{-3(c-1)} \leq e^{-3(1-1)} = 1$ for $1 \leq c \leq 1.5$ and

$$\left| y'''(c) \frac{h^3}{6} \right| \leq 19(0.1)^3(1) = 0.019.$$

(c) Using the improved Euler's method with $h = 0.1$ we obtain $y(1.5) \approx 2.080108$. With $h = 0.05$ we obtain $y(1.5) \approx 2.059166$.

(d) Since $y(1.5) = 2.053216$, the error for $h = 0.1$ is $E_{0.1} = 0.026892$, while the error for $h = 0.05$ is $E_{0.05} = 0.005950$. With global truncation error $O(h^2)$ we expect $E_{0.1}/E_{0.05} \approx 4$. We actually have $E_{0.1}/E_{0.05} = 4.52$.

Section 9.2   Runge-Kutta Methods

The terminology and concepts listed below provide an outline of the main ideas encountered in this section. These can be useful when preparing for a quiz or test.

Terminology and Concepts

- Runge-Kutta methods
- fourth-order Runge-Kutta method (RK4 method)
- truncation errors for the RK4 method

The basic skills listed below summarize the more mechanical types of problems encountered in the exercise set for this section.

Basic Skills

- use the RK4 method to approximate the solution of a first-order initial-value problem

Exercises 9.2 *Hints, Suggestions, Solutions, and Examples*

3.

x_n	y_n
1.00	5.0000
1.10	3.9724
1.20	3.2284
1.30	2.6945
1.40	2.3163
1.50	2.0533

6.

x_n	y_n
0.00	1.0000
0.10	1.1115
0.20	1.2530
0.30	1.4397
0.40	1.6961
0.50	2.0670

9.

x_n	y_n
0.00	0.5000
0.10	0.5213
0.20	0.5358
0.30	0.5443
0.40	0.5482
0.50	0.5493

12.

x_n	y_n
0.00	0.5000
0.10	0.5250
0.20	0.5498
0.30	0.5744
0.40	0.5987
0.50	0.6225

15. (a)

x_n	$h=0.05$	$h=0.1$
1.00	1.0000	1.0000
1.05	1.1112	
1.10	1.2511	1.2511
1.15	1.4348	
1.20	1.6934	1.6934
1.25	2.1047	
1.30	2.9560	2.9425
1.35	7.8981	
1.40	1.0608×10^{15}	903.0282

(b)

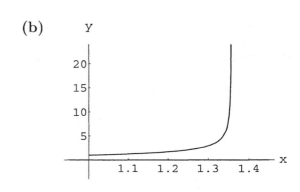

18. (a) Using $y^{(5)} = -1026e^{-3(x-1)}$ we see that the local truncation error is

$$\left| y^{(5)}(c)\frac{h^5}{120} \right| = 8.55h^5 e^{-3(c-1)}.$$

(b) Since $e^{-3(x-1)}$ is a decreasing function for $1 \le x \le 1.5$, $e^{-3(c-1)} \le e^{-3(1-1)} = 1$ for $1 \le c \le 1.5$ and

$$y^{(5)}(c)\frac{h^5}{120} \le 8.55(0.1)^5(1) = 0.0000855.$$

(c) Using the RK4 method with $h = 0.1$ we obtain $y(1.5) \approx 2.053338827$. With $h = 0.05$ we obtain $y(1.5) \approx 2.053222989$.

Section 9.3   Multistep Methods

The terminology and concepts listed below provide an outline of the main ideas encountered in this section. These can be useful when preparing for a quiz or test.

Terminology and Concepts

- predictor-corrector methods
- Adams-Bashforth-Moulton method

The basic skills listed below summarize the more mechanical types of problems encountered in the exercise set for this section.

Basic Skills

- use the Adams-Bashforth-Moulton predictor-corrector method to approximate the solution of a first-order IVP

Exercises 9.3   *Hints, Suggestions, Solutions, and Examples*

In the tables in this section "ABM" stands for "Adams-Bashforth-Moulton".

3. The first predictor is $y_4^* = 0.73318477$.

x_n	y_n	
0.0	1.00000000	init. cond.
0.2	0.73280000	RK4
0.4	0.64608032	RK4
0.6	0.65851653	RK4
0.8	0.72319464	ABM

6. The first predictor for $h = 0.2$ is $y_4^* = 3.34828434$.

x_n	h=0.2		h=0.1	
0.0	1.00000000	init. cond.	1.00000000	init. cond.
0.1			1.21017082	RK4
0.2	1.44139950	RK4	1.44140511	RK4
0.3			1.69487942	RK4
0.4	1.97190167	RK4	1.97191536	ABM
0.5			2.27400341	ABM
0.6	2.60280694	RK4	2.60283209	ABM
0.7			2.96031780	ABM
0.8	3.34860927	ABM	3.34863769	ABM
0.9			3.77026548	ABM
1.0	4.22797875	ABM	4.22801028	ABM

Section 9.4 **Higher-Order Equations and Systems**

The terminology and concepts listed below provide an outline of the main ideas encountered in this section. These can be useful when preparing for a quiz or test.

Terminology and Concepts

- express a second-order IVP as an IVP for a first-order system
- express a higher-order IVP as an IVP for a first-order system
- numerical solution of an IVP for a first-order system using Euler's method and the RK4 method

The basic skills listed below summarize the more mechanical types of problems encountered in the exercise set for this section.

Basic Skills

- solve an IVP for a first-order system consisting of two equations

Exercises 9.4   *Hints, Suggestions, Solutions, and Examples*

3. The substitution $y' = u$ leads to the system

$$y' = u, \qquad u' = 4u - 4y.$$

Using formula (4) in the text with x corresponding to t, y corresponding to x, and u corresponding to y, we obtain the table shown.

x_n	h=0.2 y_n	h=0.2 u_n	h=0.1 y_n	h=0.1 u_n
0.0	-2.0000	1.0000	-2.0000	1.0000
0.1			-1.8321	2.4427
0.2	-1.4928	4.4731	-1.4919	4.4753

6. Using $h = 0.1$, the RK4 method for a system, and a numerical solver, we obtain

t_n	h=0.2 i_{1n}	h=0.2 i_{3n}
0.0	0.0000	0.0000
0.1	2.5000	3.7500
0.2	2.8125	5.7813
0.3	2.0703	7.4023
0.4	0.6104	9.1919
0.5	-1.5619	11.4877

9.

t_n	h=0.2 x_n	h=0.2 y_n	h=0.1 x_n	h=0.1 y_n
0.0	-3.0000	5.0000	-3.0000	5.0000
0.1			-3.4790	4.6707
0.2	-3.9123	4.2857	-3.9123	4.2857

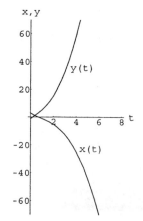

12. Solving for x' and y' we obtain the system

$$x' = \frac{1}{2}y - 3t^2 + 2t - 5$$

$$y' = -\frac{1}{2}y + 3t^2 + 2t + 5.$$

t_n	h=0.2 x_n	h=0.2 y_n	h=0.1 x_n	h=0.1 y_n
0.0	3.0000	-1.0000	3.0000	-1.0000
0.1			2.4727	-0.4527
0.2	1.9867	0.0933	1.9867	0.0933

Section 9.5

Second-Order Boundary-Value Problems

The terminology and concepts listed below provide an outline of the main ideas encountered in this section. These can be useful when preparing for a quiz or test.

Terminology and Concepts

- difference quotient
- finite difference
- forward difference
- backward difference
- central difference
- finite difference method

The basic skills listed below summarize the more mechanical types of problems encountered in the exercise set for this section.

Basic Skills

   • use the finite difference method to approximate the solution of a second-order boundary-value problem

Exercises 9.5   *Hints, Suggestions, Solutions, and Examples*

3. We identify $P(x) = 2$, $Q(x) = 1$, $f(x) = 5x$, and $h = (1-0)/5 = 0.2$. Then the finite difference equation is

$$1.2y_{i+1} - 1.96y_i + 0.8y_{i-1} = 0.04(5x_i).$$

The solution of the corresponding linear system gives

x	0.0	0.2	0.4	0.6	0.8	1.0
y	0.0000	-0.2259	-0.3356	-0.3308	-0.2167	0.0000

6. We identify $P(x) = 5$, $Q(x) = 0$, $f(x) = 4\sqrt{x}$, and $h = (2-1)/6 = 0.1667$. Then the finite difference equation is

$$1.4167y_{i+1} - 2y_i + 0.5833y_{i-1} = 0.2778(4\sqrt{x_i}).$$

The solution of the corresponding linear system gives

x	1.0000	1.1667	1.3333	1.5000	1.6667	1.8333	2.0000
y	1.0000	-0.5918	-1.1626	-1.3070	-1.2704	-1.1541	-1.0000

9. We identify $P(x) = 1 - x$, $Q(x) = x$, $f(x) = x$, and $h = (1-0)/10 = 0.1$. Then the finite difference equation is

$$[1 + 0.05(1 - x_i)]y_{i+1} + [-2 + 0.01x_i]y_i + [1 - 0.05(1 - x_i)]y_{i-1} = 0.01x_i.$$

The solution of the corresponding linear system gives

x	0.0	0.1	0.2	0.3	0.4	0.5	0.6
y	0.0000	0.2660	0.5097	0.7357	0.9471	1.1465	1.3353

0.7	0.8	0.9	1.0
1.5149	1.6855	1.8474	2.0000

12. We identify $P(r) = 2/r$, $Q(r) = 0$, $f(r) = 0$, and $h = (4-1)/6 = 0.5$. Then the finite difference equation is

$$\left(1 + \frac{0.5}{r_i}\right)u_{i+1} - 2u_i + \left(1 - \frac{0.5}{r_i}\right)u_{i-1} = 0.$$

The solution of the corresponding linear system gives

r	1.0	1.5	2.0	2.5	3.0	3.5	4.0
u	50.0000	72.2222	83.3333	90.0000	94.4444	97.6190	100.0000

Chapter 9 in Review

Hints, Suggestions, Solutions, and Examples

3.

x_n	Euler h=0.1	Euler h=0.05	Imp. Euler h=0.1	Imp. Euler h=0.05	RK4 h=0.1	RK4 h=0.05
0.50	0.5000	0.5000	0.5000	0.5000	0.5000	0.5000
0.55		0.5500		0.5512		0.5512
0.60	0.6000	0.6024	0.6048	0.6049	0.6049	0.6049
0.65		0.6573		0.6609		0.6610
0.70	0.7095	0.7144	0.7191	0.7193	0.7194	0.7194
0.75		0.7739		0.7800		0.7801
0.80	0.8283	0.8356	0.8427	0.8430	0.8431	0.8431
0.85		0.8996		0.9082		0.9083
0.90	0.9559	0.9657	0.9752	0.9755	0.9757	0.9757
0.95		1.0340		1.0451		1.0452
1.00	1.0921	1.1044	1.1163	1.1168	1.1169	1.1169

6. The first predictor is $y_3^* = 1.14822731$.

x_n	y_n	
0.0	2.00000000	init. cond.
0.1	1.65620000	RK4
0.2	1.41097281	RK4
0.3	1.24645047	RK4
0.4	1.14796764	ABM

10 Plane Autonomous Systems

Autonomous Systems

The terminology and concepts listed below provide an outline of the main ideas encountered in this section. These can be useful when preparing for a quiz or test.

Terminology and Concepts

- plane autonomous system of first-order DEs
- dynamical system of DEs
- state (response) of the system
- vector field
- critical (stationary) point
- equilibrium solution

The basic skills listed below summarize the more mechanical types of problems encountered in the exercise set for this section.

Basic Skills

- express a nonlinear second-order DE as a plane autonomous system
- find the critical points of a plane autonomous system
- find the general solution of a linear system of DEs with constant coefficients
- express a nonlinear plane autonomous system using polar coordintates
- determine if a linear system of DEs has periodic solutions

Exercises 10.1   *Hints, Suggestions, Solutions, and Examples*

3. The corresponding plane autonomous system is

$$x' = y, \quad y' = x^2 - y(1 - x^3).$$

207

If (x, y) is a critical point, $y = 0$ and so $x^2 - y(1 - x^3) = x^2 = 0$. Therefore $(0, 0)$ is the sole critical point.

6. The corresponding plane autonomous system is

$$x' = y, \quad y' = -x + \epsilon x |x|.$$

If (x, y) is a critical point, $y = 0$ and $-x + \epsilon x |x| = x(-1 + \epsilon |x|) = 0$. Hence $x = 0$, $1/\epsilon$, $-1/\epsilon$. The critical points are $(0, 0)$, $(1/\epsilon, 0)$ and $(-1/\epsilon, 0)$.

9. From $x - y = 0$ we have $y = x$. Substituting into $3x^2 - 4y = 0$ we obtain $3x^2 - 4x = x(3x - 4) = 0$. It follows that $(0, 0)$ and $(4/3, 4/3)$ are the critical points of the system.

12. Adding the two equations we obtain $10 - 15y/(y + 5) = 0$. It follows that $y = 10$, and from $-2x + y + 10 = 0$ we conclude that $x = 10$. Therefore $(10, 10)$ is the sole critical point of the system.

15. From $x(1 - x^2 - 3y^2) = 0$ we have $x = 0$ or $x^2 + 3y^2 = 1$. If $x = 0$, then substituting into $y(3 - x^2 - 3y^2)$ gives $y(3 - 3y^2) = 0$. Therefore $y = 0$, 1, -1. Likewise $x^2 = 1 - 3y^2$ yields $2y = 0$ so that $y = 0$ and $x^2 = 1 - 3(0)^2 = 1$. The critical points of the system are therefore $(0, 0)$, $(0, 1)$, $(0, -1)$, $(1, 0)$, and $(-1, 0)$.

18. **(a)** From Exercises 8.2, Problem 6, $x = c_1 + 2c_2 e^{-5t}$ and $y = 3c_1 + c_2 e^{-5t}$, which is not periodic.

(b) From $\mathbf{X}(0) = (3, 4)$ it follows that $c_1 = c_2 = 1$. Therefore $x = 1 + 2e^{-5t}$ and $y = 3 + e^{-5t}$ gives $y = \frac{1}{2}(x - 1) + 3$.

(c)

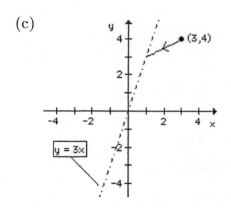

21. **(a)** From Exercises 8.2, Problem 35, $x = c_1(\sin t - \cos t)e^{4t} + c_2(-\sin t - \cos t)e^{4t}$ and $y = 2c_1(\cos t)e^{4t} + 2c_2(\sin t)e^{4t}$. Because of the presence of e^{4t}, there are no periodic solutions.

(b) From $\mathbf{X}(0) = (-1, 2)$ it follows that $c_1 = 1$ and $c_2 = 0$. Therefore $x = (\sin t - \cos t)e^{4t}$ and $y = 2(\cos t)e^{4t}$.

(c)

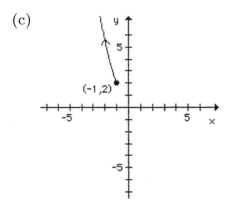

24. Switching to polar coordinates,

$$\frac{dr}{dt} = \frac{1}{r}\left(x\frac{dx}{dt} + y\frac{dy}{dt}\right) = \frac{1}{r}(xy - x^2 r^2 - xy + y^2 r^2) = r^3$$

$$\frac{d\theta}{dt} = \frac{1}{r^2}\left(-y\frac{dx}{dt} + x\frac{dy}{dt}\right) = \frac{1}{r^2}(-y^2 - xyr^2 - x^2 + xyr^2) = -1\,.$$

If we use separation of variables, it follows that

$$r = \frac{1}{\sqrt{-2t + c_1}} \quad \text{and} \quad \theta = -t + c_2.$$

Since $\mathbf{X}(0) = (4,0)$, $r = 4$ and $\theta = 0$ when $t = 0$. It follows that $c_2 = 0$ and $c_1 = \frac{1}{16}$. The final solution may be written as

$$r = \frac{4}{\sqrt{1 - 32t}}, \qquad \theta = -t.$$

Note that $r \to \infty$ as $t \to \left(\frac{1}{32}\right)$. Because $0 \le t \le \frac{1}{32}$, the curve is not a spiral.

27. The system has no critical points, so there are no periodic solutions.

30. The system has no critical points, so there are no periodic solutions.

Section 10.2   Stability of Linear Systems

The terminology and concepts listed below provide an outline of the main ideas encountered in this section. These can be useful when preparing for a quiz or test.

Terminology and Concepts

- locally stable critical point
- unstable cricital point

- phase plane

- phase portrait

- eigenvalues and eigenvectors of the coefficient matrix of a linear first-order system of DEs

- stable and unstable modes

- saddle point

- degenerate stable and unstable modes

- center

- stable and unstable spiral points

The basic skills listed below summarize the more mechanical types of problems encountered in the exercise set for this section.

Basic Skills

- given the solution of a linear system $\mathbf{X}' = \mathbf{AX}$ where $\mathbf{A}$ is a 2×2 matrix, discuss the nature of the solutions in a neighborhood of $(0,0)$

- classify the critical point $(0,0)$ of a linear system

$$x' = ax + by$$

$$y' = cx + dy$$

Exercises 10.2   *Hints, Suggestions, Solutions, and Examples*

3. (a) All solutions are unstable spirals which become unbounded as t increases.

   (b)

6. (a) All solutions become unbounded and $y = x/2$ serves as the asymptote.

210

(b)

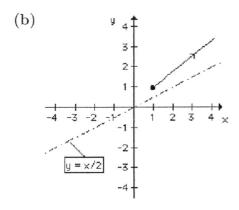

9. Since $\Delta = -41 < 0$, we conclude from Figure 10.18 that $(0,0)$ is a saddle point.

12. Since $\Delta = 1$ and $\tau = -1$, $\tau^2 - 4\Delta = -3$ and so from Figure 10.18, $(0,0)$ is a stable spiral point.

15. Since $\Delta = 0.01$ and $\tau = -0.03$, $\tau^2 - 4\Delta < 0$ and so from Figure 10.18, $(0,0)$ is a stable spiral point.

18. Note that $\Delta = 1$ and $\tau = \mu$. Therefore we need both $\tau = \mu < 0$ and $\tau^2 - 4\Delta = \mu^2 - 4 < 0$ for $(0,0)$ to be a stable spiral point. These two conditions can be written as $-2 < \mu < 0$.

21. $\mathbf{AX}_1 + \mathbf{F} = \mathbf{0}$ implies that $\mathbf{AX}_1 = -\mathbf{F}$ or $\mathbf{X}_1 = -\mathbf{A}^{-1}\mathbf{F}$. Since $\mathbf{X}_p(t) = -\mathbf{A}^{-1}\mathbf{F}$ is a particular solution, it follows from Theorem 8.6 that $\mathbf{X}(t) = \mathbf{X}_c(t) + \mathbf{X}_1$ is the general solution to $\mathbf{X}' = \mathbf{AX} + \mathbf{F}$. If $\tau < 0$ and $\Delta > 0$ then $\mathbf{X}_c(t)$ approaches $(0,0)$ by Theorem 10.1(a). It follows that $\mathbf{X}(t)$ approaches $\mathbf{X}_1$ as $t \to \infty$.

24. (a) The critical point is $\mathbf{X}_1 = (-1, -2)$.

    (b) From the graph, $\mathbf{X}_1$ appears to be a stable node or a degenerate stable node.

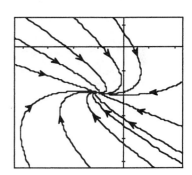

    (c) Since $\tau = -16$, $\Delta = 64$, and $\tau^2 - 4\Delta = 0$, $(0,0)$ is a degenerate stable node.

Section 10.3 Linearization and Local Stability

The terminology and concepts listed below provide an outline of the main ideas encountered in this section. These can be useful when preparing for a quiz or test.

Terminology and Concepts

- stable critical point
- asymptotically stable critical point
- unstable critical point
- linearization of a nonlinear system
- Jacobian matrix
- classification of critical points
- phase-plane method

The basic skills listed below summarize the more mechanical types of problems encountered in the exercise set for this section.

Basic Skills

- find critical points and classify each critical point of a first-order autonomous DE as either asympototically stable or unstable
- find critical points and classify each critical point of a first-order system of two DEs as a stable node, a stable spiral point, an unstable spiral point, an unstable node, or a saddle point
- find critical points and classify each critical point of a second-order DE as a stable node, a stable spiral point, an unstable spiral point, an unstable node, or a saddle point.

Exercises 10.3   *Hints, Suggestions, Solutions, and Examples*

3. The critical points are $x = 0$ and $x = n + 1$. Since $g'(x) = k(n+1) - 2kx$, $g'(0) = k(n+1) > 0$ and $g'(n+1) = -k(n+1) < 0$. Therefore $x = 0$ is unstable while $x = n + 1$ is asymptotically stable. See Theorem 10.2.

6. The only critical point is $v = mg/k$. Now $g(v) = g - (k/m)v$ and so $g'(v) = -k/m < 0$. Therefore $v = mg/k$ is an asymptotically stable critical point by Theorem 10.2.

212

9. Critical points occur at $P = a/b$, c but not at $P = 0$. Since $g'(P) = (a - bP) + (P - c)(-b)$,

$$g'(a/b) = (a/b - c)(-b) = -a + bc \quad \text{and} \quad g'(c) = a - bc.$$

Since $a < bc$, $-a + bc > 0$ and $a - bc < 0$. Therefore $P = a/b$ is unstable while $P = c$ is asymptotically stable.

12. Critical points are $(1, 0)$ and $(-1, 0)$, and

$$\mathbf{g}'(\mathbf{X}) = \begin{pmatrix} 2x & -2y \\ 0 & 2 \end{pmatrix}.$$

At $\mathbf{X} = (1, 0)$, $\tau = 4$, $\Delta = 4$, and so $\tau^2 - 4\Delta = 0$. We can conclude that $(1, 0)$ is unstable but we are unable to classify this critical point any further. At $\mathbf{X} = (-1, 0)$, $\Delta = -4 < 0$ and so $(-1, 0)$ is a saddle point.

15. Since $x^2 - y^2 = 0$, $y^2 = x^2$ and so $x^2 - 3x + 2 = (x - 1)(x - 2) = 0$. It follows that the critical points are $(1, 1)$, $(1, -1)$, $(2, 2)$, and $(2, -2)$. We next use the Jacobian

$$\mathbf{g}'(\mathbf{X}) = \begin{pmatrix} -3 & 2y \\ 2x & -2y \end{pmatrix}$$

to classify these four critical points. For $\mathbf{X} = (1, 1)$, $\tau = -5$, $\Delta = 2$, and so $\tau^2 - 4\Delta = 17 > 0$. Therefore $(1, 1)$ is a stable node. For $\mathbf{X} = (1, -1)$, $\Delta = -2 < 0$ and so $(1, -1)$ is a saddle point. For $\mathbf{X} = (2, 2)$, $\Delta = -4 < 0$ and so we have another saddle point. Finally, if $\mathbf{X} = (2, -2)$, $\tau = 1$, $\Delta = 4$, and so $\tau^2 - 4\Delta = -15 < 0$. Therefore $(2, -2)$ is an unstable spiral point.

18. We found that $(0, 0)$, $(0, 1)$, $(0, -1)$, $(1, 0)$ and $(-1, 0)$ were the critical points in Exercise 15, Section 10.1. The Jacobian is

$$\mathbf{g}'(\mathbf{X}) = \begin{pmatrix} 1 - 3x^2 - 3y^2 & -6xy \\ -2xy & 3 - x^2 - 9y^2 \end{pmatrix}.$$

For $\mathbf{X} = (0, 0)$, $\tau = 4$, $\Delta = 3$ and so $\tau^2 - 4\Delta = 4 > 0$. Therefore $(0, 0)$ is an unstable node. Both $(0, 1)$ and $(0, -1)$ give $\tau = -8$, $\Delta = 12$, and $\tau^2 - 4\Delta = 16 > 0$. These two critical points are therefore stable nodes. For $\mathbf{X} = (1, 0)$ or $(-1, 0)$, $\Delta = -4 < 0$ and so saddle points occur.

21. The corresponding plane autonomous system is

$$\theta' = y, \quad y' = (\cos\theta - \frac{1}{2})\sin\theta.$$

Since $|\theta| < \pi$, it follows that critical points are $(0, 0)$, $(\pi/3, 0)$ and $(-\pi/3, 0)$. The Jacobian matrix is

$$\mathbf{g}'(\mathbf{X}) = \begin{pmatrix} 0 & 1 \\ \cos 2\theta - \frac{1}{2}\cos\theta & 0 \end{pmatrix}$$

and so at $(0, 0)$, $\tau = 0$ and $\Delta = -1/2$. Therefore $(0, 0)$ is a saddle point. For $\mathbf{X} = (\pm\pi/3, 0)$, $\tau = 0$ and $\Delta = 3/4$. It is not possible to classify either critical point in this borderline case.

24. The corresponding plane autonomous system is

$$x' = y, \quad y' = -\frac{4x}{1 + x^2} - 2y$$

and the only critical point is $(0,0)$. Since the Jacobian matrix is

$$\mathbf{g}'(\mathbf{X}) = \begin{pmatrix} 0 & 1 \\ -4\dfrac{1 - x^2}{(1 + x^2)^2} & -2 \end{pmatrix},$$

$\tau = -2$, $\Delta = 4$, $\tau^2 - 4\Delta = -12$, and so $(0,0)$ is a stable spiral point.

27. The corresponding plane autonomous system is

$$x' = y, \quad y' = -\frac{(\beta + \alpha^2 y^2)x}{1 + \alpha^2 x^2}$$

and the Jacobian matrix is

$$\mathbf{g}'(\mathbf{X}) = \begin{pmatrix} 0 & 1 \\ \dfrac{(\beta + \alpha y^2)(\alpha^2 x^2 - 1)}{(1 + \alpha^2 x^2)^2} & \dfrac{-2\alpha^2 yx}{1 + \alpha^2 x^2} \end{pmatrix}.$$

For $\mathbf{X} = (0,0)$, $\tau = 0$ and $\Delta = \beta$. Since $\beta < 0$, we can conclude that $(0,0)$ is a saddle point.

30. (a) The corresponding plane autonomous system is

$$x' = y, \quad y' = \epsilon\left(y - \frac{1}{3}y^3\right) - x$$

and so the only critical point is $(0,0)$. Since the Jacobian matrix is

$$\mathbf{g}'(\mathbf{X}) = \begin{pmatrix} 0 & 1 \\ -1 & \epsilon(1 - y^2) \end{pmatrix},$$

$\tau = \epsilon$, $\Delta = 1$, and so $\tau^2 - 4\Delta = \epsilon^2 - 4$ at the critical point $(0,0)$.

(b) When $\tau = \epsilon > 0$, $(0,0)$ is an unstable critical point.

(c) When $\epsilon < 0$ and $\tau^2 - 4\Delta = \epsilon^2 - 4 < 0$, $(0,0)$ is a stable spiral point. These two requirements can be written as $-2 < \epsilon < 0$.

(d) When $\epsilon = 0$, $x'' + x = 0$ and so $x = c_1 \cos t + c_2 \sin t$. Therefore all solutions are periodic (with period 2π) and so $(0,0)$ is a center.

33. (a) $x' = 2xy = 0$ implies that either $x = 0$ or $y = 0$. If $x = 0$, then from $1 - x^2 + y^2 = 0$, $y^2 = -1$ and there are no real solutions. If $y = 0$, $1 - x^2 = 0$ and so $(1,0)$ and $(-1,0)$ are critical points. The Jacobian matrix is

$$\mathbf{g}'(\mathbf{X}) = \begin{pmatrix} 2y & 2x \\ -2x & 2y \end{pmatrix}$$

and so $\tau = 0$ and $\Delta = 4$ at either $\mathbf{X} = (1,0)$ or $(-1,0)$. We obtain no information about these critical points in this borderline case.

(b) The DE is

$$\frac{dy}{dx} = \frac{y'}{x'} = \frac{1 - x^2 + y^2}{2xy}$$

or

$$2xy \frac{dy}{dx} = 1 - x^2 + y^2.$$

Letting $\mu = y^2/x$, it follows that $d\mu/dx = (1/x^2) - 1$ and so $\mu = -(1/x) - x + 2c$. Therefore $y^2/x = -(1/x) - x + 2c$ which can be put in the form $(x - c)^2 + y^2 = c^2 - 1$. The solution curves are shown and so both $(1, 0)$ and $(-1, 0)$ are centers.

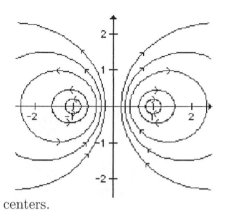

36. The corresponding plane autonomous system is

$$x' = y, \quad y' = \epsilon x^2 - x + 1$$

and so the critical points must satisfy $y = 0$ and

$$x = \frac{1 \pm \sqrt{1 - 4\epsilon}}{2\epsilon}.$$

Therefore we must require that $\epsilon \le \frac{1}{4}$ for real solutions to exist. We will use the Jacobian matrix

$$\mathbf{g}'(\mathbf{X}) = \begin{pmatrix} 0 & 1 \\ 2\epsilon x - 1 & 0 \end{pmatrix}$$

to attempt to classify $((1 \pm \sqrt{1 - 4\epsilon})/2\epsilon, 0)$ when $\epsilon \le 1/4$. Note that $\tau = 0$ and $\Delta = \mp\sqrt{1 - 4\epsilon}$. For $\mathbf{X} = ((1 + \sqrt{1 - 4\epsilon})/2\epsilon, 0)$ and $\epsilon < 1/4$, $\Delta < 0$ and so a saddle point occurs. For $\mathbf{X} = ((1 - \sqrt{1 - 4\epsilon})/2\epsilon, 0)$, $\Delta \ge 0$ and we are not able to classify this critical point using linearization.

39. (a) Letting $x = \theta$ and $y = x'$ we obtain the system $x' = y$ and $y' = 1/2 - \sin x$. Since $\sin \pi/6 = \sin 5\pi/6 = 1/2$ we see that $(\pi/6, 0)$ and $(5\pi/6, 0)$ are critical points of the system.

(b) The Jacobian matrix is

$$\mathbf{g}'(\mathbf{X}) = \begin{pmatrix} 0 & 1 \\ -\cos x & 0 \end{pmatrix}$$

and so

$$\mathbf{A}_1 = \mathbf{g}' = ((\pi/6, 0)) = \begin{pmatrix} 0 & 1 \\ -\sqrt{3}/2 & 0 \end{pmatrix} \quad \text{and} \quad \mathbf{A}_2 = \mathbf{g}' = ((5\pi/6, 0)) = \begin{pmatrix} 0 & 1 \\ \sqrt{3}/2 & 0 \end{pmatrix}.$$

Since $\det \mathbf{A}_1 > 0$ and the trace of $\mathbf{A}_1$ is 0, no conclusion can be drawn regarding the critical point $(\pi/6, 0)$. Since $\det \mathbf{A}_2 < 0$, we see that $(5\pi/6, 0)$ is a saddle point.

(c) From the system in part (a) we obtain the first-order DE

$$\frac{dy}{dx} = \frac{1/2 - \sin x}{y}.$$

Separating variables and integrating we obtain

$$\int y\,dy = \int \left(\frac{1}{2} - \sin x\right) dx$$

and

$$\frac{1}{2}y^2 = \frac{1}{2}x + \cos x + c_1$$

or

$$y^2 = x + 2\cos x + c_2.$$

For x_0 near $\pi/6$, if $\mathbf{X}(0) = (x_0, 0)$ then $c_2 = -x_0 - 2\cos x_0$ and $y^2 = x + 2\cos x - x_0 - 2\cos x_0$. Thus, there are two values of y for each x in a sufficiently small interval around $\pi/6$. Therefore $(\pi/6, 0)$ is a center.

Section 10.4   Autonomous Systems as Mathematical Models

The terminology and concepts listed below provide an outline of the main ideas encountered in this section. These can be useful when preparing for a quiz or test.

Terminology and Concepts

- nonlinear pendulum
- nonlinear oscillations
- Lotka-Volterra preditor-prey model
- Lotka-Volterra competition model

Exercises 10.4   *Hints, Suggestions, Solutions, and Examples*

3. The corresponding plane autonomous system is

$$x' = y, \quad y' = -g\,\frac{f'(x)}{1 + [f'(x)]^2} - \frac{\beta}{m}\,y$$

and

$$\frac{\partial}{\partial x}\left(-g\,\frac{f'(x)}{1 + [f'(x)]^2} - \frac{\beta}{m}\,y\right) = -g\,\frac{(1 + [f'(x)]^2)f''(x) - f'(x)2f'(x)f''(x)}{(1 + [f'(x)]^2)^2}.$$

If $\mathbf{X}_1 = (x_1, y_1)$ is a critical point, $y_1 = 0$ and $f'(x_1) = 0$. The Jacobian at this critical point is therefore

$$\mathbf{g}'(\mathbf{X}_1) = \begin{pmatrix} 0 & 1 \\ -gf''(x_1) & -\dfrac{\beta}{m} \end{pmatrix}.$$

6. (a) If $f(x) = \cosh x$, $f'(x) = \sinh x$ and $[f'(x)]^2 + 1 = \sinh^2 x + 1 = \cosh^2 x$. Therefore

$$\frac{dy}{dx} = \frac{y'}{x'} = -g\frac{\sinh x}{\cosh^2 x}\frac{1}{y}.$$

We can separate variables to show that $y^2 = 2g/\cosh x + c$. But $x(0) = x_0$ and $y(0) = x'(0) = v_0$. Therefore $c = v_0^2 - (2g/\cosh x_0)$ and so

$$y^2 = \frac{2g}{\cosh x} - \frac{2g}{\cosh x_0} + v_0^2.$$

Now

$$\frac{2g}{\cosh x} - \frac{2g}{\cosh x_0} + v_0^2 \geq 0 \quad \text{if and only if} \quad \cosh x \leq \frac{2g\cosh x_0}{2g - v_0^2\cosh x_0}$$

and the solution to this inequality is an interval $[-a, a]$. Therefore each x in $(-a, a)$ has two corresponding values of y and so the solution is periodic.

(b) Since $z = \cosh x$, the maximum height occurs at the largest value of x on the cycle. From (a), $x_{\max} = a$ where $\cosh a = 2g\cosh x_0/(2g - v_0^2\cosh x_0)$. Therefore

$$z_{\max} = \frac{2g\cosh x_0}{2g - v_0^2\cosh x_0}.$$

9. (a) In the Lotka-Volterra Model the average number of predators is d/c and the average number of prey is a/b. But

$$x' = -ax + bxy - \epsilon_1 x = -(a + \epsilon_1)x + bxy$$

$$y' = -cxy + dy - \epsilon_2 y = -cxy + (d - \epsilon_2)y$$

and so the new critical point in the first quadrant is $(d/c - \epsilon_2/c, a/b + \epsilon_1/b)$.

(b) The average number of predators $d/c - \epsilon_2/c$ has decreased while the average number of prey $a/b + \epsilon_1/b$ has increased. The fishery science model is consistent with Volterra's principle.

12. $\Delta = r_1 r_2$, $\tau = r_1 + r_2$ and $\tau^2 - 4\Delta = (r_1 + r_2)^2 - 4r_1 r_2 = (r_1 - r_2)^2$. Therefore when $r_1 \neq r_2$, $(0, 0)$ is an unstable node.

15. $K_1/\alpha_{12} < K_2 < K_1\alpha_{21}$ and so $\alpha_{12}\alpha_{21} > 1$. Therefore $\Delta = (1 - \alpha_{12}\alpha_{21})\hat{x}\hat{y}\,r_1 r_2/K_1 K_2 < 0$ and so $(\hat{x}, \hat{y})$ is a saddle point.

3. a center or a saddle point

6. True

9. The system is linear and we identify $\Delta = -\alpha$ and $\tau = \alpha + 1$. Since a critical point will be a center when $\Delta > 0$ and $\tau = 0$ we see that for $\alpha = -1$ critical points will be centers and solutions will be periodic. Note also that when $\alpha = -1$ the system is

$$x' = -x - 2y$$

$$y' = x + y,$$

which does have an isolated critical point at $(0,0)$.

12. (a) If $\mathbf{X}(0) = \mathbf{X}_0$ lies on the line $y = -2x$, then $\mathbf{X}(t)$ approaches $(0,0)$ along this line. For all other initial conditions, $\mathbf{X}(t)$ approaches $(0,0)$ from the direction determined by the line $y = x$.

(b) If $\mathbf{X}(0) = \mathbf{X}_0$ lies on the line $y = -x$, then $\mathbf{X}(t)$ approaches $(0,0)$ along this line. For all other initial conditions, $\mathbf{X}(t)$ becomes unbounded and $y = 2x$ serves as an asymptote.

15. From $x = r\cos\theta$, $y = r\sin\theta$ we have

$$\frac{dx}{dt} = -r\sin\theta\,\frac{d\theta}{dt} + \frac{dr}{dt}\cos\theta$$

$$\frac{dy}{dt} = r\cos\theta\,\frac{d\theta}{dt} + \frac{dr}{dt}\sin\theta.$$

Then $r' = \alpha r$, $\theta' = 1$ gives

$$\frac{dx}{dt} = -r\sin\theta + \alpha r\cos\theta$$

$$\frac{dy}{dt} = r\cos\theta + \alpha r\sin\theta.$$

We see that $r = 0$, which corresponds to $\mathbf{X} = (0,0)$, is a critical point. Solving $r' = \alpha r$ we have $r = c_1 e^{\alpha t}$. Thus, when $\alpha < 0$, $\lim_{t\to\infty} r(t) = 0$ and $(0,0)$ is a stable critical point. When $\alpha = 0$, $r' = 0$ and $r = c_1$. In this case $(0,0)$ is a center, which is stable. Therefore, $(0,0)$ is a stable critical point for the system when $\alpha \leq 0$.

18. Using the phase-plane method we obtain

$$\frac{dy}{dx} = \frac{y'}{x'} = \frac{-2x\sqrt{y^2+1}}{y}.$$

We can separate variables to show that $\sqrt{y^2 + 1} = -x^2 + c$. But $x(0) = x_0$ and $y(0) = x'(0) = 0$. It follows that $c = 1 + x_0^2$ so that $y^2 = (1 + x_0^2 - x^2)^2 - 1$. Note that $1 + x_0^2 - x^2 > 1$ for $-x_0 < x < x_0$ and $y = 0$ for $x = \pm x_0$. Each x with $-x_0 < x < x_0$ has two corresponding values of y and so the solution $\mathbf{X}(t)$ with $\mathbf{X}(0) = (x_0, 0)$ is periodic.

11 Orthogonal Functions and Fourier Series

Orthogonal Functions

The terminology and concepts listed below provide an outline of the main ideas encountered in this section. These can be useful when preparing for a quiz or test.

Terminology and Concepts

- inner product of two vectors in 3-space
- inner product of two functions on an interval
- orthogonal functions and orthogonal sets
- norm of a function
- orthonormal set
- orthogonal series expansion of a function
- orthogonality with respect to a weight function

The basic skills listed below summarize the more mechanical types of problems encountered in the exercise set for this section.

Basic Skills

- show that a set of functions is orthogonal on a given interval
- show that a set of functions is orthogonal with respect to a given weight function on an interval

Exercises 11.1 *Hints, Suggestions, Solutions, and Examples*

3. $\int_0^2 e^x(xe^{-x} - e^{-x})dx = \int_0^2 (x-1)dx = \left(\frac{1}{2}x^2 - x\right)\Big|_0^2 = 0$

6. $\int_{\pi/4}^{5\pi/4} e^x \sin x\, dx = \left(\frac{1}{2}e^x \sin x - \frac{1}{2}e^x \cos x\right)\Big|_{\pi/4}^{5\pi/4} = 0$

9. For $m \neq n$

$$\int_0^\pi \sin nx \sin mx \, dx = \frac{1}{2} \int_0^\pi \Big(\cos(n-m)x - \cos(n+m)x \Big) \, dx$$

$$= \frac{1}{2(n-m)} \sin(n-m)x \, \Big|_0^\pi - \frac{1}{2(n+m)} \sin(n+m)x \, \Big|_0^\pi$$

$$= 0.$$

For $m = n$

$$\int_0^\pi \sin^2 nx \, dx = \int_0^\pi \left(\frac{1}{2} - \frac{1}{2} \cos 2nx \right) dx = \frac{1}{2} x \, \Big|_0^\pi - \frac{1}{4n} \sin 2nx \, \Big|_0^\pi = \frac{\pi}{2}$$

so that

$$\| \sin nx \| = \sqrt{\frac{\pi}{2}}.$$

12. For $m \neq n$, we use Problems 11 and 10:

$$\int_{-p}^p \cos \frac{n\pi}{p} x \cos \frac{m\pi}{p} x \, dx = 2 \int_0^p \cos \frac{n\pi}{p} x \cos \frac{m\pi}{p} x \, dx = 0$$

$$\int_{-p}^p \sin \frac{n\pi}{p} x \sin \frac{m\pi}{p} x \, dx = 2 \int_0^p \sin \frac{n\pi}{p} x \sin \frac{m\pi}{p} x \, dx = 0.$$

Also

$$\int_{-p}^p \sin \frac{n\pi}{p} x \cos \frac{m\pi}{p} x \, dx = \frac{1}{2} \int_{-p}^p \left(\sin \frac{(n-m)\pi}{p} x + \sin \frac{(n+m)\pi}{p} x \right) dx = 0,$$

$$\int_{-p}^p 1 \cdot \cos \frac{n\pi}{p} x \, dx = \frac{p}{n\pi} \sin \frac{n\pi}{p} x \, \Big|_{-p}^p = 0,$$

$$\int_{-p}^p 1 \cdot \sin \frac{n\pi}{p} x \, dx = -\frac{p}{n\pi} \cos \frac{n\pi}{p} x \, \Big|_{-p}^p = 0,$$

and

$$\int_{-p}^p \sin \frac{n\pi}{p} x \cos \frac{n\pi}{p} x \, dx = \int_{-p}^p \frac{1}{2} \sin \frac{2n\pi}{p} x \, dx = -\frac{p}{4n\pi} \cos \frac{2n\pi}{p} x \, \Big|_{-p}^p = 0.$$

For $m = n$

$$\int_{-p}^p \cos^2 \frac{n\pi}{p} x \, dx = \int_{-p}^p \left(\frac{1}{2} + \frac{1}{2} \cos \frac{2n\pi}{p} x \right) dx = p,$$

$$\int_{-p}^p \sin^2 \frac{n\pi}{p} x \, dx = \int_{-p}^p \left(\frac{1}{2} - \frac{1}{2} \cos \frac{2n\pi}{p} x \right) dx = p,$$

and

$$\int_{-p}^p 1^2 dx = 2p$$

so that

$$\|1\| = \sqrt{2p}, \quad \left\|\cos\frac{n\pi}{p}x\right\| = \sqrt{p}, \quad \text{and} \quad \left\|\sin\frac{n\pi}{p}x\right\| = \sqrt{p}.$$

15. By orthogonality $\int_a^b \phi_0(x)\phi_n(x)dx = 0$ for $n = 1, 2, 3, \ldots$; that is, $\int_a^b \phi_n(x)dx = 0$ for $n = 1, 2, 3, \ldots$.

18. Setting

$$0 = \int_{-2}^{2} f_3(x)f_1(x)\,dx = \int_{-2}^{2}\left(x^2 + c_1 x^3 + c_2 x^4\right)dx = \frac{16}{3} + \frac{64}{5}c_2$$

and

$$0 = \int_{-2}^{2} f_3(x)f_2(x)\,dx = \int_{-2}^{2}\left(x^3 + c_1 x^4 + c_2 x^5\right)dx = \frac{64}{5}c_1$$

we obtain $c_1 = 0$ and $c_2 = -5/12$.

Section 11.2   Fourier Series

The terminology and concepts listed below provide an outline of the main ideas encountered in this section. These can be useful when preparing for a quiz or test.

Terminology and Concepts

- Fourier series of a function
- Fourier coefficients of a function
- conditions under which a Fourier series converges
- convergence at a point of finite discontinuity
- fundamental period of a Fourier series
- periodic extension of a function
- sequence of partial sums of a Fourier series

The basic skills listed below summarize the more mechanical types of problems encountered in the exercise set for this section.

Basic Skills

- find the Fourier series of a function on a given interval

3. $a_0 = \displaystyle\int_{-1}^{1} f(x)\,dx = \int_{-1}^{0} 1\,dx + \int_{0}^{1} x\,dx = \frac{3}{2}$

$a_n = \displaystyle\int_{-1}^{1} f(x)\cos n\pi x\,dx = \int_{-1}^{0} \cos n\pi x\,dx + \int_{0}^{1} x\cos n\pi x\,dx = \frac{1}{n^2\pi^2}[(-1)^n - 1]$,

$b_n = \displaystyle\int_{-1}^{1} f(x)\sin n\pi x\,dx = \int_{-1}^{0} \sin n\pi x\,dx + \int_{0}^{1} x\sin n\pi x\,dx = -\frac{1}{n\pi}$

$f(x) = \dfrac{3}{4} + \displaystyle\sum_{n=1}^{\infty} \left[\frac{(-1)^n - 1}{n^2\pi^2}\cos n\pi x - \frac{1}{n\pi}\sin n\pi x \right]$

6. $a_0 = \dfrac{1}{\pi}\displaystyle\int_{-\pi}^{\pi} f(x)\,dx = \frac{1}{\pi}\int_{-\pi}^{0} \pi^2\,dx + \frac{1}{\pi}\int_{0}^{\pi} \left(\pi^2 - x^2\right)dx = \frac{5}{3}\pi^2$

$a_n = \dfrac{1}{\pi}\displaystyle\int_{-\pi}^{\pi} f(x)\cos nx\,dx = \frac{1}{\pi}\int_{-\pi}^{0} \pi^2\cos nx\,dx + \frac{1}{\pi}\int_{0}^{\pi}\left(\pi^2 - x^2\right)\cos nx\,dx$

$= \dfrac{1}{\pi}\left(\dfrac{\pi^2 - x^2}{n}\sin nx \Big|_0^\pi + \dfrac{2}{n}\int_0^\pi x\sin nx\,dx \right) = \dfrac{2}{n^2}(-1)^{n+1}$

$b_n = \dfrac{1}{\pi}\displaystyle\int_{-\pi}^{\pi} f(x)\sin nx\,dx = \frac{1}{\pi}\int_{-\pi}^{0} \pi^2\sin nx\,dx + \frac{1}{\pi}\int_{0}^{\pi}\left(\pi^2 - x^2\right)\sin nx\,dx$

$= \dfrac{\pi}{n}[(-1)^n - 1] + \dfrac{1}{\pi}\left(\dfrac{x^2 - \pi^2}{n}\cos nx \Big|_0^\pi - \dfrac{2}{n}\int_0^\pi x\cos nx\,dx \right) = \dfrac{\pi}{n}(-1)^n + \dfrac{2}{n^3\pi}[1 - (-1)^n]$

$f(x) = \dfrac{5\pi^2}{6} + \displaystyle\sum_{n=1}^{\infty} \left[\frac{2}{n^2}(-1)^{n+1}\cos nx + \left(\frac{\pi}{n}(-1)^n + \frac{2[1 - (-1)^n]}{n^3\pi} \right)\sin nx \right]$

9. $a_0 = \dfrac{1}{\pi}\displaystyle\int_{-\pi}^{\pi} f(x)\,dx = \frac{1}{\pi}\int_{0}^{\pi} \sin x\,dx = \frac{2}{\pi}$

$a_n = \dfrac{1}{\pi}\displaystyle\int_{-\pi}^{\pi} f(x)\cos nx\,dx = \frac{1}{\pi}\int_{0}^{\pi} \sin x\,\cos nx\,dx = \frac{1}{2\pi}\int_{0}^{\pi}\left(\sin(1+n)x + \sin(1-n)x\right)dx$

$= \dfrac{1 + (-1)^n}{\pi(1 - n^2)}$ for $n = 2, 3, 4, \ldots$

$a_1 = \dfrac{1}{2\pi}\displaystyle\int_{0}^{\pi} \sin 2x\,dx = 0$

$b_n = \dfrac{1}{\pi}\displaystyle\int_{-\pi}^{\pi} f(x)\sin nx\,dx = \frac{1}{\pi}\int_{0}^{\pi} \sin x\,\sin nx\,dx$

$= \dfrac{1}{2\pi}\displaystyle\int_{0}^{\pi}\left(\cos(1-n)x - \cos(1+n)x\right)dx = 0$ for $n = 2, 3, 4, \ldots$

$b_1 = \dfrac{1}{2\pi}\displaystyle\int_{0}^{\pi}(1 - \cos 2x)\,dx = \frac{1}{2}$

$$f(x) = \frac{1}{\pi} + \frac{1}{2}\sin x + \sum_{n=2}^{\infty} \frac{1+(-1)^n}{\pi(1-n^2)}\cos nx$$

12. $a_0 = \frac{1}{2}\int_{-2}^{2} f(x)\,dx = \frac{1}{2}\left(\int_{0}^{1} x\,dx + \int_{1}^{2} 1\,dx\right) = \frac{3}{4}$

$a_n = \frac{1}{2}\int_{-2}^{2} f(x)\cos\frac{n\pi}{2}x\,dx = \frac{1}{2}\left(\int_{0}^{1} x\cos\frac{n\pi}{2}x\,dx + \int_{1}^{2}\cos\frac{n\pi}{2}x\,dx\right) = \frac{2}{n^2\pi^2}\left(\cos\frac{n\pi}{2}-1\right)$

$b_n = \frac{1}{2}\int_{-2}^{2} f(x)\sin\frac{n\pi}{2}x\,dx = \frac{1}{2}\left(\int_{0}^{1} x\sin\frac{n\pi}{2}x\,dx + \int_{1}^{2}\sin\frac{n\pi}{2}x\,dx\right)$

$ = \frac{2}{n^2\pi^2}\left(\sin\frac{n\pi}{2} + \frac{n\pi}{2}(-1)^{n+1}\right)$

$f(x) = \frac{3}{8} + \sum_{n=1}^{\infty}\left[\frac{2}{n^2\pi^2}\left(\cos\frac{n\pi}{2}-1\right)\cos\frac{n\pi}{2}x + \frac{2}{n^2\pi^2}\left(\sin\frac{n\pi}{2} + \frac{n\pi}{2}(-1)^{n+1}\right)\sin\frac{n\pi}{2}x\right]$

15. $a_0 = \frac{1}{\pi}\int_{-\pi}^{\pi} f(x)\,dx = \frac{1}{\pi}\int_{-\pi}^{\pi} e^x\,dx = \frac{1}{\pi}(e^{\pi} - e^{-\pi})$

$a_n = \frac{1}{\pi}\int_{-\pi}^{\pi} f(x)\cos nx\,dx = \frac{(-1)^n(e^{\pi} - e^{-\pi})}{\pi(1+n^2)}$

$b_n = \frac{1}{\pi}\int_{-\pi}^{\pi} f(x)\sin nx\,dx = \frac{1}{\pi}\int_{-\pi}^{\pi} e^x\sin nx\,dx = \frac{(-1)^n n(e^{-\pi} - e^{\pi})}{\pi(1+n^2)}$

$f(x) = \frac{e^{\pi} - e^{-\pi}}{2\pi} + \sum_{n=1}^{\infty}\left[\frac{(-1)^n(e^{\pi} - e^{-\pi})}{\pi(1+n^2)}\cos nx + \frac{(-1)^n n(e^{-\pi} - e^{\pi})}{\pi(1+n^2)}\sin nx\right]$

17. (*Hint*) Consider the values to which the Fourier series in Problem 5 converges when $x = 0$ and $x = \pi$.

18. From Problem 17

$$\frac{\pi^2}{8} = \frac{1}{2}\left(\frac{\pi^2}{6} + \frac{\pi^2}{12}\right) = \frac{1}{2}\left(2 + \frac{2}{3^2} + \frac{2}{5^2} + \cdots\right) = 1 + \frac{1}{3^2} + \frac{1}{5^2} + \cdots.$$

21. (a) Letting $c_0 = a_0/2$, $c_n = (a_n - ib_n)/2$, and $c_{-n} = (a_n + ib_n)/2$ we have

$$f(x) = \frac{a_0}{2} + \sum_{n=1}^{\infty} \left(a_n \cos \frac{n\pi}{p}x + b_n \sin \frac{n\pi}{p}x \right)$$

$$= c_0 + \sum_{n=1}^{\infty} \left(a_n \frac{e^{in\pi x/p} + e^{-in\pi x/p}}{2} + b_n \frac{e^{in\pi x/p} - e^{-in\pi x/p}}{2i} \right)$$

$$= c_0 + \sum_{n=1}^{\infty} \left(a_n \frac{e^{in\pi x/p} + e^{-in\pi x/p}}{2} - b_n \frac{ie^{in\pi x/p} - ie^{-in\pi x/p}}{2} \right)$$

$$= c_0 + \sum_{n=1}^{\infty} \left(\frac{a_n - ib_n}{2} e^{in\pi x/p} + \frac{a_n + ib_n}{2} e^{-in\pi x/p} \right)$$

$$= c_0 + \sum_{n=1}^{\infty} \left(c_n e^{in\pi x/p} + c_{-n} e^{i(-n)\pi x/p} \right) = \sum_{n=-\infty}^{\infty} c_n e^{in\pi x/p}.$$

(b) From part(a) we have

$$c_n = \frac{1}{2}(a_n - ib_n) = \frac{1}{2p} \int_{-p}^{p} f(x) \left(\cos \frac{n\pi}{p}x - i \sin \frac{n\pi}{p}x \right) dx = \frac{1}{2p} \int_{-p}^{p} f(x) e^{-in\pi x/p} dx$$

and

$$c_{-n} = \frac{1}{2}(a_n + ib_n) = \frac{1}{2p} \int_{-p}^{p} f(x) \left(\cos \frac{n\pi}{p}x + i \sin \frac{n\pi}{p}x \right) dx = \frac{1}{2p} \int_{-p}^{p} f(x) e^{in\pi x/p} dx$$

for $n = 1, 2, 3, \ldots$. Thus, for $n = \pm 1, \pm 2, \pm 3, \ldots$,

$$c_n = \frac{1}{2p} \int_{-p}^{p} f(x) e^{-in\pi x/p} dx.$$

When $n = 0$ the above formula gives

$$c_0 = \frac{1}{2p} \int_{-p}^{p} f(x) dx,$$

which is $a_0/2$ where a_0 is (9) in the text. Therefore

$$c_n = \frac{1}{2p} \int_{-p}^{p} f(x) e^{-in\pi x/p} dx, \quad n = 0, \pm 1, \pm 2, \ldots.$$

Section 11.3

Fourier Cosine and Sine Series

The terminology and concepts listed below provide an outline of the main ideas encountered in this section. These can be useful when preparing for a quiz or test.

Terminology and Concepts

- properties of even and odd functions
- cosine series of an even function
- sine series of an odd function
- Gibbs phenomenon
- half-range expansions

The basic skills listed below summarize the more mechanical types of problems encountered in the exercise set for this section.

Basic Skills

- determine if a given function is even or odd
- expand a given function in an appropriate cosine or sine series
- find half-range cosine and sine expansions of a given function
- expand a function defined on an interval of the form (a, b) in a Fourier series
- use a Fourier series to find a particular solution of an ordinary DE with a periodic driving force

Exercises 11.3 *Hints, Suggestions, Solutions, and Examples*

3. Since $f(-x) = (-x)^2 - x = x^2 - x$, $f(x)$ is neither even nor odd.

6. Since $f(-x) = e^{-x} - e^x = -f(x)$, $f(x)$ is an odd function.

9. Since $f(x)$ is not defined for $x < 0$, it is neither even nor odd.

12. Since $f(x)$ is an even function, we expand in a cosine series:

$$a_0 = \int_1^2 1\,dx = 1$$

$$a_n = \int_1^2 \cos\frac{n\pi}{2}x\,dx = -\frac{2}{n\pi}\sin\frac{n\pi}{2}.$$

Thus

$$f(x) = \frac{1}{2} + \sum_{n=1}^{\infty} \frac{-2}{n\pi} \sin\frac{n\pi}{2} \cos\frac{n\pi}{2}x.$$

15. Since $f(x)$ is an even function, we expand in a cosine series:

$$a_0 = 2\int_0^1 x^2\,dx = \frac{2}{3}$$

$$a_n = 2\int_0^1 x^2 \cos n\pi x\,dx = 2\left(\frac{x^2}{n\pi}\sin n\pi x\,\bigg|_0^1 - \frac{2}{n\pi}\int_0^1 x\sin n\pi x\,dx\right) = \frac{4}{n^2\pi^2}(-1)^n.$$

Thus

$$f(x) = \frac{1}{3} + \sum_{n=1}^{\infty} \frac{4}{n^2\pi^2}(-1)^n \cos n\pi x.$$

18. Since $f(x)$ is an odd function, we expand in a sine series:

$$b_n = \frac{2}{\pi}\int_0^{\pi} x^3 \sin nx\,dx = \frac{2}{\pi}\left(-\frac{x^3}{n}\cos nx\,\bigg|_0^{\pi} + \frac{3}{n}\int_0^{\pi} x^2 \cos nx\,dx\right)$$

$$= \frac{2\pi^2}{n}(-1)^{n+1} - \frac{12}{n^2\pi}\int_0^{\pi} x\sin nx\,dx$$

$$= \frac{2\pi^2}{n}(-1)^{n+1} - \frac{12}{n^2\pi}\left(-\frac{x}{n}\cos nx\,\bigg|_0^{\pi} + \frac{1}{n}\int_0^{\pi} \cos nx\,dx\right) = \frac{2\pi^2}{n}(-1)^{n+1} + \frac{12}{n^3}(-1)^n.$$

Thus

$$f(x) = \sum_{n=1}^{\infty}\left(\frac{2\pi^2}{n}(-1)^{n+1} + \frac{12}{n^3}(-1)^n\right)\sin nx.$$

21. Since $f(x)$ is an even function, we expand in a cosine series:

$$a_0 = \int_0^1 x\,dx + \int_1^2 1\,dx = \frac{3}{2}$$

$$a_n = \int_0^1 x\cos\frac{n\pi}{2}x\,dx + \int_1^2 \cos\frac{n\pi}{2}x\,dx = \frac{4}{n^2\pi^2}\left(\cos\frac{n\pi}{2} - 1\right).$$

Thus

$$f(x) = \frac{3}{4} + \sum_{n=1}^{\infty} \frac{4}{n^2\pi^2}\left(\cos\frac{n\pi}{2} - 1\right)\cos\frac{n\pi}{2}x.$$

24. Since $f(x)$ is an even function, we expand in a cosine series. [See the solution of Problem 10 in Exercise 11.2 for the computation of the integrals.]

$$a_0 = \frac{2}{\pi/2}\int_0^{\pi/2} \cos x\,dx = \frac{4}{\pi}$$

$$a_n = \frac{2}{\pi/2}\int_0^{\pi/2} \cos x\cos\frac{n\pi}{\pi/2}x\,dx = \frac{4(-1)^{n+1}}{\pi(4n^2 - 1)}$$

Thus

$$f(x) = \frac{2}{\pi} + \sum_{n=1}^{\infty} \frac{4(-1)^{n+1}}{\pi(4n^2 - 1)} \cos 2nx.$$

27. $a_0 = \dfrac{4}{\pi} \displaystyle\int_0^{\pi/2} \cos x \, dx = \dfrac{4}{\pi}$

$a_n = \dfrac{4}{\pi} \displaystyle\int_0^{\pi/2} \cos x \cos 2nx \, dx = \dfrac{2}{\pi} \displaystyle\int_0^{\pi/2} [\cos(2n+1)x + \cos(2n-1)x] \, dx = \dfrac{4(-1)^n}{\pi(1-4n^2)}$

$b_n = \dfrac{4}{\pi} \displaystyle\int_0^{\pi/2} \cos x \sin 2nx \, dx = \dfrac{2}{\pi} \displaystyle\int_0^{\pi/2} [\sin(2n+1)x + \sin(2n-1)x] \, dx = \dfrac{8n}{\pi(4n^2-1)}$

$f(x) = \dfrac{2}{\pi} + \displaystyle\sum_{n=1}^{\infty} \dfrac{4(-1)^n}{\pi(1-4n^2)} \cos 2nx$

$f(x) = \displaystyle\sum_{n=1}^{\infty} \dfrac{8n}{\pi(4n^2-1)} \sin 2nx$

30. $a_0 = \dfrac{1}{\pi} \displaystyle\int_\pi^{2\pi} (x-\pi) \, dx = \dfrac{\pi}{2}$

$a_n = \dfrac{1}{\pi} \displaystyle\int_\pi^{2\pi} (x-\pi) \cos \dfrac{n}{2} x \, dx = \dfrac{4}{n^2\pi} \left((-1)^n - \cos \dfrac{n\pi}{2} \right)$

$b_n = \dfrac{1}{\pi} \displaystyle\int_\pi^{2\pi} (x-\pi) \sin \dfrac{n}{2} x \, dx = \dfrac{2}{n}(-1)^{n+1} - \dfrac{4}{n^2\pi} \sin \dfrac{n\pi}{2}$

$f(x) = \dfrac{\pi}{4} + \displaystyle\sum_{n=1}^{\infty} \dfrac{4}{n^2\pi} \left((-1)^n - \cos \dfrac{n\pi}{2} \right) \cos \dfrac{n}{2} x$

$f(x) = \displaystyle\sum_{n=1}^{\infty} \left(\dfrac{2}{n}(-1)^{n+1} - \dfrac{4}{n^2\pi} \sin \dfrac{n\pi}{2} \right) \sin \dfrac{n}{2} x$

33. $a_0 = 2 \displaystyle\int_0^1 (x^2 + x) \, dx = \dfrac{5}{3}$

$a_n = 2 \displaystyle\int_0^1 (x^2+x) \cos n\pi x \, dx = \dfrac{2(x^2+x)}{n\pi} \sin n\pi x \bigg|_0^1 - \dfrac{2}{n\pi} \displaystyle\int_0^1 (2x+1) \sin n\pi x \, dx = \dfrac{2}{n^2\pi^2}[3(-1)^n - 1]$

$b_n = 2 \displaystyle\int_0^1 (x^2+x) \sin n\pi x \, dx = -\dfrac{2(x^2+x)}{n\pi} \cos n\pi x \bigg|_0^1 + \dfrac{2}{n\pi} \displaystyle\int_0^1 (2x+1) \cos n\pi x \, dx$

$= \dfrac{4}{n\pi}(-1)^{n+1} + \dfrac{4}{n^3\pi^3}[(-1)^n - 1]$

$f(x) = \dfrac{5}{6} + \displaystyle\sum_{n=1}^{\infty} \dfrac{2}{n^2\pi^2}[3(-1)^n - 1] \cos n\pi x$

$f(x) = \displaystyle\sum_{n=1}^{\infty} \left(\dfrac{4}{n\pi}(-1)^{n+1} + \dfrac{4}{n^3\pi^3}[(-1)^n - 1] \right) \sin n\pi x$

36. $a_0 = \dfrac{2}{\pi} \displaystyle\int_0^\pi x\, dx = \pi$

$a_n = \dfrac{2}{\pi} \displaystyle\int_0^\pi x \cos 2nx\, dx = 0$

$b_n = \dfrac{2}{\pi} \displaystyle\int_0^\pi x \sin 2nx\, dx = -\dfrac{1}{n}$

$f(x) = \dfrac{\pi}{2} - \displaystyle\sum_{n=1}^\infty \dfrac{1}{n} \sin 2nx$

39. We have

$$b_n = \frac{2}{\pi} \int_0^\pi 5 \sin nt\, dt = \frac{10}{n\pi}[1 - (-1)^n]$$

so that

$$f(t) = \sum_{n=1}^\infty \frac{10[1 - (-1)^n]}{n\pi} \sin nt.$$

Substituting the assumption $x_p(t) = \sum_{n=1}^\infty B_n \sin nt$ into the DE then gives

$$x_p'' + 10x_p = \sum_{n=1}^\infty B_n(10 - n^2) \sin nt = \sum_{n=1}^\infty \frac{10[1 - (-1)^n]}{n\pi} \sin nt$$

and so $B_n = 10[1 - (-1)^n]/n\pi(10 - n^2)$. Thus

$$x_p(t) = \frac{10}{\pi} \sum_{n=1}^\infty \frac{1 - (-1)^n}{n(10 - n^2)} \sin nt.$$

42. We have

$$a_0 = \frac{2}{1/2} \int_0^{1/2} t\, dt = \frac{1}{2}$$

$$a_n = \frac{2}{1/2} \int_0^{1/2} t \cos 2n\pi t\, dt = \frac{1}{n^2\pi^2}[(-1)^n - 1]$$

so that

$$f(t) = \frac{1}{4} + \sum_{n=1}^\infty \frac{(-1)^n - 1}{n^2\pi^2} \cos 2n\pi t.$$

Substituting the assumption

$$x_p(t) = \frac{A_0}{2} + \sum_{n=1}^\infty A_n \cos 2n\pi t$$

into the DE then gives

$$\frac{1}{4} x_p'' + 12x_p = 6A_0 + \sum_{n=1}^\infty A_n(12 - n^2\pi^2) \cos 2n\pi t = \frac{1}{4} + \sum_{n=1}^\infty \frac{(-1)^n - 1}{n^2\pi^2} \cos 2n\pi t$$

and $A_0 = 1/24$, $A_n = [(-1)^n - 1]/n^2\pi^2(12 - n^2\pi^2)$. Thus

$$x_p(t) = \frac{1}{48} + \frac{1}{\pi^2} \sum_{n=1}^\infty \frac{(-1)^n - 1}{n^2(12 - n^2\pi^2)} \cos 2n\pi t.$$

45. (a) We have

$$b_n = \frac{2}{L}\int_0^L \frac{w_0 x}{L}\sin\frac{n\pi}{L}x\,dx = \frac{2w_0}{n\pi}(-1)^{n+1}$$

so that

$$w(x) = \sum_{n=1}^{\infty}\frac{2w_0}{n\pi}(-1)^{n+1}\sin\frac{n\pi}{L}x.$$

(b) If we assume $y_p(x) = \sum_{n=1}^{\infty}B_n\sin(n\pi x/L)$ then

$$y_p^{(4)} = \sum_{n=1}^{\infty}\frac{n^4\pi^4}{L^4}B_n\sin\frac{n\pi}{L}x$$

and so the DE $EIy_p^{(4)} = w(x)$ gives

$$B_n = \frac{2w_0(-1)^{n+1}L^4}{EIn^5\pi^5}.$$

Thus

$$y_p(x) = \frac{2w_0 L^4}{EI\pi^5}\sum_{n=1}^{\infty}\frac{(-1)^{n+1}}{n^5}\sin\frac{n\pi}{L}x.$$

Section 11.4   Sturm-Liouville Problem

The terminology and concepts listed below provide an outline of the main ideas encountered in this section. These can be useful when preparing for a quiz or test.

Terminology and Concepts

- eigenvalues and eigenfunctions of a BVP
- regular Sturm-Liouville problem
- homogeneous boundary conditions
- separated boundary conditions
- singular Sturm-Liouville problem
- periodic boundary-value problem
- self-adjoint form of a linear second-order DE

The basic skills listed below summarize the more mechanical types of problems encountered in the exercise set for this section.

Basic Skills

- for a given BVP find the equation that defines the eigenvalues and find the corresponding eigenfunctions
- put a second-order DE into self-adjoint form and give an orthogonality relation

Exercises 11.4 *Hints, Suggestions, Solutions, and Examples*

3. For $\lambda = 0$ the solution of $y'' = 0$ is $y = c_1 x + c_2$. The condition $y'(0) = 0$ implies $c_1 = 0$, so $\lambda = 0$ is an eigenvalue with corresponding eigenfunction 1.

For $\lambda = -\alpha^2 < 0$ we have $y = c_1 \cosh \alpha x + c_2 \sinh \alpha x$ and $y' = c_1 \alpha \sinh \alpha x + c_2 \alpha \cosh \alpha x$. The condition $y'(0) = 0$ implies $c_2 = 0$ and so $y = c_1 \cosh \alpha x$. Now the condition $y'(L) = 0$ implies $c_1 = 0$. Thus $y = 0$ and there are no negative eigenvalues.

For $\lambda = \alpha^2 > 0$ we have $y = c_1 \cos \alpha x + c_2 \sin \alpha x$ and $y' = -c_1 \alpha \sin \alpha x + c_2 \alpha \cos \alpha x$. The condition $y'(0) = 0$ implies $c_2 = 0$ and so $y = c_1 \cos \alpha x$. Now the condition $y'(L) = 0$ implies $-c_1 \alpha \sin \alpha L = 0$. For $c_1 \neq 0$ this condition will hold when $\alpha L = n\pi$ or $\lambda = \alpha^2 = n^2 \pi^2 / L^2$, where $n = 1, 2, 3, \ldots$. These are the positive eigenvalues with corresponding eigenfunctions $\cos(n\pi x/L)$, $n = 1, 2, 3, \ldots$.

4. (*Hint*) For possible negative eigenvalues start with $\lambda = -\alpha^2 < 0$. For possible positive eigenvalues start with $\lambda = \alpha^2 > 0$.

6. The eigenfunctions are $\sin \alpha_n x$ where $\tan \alpha_n = -\alpha_n$. Thus

$$\| \sin \alpha_n x \|^2 = \int_0^1 \sin^2 \alpha_n x \, dx = \frac{1}{2} \int_0^1 (1 - \cos 2\alpha_n x) \, dx$$

$$= \frac{1}{2} \left(x - \frac{1}{2\alpha_n} \sin 2\alpha_n x \right) \Big|_0^1 = \frac{1}{2} \left(1 - \frac{1}{2\alpha_n} \sin 2\alpha_n \right)$$

$$= \frac{1}{2} \left[1 - \frac{1}{2\alpha_n} (2 \sin \alpha_n \cos \alpha_n) \right]$$

$$= \frac{1}{2} \left[1 - \frac{1}{\alpha_n} \tan \alpha_n \cos \alpha_n \cos \alpha_n \right]$$

$$= \frac{1}{2} \left[1 - \frac{1}{\alpha_n} \left(-\alpha_n \cos^2 \alpha_n \right) \right] = \frac{1}{2} \left(1 + \cos^2 \alpha_n \right).$$

9. To obtain the self-adjoint form we note that an integrating factor is $(1/x)e^{\int (1-x)dx/x} = e^{-x}$. Thus, the DE is

$$xe^{-x} y'' + (1-x)e^{-x} y' + n e^{-x} y = 0$$

and the self-adjoint form is

$$\frac{d}{dx}\left[xe^{-x}y'\right] + ne^{-x}y = 0.$$

Identifying the weight function $p(x) = e^{-x}$ and noting that since $r(x) = xe^{-x}$, $r(0) = 0$ and $\lim_{x\to\infty} r(x) = 0$, we have the orthogonality relation

$$\int_0^\infty e^{-x}L_m(x)L_n(x)\,dx = 0,\ m \neq n.$$

12. **(a)** Letting $\lambda = \alpha^2$ the DE becomes $x^2y'' + xy' + (\alpha^2x^2 - 1)y = 0$. This is the parametric Bessel equation with $\nu = 1$. The general solution is

$$y = c_1 J_1(\alpha x) + c_2 Y_1(\alpha x).$$

Since Y is unbounded at 0 we must have $c_2 = 0$, so that $y = c_1 J_1(\alpha x)$. The condition $J_1(3\alpha) = 0$ defines the eigenvalues $\lambda_n = \alpha_n^2$ for $n = 1, 2, 3, \ldots$. The corresponding eigenfunctions are $J_1(\alpha_n x)$.

(b) Using a CAS or Table 6.1 in the text to solve $J_1(3\alpha) = 0$ we find $3\alpha_1 = 3.8317$, $3\alpha_2 = 7.0156$, $3\alpha_3 = 10.1735$, and $3\alpha_4 = 13.3237$. The corresponding eigenvalues are $\lambda_1 = \alpha_1^2 = 1.6313$, $\lambda_2 = \alpha_2^2 = 5.4687$, $\lambda_3 = \alpha_3^2 = 11.4999$, and $\lambda_4 = \alpha_4^2 = 19.7245$.

Section 11.5   Bessel and Legendre Series

The terminology and concepts listed below provide an outline of the main ideas encountered in this section. These can be useful when preparing for a quiz or test.

Terminology and Concepts

- Bessel functions

- Fourier-Bessel series

- differential recurrence relations

- convergence of a Fourier-Bessel series

- Fourier-Legendre series

- convergence of a Fourier-Legendre series

The basic skills listed below summarize the more mechanical types of problems encountered in the exercise set for this section.

Basic Skills

- expand a function in a Fourier-Bessel series using Bessel functions of the same order as in the boundary condition
- expand a function in a Fourier-Legendre series

Exercises 11.5   *Hints, Suggestions, Solutions, and Examples*

3. The boundary condition indicates that we use (15) and (16) in the text. With $b = 2$ we obtain

$$c_i = \frac{2}{4J_1^2(2\alpha_i)} \int_0^2 x J_0(\alpha_i x)\, dx$$

$$\boxed{t = \alpha_i x \qquad dt = \alpha_i\, dx}$$

$$= \frac{1}{2J_1^2(2\alpha_i)} \cdot \frac{1}{\alpha_i^2} \int_0^{2\alpha_i} t J_0(t)\, dt$$

$$= \frac{1}{2\alpha_i^2 J_1^2(2\alpha_i)} \int_0^{2\alpha_i} \frac{d}{dt}[t J_1(t)]\, dt \qquad \text{[From (5) in the text]}$$

$$= \frac{1}{2\alpha_i^2 J_1^2(2\alpha_i)} t J_1(t) \Big|_0^{2\alpha_i}$$

$$= \frac{1}{\alpha_i J_1(2\alpha_i)}\,.$$

Thus

$$f(x) = \sum_{i=1}^{\infty} \frac{1}{\alpha_i J_1(2\alpha_i)} J_0(\alpha_i x).$$

6. Writing the boundary condition in the form

$$2J_0(2\alpha) + 2\alpha J_0'(2\alpha) = 0$$

we identify $b = 2$ and $h = 2$. Using (17) and (18) in the text we obtain

$$c_i = \frac{2\alpha_i^2}{(4\alpha_i^2 + 4)J_0^2(2\alpha_i)} \int_0^2 x J_0(\alpha_i x)\, dx$$

$$\boxed{t = \alpha_i x \qquad dt = \alpha_i\, dx}$$

$$= \frac{\alpha_i^2}{2(\alpha_i^2 + 1)J_0^2(2\alpha_i)} \cdot \frac{1}{\alpha_i^2} \int_0^{2\alpha_i} t J_0(t)\, dt$$

$$= \frac{1}{2(\alpha_i^2 + 1)J_0^2(2\alpha_i)} \int_0^{2\alpha_i} \frac{d}{dt}[t J_1(t)]\, dt \qquad \text{[From (5) in the text]}$$

233

$$= \frac{1}{2(\alpha_i^2 + 1)J_0^2(2\alpha_i)} t J_1(t) \Big|_0^{2\alpha_i}$$

$$= \frac{\alpha_i J_1(2\alpha_i)}{(\alpha_i^2 + 1)J_0^2(2\alpha_i)}.$$

Thus

$$f(x) = \sum_{i=1}^{\infty} \frac{\alpha_i J_1(2\alpha_i)}{(\alpha_i^2 + 1)J_0^2(2\alpha_i)} J_0(\alpha_i x).$$

9. The boundary condition indicates that we use (19) and (20) in the text. With $b = 3$ we obtain

$$c_1 = \frac{2}{9} \int_0^3 x x^2 \, dx = \frac{2}{9} \frac{x^4}{4} \Big|_0^3 = \frac{9}{2},$$

$$c_i = \frac{2}{9 J_0^2(3\alpha_i)} \int_0^3 x J_0(\alpha_i x) x^2 \, dx$$

$$\boxed{t = \alpha_i x \qquad dt = \alpha_i \, dx}$$

$$= \frac{2}{9 J_0^2(3\alpha_i)} \cdot \frac{1}{\alpha_i^4} \int_0^{3\alpha_i} t^3 J_0(t) \, dt$$

$$= \frac{2}{9\alpha_i^4 J_0^2(3\alpha_i)} \int_0^{3\alpha_i} t^2 \frac{d}{dt} [t J_1(t)] \, dt$$

$$\boxed{\begin{array}{ll} u = t^2 & dv = \frac{d}{dt}[t J_1(t)] \, dt \\ du = 2t \, dt & v = t J_1(t) \end{array}}$$

$$= \frac{2}{9\alpha_i^4 J_0^2(3\alpha_i)} \left(t^3 J_1(t) \Big|_0^{3\alpha_i} - 2 \int_0^{3\alpha_i} t^2 J_1(t) \, dt \right).$$

With $n = 0$ in equation (6) in the text we have $J_0'(x) = -J_1(x)$, so the boundary condition $J_0'(3\alpha_i) = 0$ implies $J_1(3\alpha_i) = 0$. Then

$$c_i = \frac{2}{9\alpha_i^4 J_0^2(3\alpha_i)} \left(-2 \int_0^{3\alpha_i} \frac{d}{dt} [t^2 J_2(t)] \, dt \right) = \frac{2}{9\alpha_i^4 J_0^2(3\alpha_i)} \left(-2t^2 J_2(t) \Big|_0^{3\alpha_i} \right)$$

$$= \frac{2}{9\alpha_i^4 J_0^2(3\alpha_i)} \left[-18\alpha_i^2 J_2(3\alpha_i) \right] = \frac{-4 J_2(3\alpha_i)}{\alpha_i^2 J_0^2(3\alpha_i)}.$$

Thus

$$f(x) = \frac{9}{2} - 4 \sum_{i=1}^{\infty} \frac{J_2(3\alpha_i)}{\alpha_i^2 J_0^2(3\alpha_i)} J_0(\alpha_i x).$$

234

15. We compute

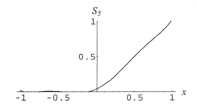

$$c_0 = \frac{1}{2}\int_0^1 xP_0(x)\,dx = \frac{1}{2}\int_0^1 x\,dx = \frac{1}{4}$$

$$c_1 = \frac{3}{2}\int_0^1 xP_1(x)\,dx = \frac{3}{2}\int_0^1 x^2\,dx = \frac{1}{2}$$

$$c_2 = \frac{5}{2}\int_0^1 xP_2(x)\,dx = \frac{5}{2}\int_0^1 \frac{1}{2}(3x^3 - x)dx = \frac{5}{16}$$

$$c_3 = \frac{7}{2}\int_0^1 xP_3(x)\,dx = \frac{7}{2}\int_0^1 \frac{1}{2}(5x^4 - 3x^2)dx = 0$$

$$c_4 = \frac{9}{2}\int_0^1 xP_4(x)\,dx = \frac{9}{2}\int_0^1 \frac{1}{8}(35x^5 - 30x^3 + 3x)dx = -\frac{3}{32}$$

$$c_5 = \frac{11}{2}\int_0^1 xP_5(x)\,dx = \frac{11}{2}\int_0^1 \frac{1}{8}(63x^6 - 70x^4 + 15x^2)dx = 0$$

$$c_6 = \frac{13}{2}\int_0^1 xP_6(x)\,dx = \frac{13}{2}\int_0^1 \frac{1}{16}(231x^7 - 315x^5 + 105x^3 - 5x)dx = \frac{13}{256}.$$

Thus

$$f(x) = \frac{1}{4}P_0(x) + \frac{1}{2}P_1(x) + \frac{5}{16}P_2(x) - \frac{3}{32}P_4(x) + \frac{13}{256}P_6(x) + \cdots.$$

The figure above is the graph of $S_5(x) = \frac{1}{4}P_0(x) + \frac{1}{2}P_1(x) + \frac{5}{16}P_2(x) - \frac{3}{32}P_4(x) + \frac{13}{256}P_6(x)$.

18. From Problem 17 we have

$$P_2(\cos\theta) = \frac{1}{4}(3\cos 2\theta + 1)$$

or

$$\cos 2\theta = \frac{4}{3}P_2(\cos\theta) - \frac{1}{3}.$$

Then, using $P_0(\cos\theta) = 1$,

$$F(\theta) = 1 - \cos 2\theta = 1 - \left[\frac{4}{3}P_2(\cos\theta) - \frac{1}{3}\right]$$

$$= \frac{4}{3} - \frac{4}{3}P_2(\cos\theta) = \frac{4}{3}P_0(\cos\theta) - \frac{4}{3}P_2(\cos\theta).$$

21. From (26) in Problem 19 in the text we find

$$c_0 = \int_0^1 x P_0(x)\, dx = \int_0^1 x\, dx = \frac{1}{2},$$

$$c_2 = 5 \int_0^1 x P_2(x)\, dx = 5 \int_0^1 \frac{1}{2}(3x^3 - x)dx = \frac{5}{8},$$

$$c_4 = 9 \int_0^1 x P_4(x)\, dx = 9 \int_0^1 \frac{1}{8}(35x^5 - 30x^3 + 3x)dx = -\frac{3}{16},$$

and

$$c_6 = 13 \int_0^1 x P_6(x)\, dx = 13 \int_0^1 \frac{1}{16}(231x^7 - 315x^5 + 105x^3 - 5x)dx = \frac{13}{128}.$$

Hence, from (25) in the text,

$$f(x) = \frac{1}{2} P_0(x) + \frac{5}{8} P_2(x) - \frac{3}{16} P_4(x) + \frac{13}{128} P_6 + \cdots.$$

On the interval $-1 < x < 1$ this series represents the function $f(x) = |x|$.

Chapter 11 in Review

Hints, Suggestions, Solutions, and Examples

3. Cosine, since f is even

6. Periodically extending the function we see that at $x = -1$ the function converges to $\frac{1}{2}(-1+0) = -\frac{1}{2}$; at $x = 0$ it converges to $\frac{1}{2}(0+1) = \frac{1}{2}$, and at $x = 1$ it converges to $\frac{1}{2}(-1+0) = -\frac{1}{2}$.

9. Since the coefficient of y in the DE is n^2, the weight function is the integrating factor

$$\frac{1}{a(x)} e^{\int (b/a)dx} = \frac{1}{1-x^2} e^{\int -\frac{x}{1-x^2}\, dx} = \frac{1}{1-x^2} e^{\frac{1}{2}\ln(1-x^2)} = \frac{\sqrt{1-x^2}}{1-x^2} = \frac{1}{\sqrt{1-x^2}}$$

on the interval $[-1, 1]$. The orthogonality relation is

$$\int_{-1}^1 \frac{1}{\sqrt{1-x^2}} T_m(x)T_n(x)\, dx = 0, \quad m \neq n.$$

12. (a) For $m \neq n$

$$\int_0^L \sin\frac{(2n+1)\pi}{2L} x \sin\frac{(2m+1)\pi}{2L} x\, dx = \frac{1}{2}\int_0^L \left(\cos\frac{n-m}{L}\pi x - \cos\frac{n+m+1}{L}\pi x\right) dx = 0.$$

(b) From

$$\int_0^L \sin^2\frac{(2n+1)\pi}{2L} x\, dx = \int_0^L \left(\frac{1}{2} - \frac{1}{2}\cos\frac{(2n+1)\pi}{L} x\right) dx = \frac{L}{2}$$

236

we see that

$$\left\| \sin \frac{(2n+1)\pi}{2L} x \right\| = \sqrt{\frac{L}{2}}.$$

15. **(a)** Since

$$a_0 = 2 \int_0^1 e^{-x} dx = 2(1 - e^{-1})$$

and

$$a_n = 2 \int_{-1}^1 e^{-x} \cos n\pi x \, dx = \frac{2}{1 + n^2\pi^2}[1 - (-1)^n e^{-1}]$$

for $n = 1, 2, 3, \ldots$ we have

$$f(x) = 1 - e^{-1} + 2 \sum_{n=1}^{\infty} \frac{1 - (-1)^n e^{-1}}{1 + n^2\pi^2} \cos n\pi x.$$

(b) Since

$$b_n = 2 \int_0^1 e^{-x} \sin n\pi x \, dx = \frac{2n\pi}{1 + n^2\pi^2}[1 - (-1)^n e^{-1}]$$

for $n = 1, 2, 3, \ldots$ we have

$$f(x) = \sum_{n=1}^{\infty} \frac{2n\pi}{1 + n^2\pi^2}[1 - (-1)^n e^{-1}] \sin n\pi x.$$

18. Expanding in a full Fourier series we have

$$a_0 = \frac{1}{2}\left(\int_0^2 x\, dx + \int_2^4 2\, dx \right) = 3$$

$$a_n = \frac{1}{2}\left(\int_0^2 x \cos \frac{n\pi x}{2}\, dx + \int_2^4 2 \cos \frac{n\pi x}{2}\, dx \right) = 2\frac{(-1)^n - 1}{n^2\pi^2}$$

$$b_n = \frac{1}{2}\left(\int_0^2 x \sin \frac{n\pi x}{2}\, dx + \int_2^4 2 \sin \frac{n\pi x}{2}\, dx \right) = 4\frac{-1}{n\pi}$$

so

$$f(x) = \frac{3}{2} + 2 \sum_{n=1}^{\infty} \left(\frac{(-1)^n - 1}{n^2\pi^2} \cos \frac{n\pi x}{2} - \frac{2}{n\pi} \sin \frac{n\pi x}{2} \right).$$

21. The boundary condition indicates that we use (15) and (16) of Section 11.5 in the text. With $b = 4$ we obtain

$$c_i = \frac{2}{16 J_1^2(4\alpha_i)} \int_0^4 x J_0(\alpha_i x) f(x)\, dx$$

$$= \frac{1}{8 J_1^2(4\alpha_i)} \int_0^2 x J_0(\alpha_i x)\, dx$$

$$\boxed{t = \alpha_i x \qquad dt = \alpha_i\, dx}$$

$$= \frac{1}{8J_1^2(4\alpha_i)} \cdot \frac{1}{\alpha_i^2} \int_0^{2\alpha_i} t J_0(t)\, dt$$

$$= \frac{1}{8J_1^2(4\alpha_i)} \int_0^{2\alpha_i} \frac{d}{dt}[t J_1(t)]\, dt \qquad \text{[From (5) in 11.5 in the text]}$$

$$= \frac{1}{8J_1^2(4\alpha_i)} t J_1(t) \Big|_0^{2\alpha_i} = \frac{J_1(2\alpha_i)}{4\alpha_i J_1^2(4\alpha_i)}.$$

Thus

$$f(x) = \frac{1}{4} \sum_{i=1}^{\infty} \frac{J_1(2\alpha_i)}{\alpha_i J_1^2(4\alpha_i)} J_0(\alpha_i x).$$

12 Boundary-Value Problems in Rectangular Coordinates

Separable Partial Differential Equations

The terminology and concepts listed below provide an outline of the main ideas encountered in this section. These can be useful when preparing for a quiz or test.

Terminology and Concepts

- general form of a linear second-order PDE
- method of separation of variables for finding a solution of a PDE
- Superposition Principle for solutions of a homogeneous linear PDE
- hyperbolic PDE
- parabolic PDE
- elliptic PDE

The basic skills listed below summarize the more mechanical types of problems encountered in the exercise set for this section.

Basic Skills

- use separation of variables to find a product solution for a PDE
- classify a PDE as hyperbolic, parabolic, or elliptic

Exercises 12.1   *Hints, Suggestions, Solutions, and Examples*

3. Substituting $u(x, y) = X(x)Y(y)$ into the PDE yields $X'Y + XY' = XY$. Separating variables and using the separation constant $-\lambda$ we obtain

$$\frac{X'}{X} = \frac{Y - Y'}{Y} = -\lambda.$$

Then

$$X' + \lambda X = 0 \quad \text{and} \quad Y' - (1 + \lambda)Y = 0$$

so that

$$X = c_1 e^{-\lambda x} \quad \text{and} \quad Y = c_2 e^{(1+\lambda)y}.$$

A particular product solution of the PDE is

$$u = XY = c_3 e^{y + \lambda(y-x)}.$$

6. Substituting $u(x, y) = X(x)Y(y)$ into the PDE yields $yX'Y + xXY' = 0$. Separating variables and using the separation constant $-\lambda$ we obtain

$$\frac{X'}{xX} = -\frac{Y'}{yY} = -\lambda.$$

When $\lambda \neq 0$

$$X' + \lambda x X = 0 \quad \text{and} \quad Y' - \lambda y Y = 0$$

so that

$$X = c_1 e^{\lambda x^2/2} \quad \text{and} \quad Y = c_2 e^{-\lambda y^2/2}.$$

A particular product solution of the PDE is

$$u = XY = c_3 e^{\lambda(x^2 - y^2)/2}.$$

When $\lambda = 0$ the DEs become $X' = 0$ and $Y' = 0$, so in this case $X = c_4$, $Y = c_5$, and $u = XY = c_6$.

9. Substituting $u(x, t) = X(x)T(t)$ into the PDE yields $kX''T - XT = XT'$. Separating variables and using the separation constant $-\lambda$ we obtain

$$\frac{kX'' - X}{X} = \frac{T'}{T} = -\lambda.$$

Then

$$X'' + \frac{\lambda - 1}{k}X = 0 \quad \text{and} \quad T' + \lambda T = 0.$$

The second DE implies $T(t) = c_1 e^{-\lambda t}$. For the first DE we consider three cases:

I. If $(\lambda - 1)/k = 0$ then $\lambda = 1$, $X'' = 0$, and $X(x) = c_2 x + c_3$, so

$$u = XT = e^{-t}(A_1 x + A_2).$$

II. If $(\lambda - 1)/k = -\alpha^2 < 0$, then $\lambda = 1 - k\alpha^2$, $X'' - \alpha^2 X = 0$, and $X(x) = c_4 \cosh \alpha x + c_5 \sinh \alpha x$, so

$$u = XT = (A_3 \cosh \alpha x + A_4 \sinh \alpha x)e^{-(1-k\alpha^2)t}.$$

III. If $(\lambda - 1)/k = \alpha^2 > 0$, then $\lambda = 1 + \lambda \alpha^2$, $X'' + \alpha^2 X = 0$, and $X(x) = c_6 \cos \alpha x + c_7 \sin \alpha x$, so

$$u = XT = (A_5 \cos \alpha x + A_6 \sin \alpha x)e^{-(1+\lambda \alpha^2)t}.$$

12. Substituting $u(x,t) = X(x)T(t)$ into the PDE yields $a^2 X''T = XT'' + 2kXT'$. Separating variables and using the separation constant $-\lambda$ we obtain

$$\frac{X''}{X} = \frac{T'' + 2kT'}{a^2 T} = -\lambda.$$

Then

$$X'' + \lambda X = 0 \qquad \text{and} \qquad T'' + 2kT' + a^2 \lambda T = 0.$$

We consider three cases:

I. If $\lambda = 0$ then $X'' = 0$ and $X(x) = c_1 x + c_2$. Also, $T'' + 2kT' = 0$ and $T(t) = c_3 + c_4 e^{-2kt}$, so

$$u = XT = (c_1 x + c_2)(c_3 + c_4 e^{-2kt}).$$

II. If $\lambda = -\alpha^2 < 0$, then $X'' - \alpha^2 X = 0$, and $X(x) = c_5 \cosh \alpha x + c_6 \sinh \alpha x$. The auxiliary equation of $T'' + 2kT' - \alpha^2 a^2 T = 0$ is $m^2 + 2km - \alpha^2 a^2 = 0$. Solving for m we obtain $m = -k \pm \sqrt{k^2 + \alpha^2 a^2}$, so $T(t) = c_7 e^{(-k+\sqrt{k^2+\alpha^2 a^2})t} + c_8 e^{(-k-\sqrt{k^2+\alpha^2 a^2})t}$. Then

$$u = XT = (c_5 \cosh \alpha x + c_6 \sinh \alpha x)(c_7 e^{(-k+\sqrt{k^2+\alpha^2 a^2})t} + c_8 e^{(-k-\sqrt{k^2+\alpha^2 a^2})t}.$$

III. If $\lambda = \alpha^2 > 0$, then $X'' + \alpha^2 X = 0$, and $X(x) = c_9 \cos \alpha x + c_{10} \sin \alpha x$. The auxiliary equation of $T'' + 2kT' + \alpha^2 a^2 T = 0$ is $m^2 + 2km + \alpha^2 a^2 = 0$. Solving for m we obtain $m = -k \pm \sqrt{k^2 - \alpha^2 a^2}$. We consider three possibilities for the discriminant $k^2 - \alpha^2 a^2$:

(i) If $k^2 - \alpha^2 a^2 = 0$ then $T(t) = c_{11} e^{-kt} + c_{12} t e^{-kt}$ and

$$u = XT = (c_9 \cos \alpha x + c_{10} \sin \alpha x)(c_{11} e^{-kt} + c_{12} t e^{-kt}).$$

From $k^2 - \alpha^2 a^2 = 0$ we have $\alpha = k/a$ so the solution can be written

$$u = XT = (c_9 \cos kx/a + c_{10} \sin kx/a)(c_{11} e^{-kt} + c_{12} t e^{-kt}).$$

(ii) If $k^2 - \alpha^2 a^2 < 0$ then $T(t) = e^{-kt}\left(c_{13} \cos \sqrt{\alpha^2 a^2 - k^2}\, t + c_{14} \sin \sqrt{\alpha^2 a^2 - k^2}\, t\right)$ and

$$u = XT = (c_9 \cos \alpha x + c_{10} \sin \alpha x)e^{-kt}\left(c_{13} \cos \sqrt{\alpha^2 a^2 - k^2}\, t + c_{14} \sin \sqrt{\alpha^2 a^2 - k^2}\, t\right).$$

(iii) If $k^2 - \alpha^2 a^2 > 0$ then $T(t) = c_{15} e^{(-k+\sqrt{k^2-\alpha^2 a^2})t} + c_{16} e^{(-k-\sqrt{k^2-\alpha^2 a^2})t}$ and

$$u = XT = (c_9 \cos \alpha x + c_{10} \sin \alpha x)\left(c_{15} e^{(-k+\sqrt{k^2-\alpha^2 a^2})t} + c_{16} e^{(-k-\sqrt{k^2-\alpha^2 a^2})t}\right).$$

15. Substituting $u(x,y) = X(x)Y(y)$ into the PDE yields $X''Y + XY'' = XY$. Separating variables and using the separation constant $-\lambda$ we obtain

$$\frac{X''}{X} = \frac{Y - Y''}{Y} = -\lambda.$$

Then

$$X'' + \lambda X = 0 \qquad \text{and} \qquad Y'' - (1 + \lambda)Y = 0.$$

We consider three cases:

I. If $\lambda = 0$ then $X'' = 0$ and $X(x) = c_1 x + c_2$. Also $Y'' - Y = 0$ and $Y(y) = c_3 \cosh y + c_4 \sinh y$ so

$$u = XY = (c_1 x + c_2)(c_3 \cosh y + c_4 \sinh y).$$

II. If $\lambda = -\alpha^2 < 0$ then $X'' - \alpha^2 X = 0$ and $Y'' + (\alpha^2 - 1)Y = 0$. The solution of the first differential equation is $X(x) = c_5 \cosh \alpha x + c_6 \sinh \alpha x$. The solution of the second DE depends on the nature of $\alpha^2 - 1$. We consider three cases:

(i) If $\alpha^2 - 1 = 0$, or $\alpha^2 = 1$, then $Y(y) = c_7 y + c_8$ nad

$$u = XY = (c_5 \cosh \alpha x + c_6 \sinh \alpha x)(c_7 y + c_8).$$

(ii) If $\alpha^2 - 1 < 0$, or $0 < \alpha^2 < 1$, then $Y(y) = c_9 \cosh \sqrt{1 - \alpha^2}\, y + c_{10} \sinh \sqrt{1 - \alpha^2}\, y$ and

$$u = XY = (c_5 \cosh \alpha x + c_6 \sinh \alpha x) \left(c_9 \cosh \sqrt{1 - \alpha^2}\, y + c_{10} \sinh \sqrt{1 - \alpha^2}\, y \right).$$

(iii) If $\alpha^2 - 1 > 0$, or $\alpha^2 > 1$, then $Y(y) = c_{11} \cos \sqrt{\alpha^2 - 1}\, y + c_{12} \sin \sqrt{\alpha^2 - 1}\, y$ and

$$u = XY = (c_5 \cosh \alpha x + c_6 \sinh \alpha x) \left(c_{11} \cos \sqrt{\alpha^2 - 1}\, y + c_{12} \sin \sqrt{\alpha^2 - 1}\, y \right).$$

III. If $\lambda = \alpha^2 > 0$, then $X'' + \alpha^2 X = 0$ and $X(x) = c_{13} \cos \alpha x + c_{14} \sin \alpha x$. Also,
$Y'' - (1 + \alpha^2)Y = 0$ and $Y(y) = c_{15} \cosh \sqrt{1 + \alpha^2}\, y + c_{16} \sinh \sqrt{1 + \alpha^2}\, y$ so

$$u = XY = (c_{13} \cos \alpha x + c_{14} \sin \alpha x) \left(c_{15} \cosh \sqrt{1 + \alpha^2}\, y + c_{16} \sinh \sqrt{1 + \alpha^2}\, y \right).$$

18. Identifying $A = 3$, $B = 5$, and $C = 1$, we compute $B^2 - 4AC = 13 > 0$. The equation is hyperbolic.

21. Identifying $A = 1$, $B = -9$, and $C = 0$, we compute $B^2 - 4AC = 81 > 0$. The equation is hyperbolic.

24. Identifying $A = 1$, $B = 0$, and $C = 1$, we compute $B^2 - 4AC = -4 < 0$. The equation is elliptic.

27. Substituting $u(r, t) = R(r)T(t)$ into the PDE yields

$$k \left(R''T + \frac{1}{r}R'T \right) = RT'.$$

Separating variables and using the separation constant $-\lambda$ we obtain

$$\frac{rR'' + R'}{rR} = \frac{T'}{kT} = -\lambda.$$

Then

$$rR'' + R' + \lambda r R = 0 \qquad \text{and} \qquad T' + \lambda k T = 0.$$

Letting $\lambda = \alpha^2$ and writing the first equation as $r^2 R'' + r R' = \alpha^2 r^2 R = 0$ we see that it is a parametric Bessel equation of order 0. As discussed in Chapter 6 of the text, it has solution

$R(r) = c_1 J_0(\alpha r) + c_2 Y_0(\alpha r)$. Since a solution of $T' + \alpha^2 kT$ is $T(t) = e^{-k\alpha^2 t}$, we see that a solution of the PDE is

$$u = RT = e^{-k\alpha^2 t}[c_1 J_0(\alpha r) + c_2 Y_0(\alpha r)].$$

30. We identify $A = xy + 1$, $B = x + 2y$, and $C = 1$. Then $B^2 - 4AC = x^2 + 4y^2 - 4$. The equation $x^2 + 4y^2 = 4$ defines an ellipse. The PDE is hyperbolic outside the ellipse, parabolic on the ellipse, and elliptic inside the ellipse.

Section 12.2   Classical PDEs and Boundary-Value Problems

The terminology and concepts listed below provide an outline of the main ideas encountered in this section. These can be useful when preparing for a quiz or test.

Terminology and Concepts

- one-dimensional heat equation
- one-dimensional wave equation
- two-dimensional form of Laplace's equation
- Laplacian
- difference between initial conditions and boundary conditions
- Dirichlet condition
- Newmann condition
- Robin condition
- boundary-value problem

The basic skills listed below summarize the more mechanical types of problems encountered in the exercise set for this section.

Basic Skills

- set up a boundary-value problem involving the heat equation, the wave equation, or Laplace's equation given a set of initial and boundary conditions

3. $k\dfrac{\partial^2 u}{\partial x^2} = \dfrac{\partial u}{\partial t}, \quad 0 < x < L, \ t > 0$

$u(0,t) = 100, \quad \dfrac{\partial u}{\partial x}\bigg|_{x=L} = -hu(L,t), \quad t > 0$

$u(x,0) = f(x), \quad 0 < x < L$

6. $a^2\dfrac{\partial^2 u}{\partial x^2} = \dfrac{\partial^2 u}{\partial t^2}, \quad 0 < x < L, \ t > 0$

$u(0,t) = 0, \quad u(L,t) = 0, \quad t > 0$

$u(x,0) = 0, \quad \dfrac{\partial u}{\partial t}\bigg|_{t=0} = \sin\dfrac{\pi x}{L}, \quad 0 < x < L$

9. $\dfrac{\partial^2 u}{\partial x^2} + \dfrac{\partial^2 u}{\partial y^2} = 0, \quad 0 < x < 4, \ 0 < y < 2$

$\dfrac{\partial u}{\partial x}\bigg|_{x=0} = 0, \quad u(4,y) = f(y), \quad 0 < y < 2$

$\dfrac{\partial u}{\partial y}\bigg|_{y=0} = 0, \quad u(x,2) = 0, \quad 0 < x < 4$

Section 12.3   Heat Equation

The terminology and concepts listed below provide an outline of the main ideas encountered in this section. These can be useful when preparing for a quiz or test.

Terminology and Concepts

- solution of the heat equation

The basic skills listed below summarize the more mechanical types of problems encountered in the exercise set for this section.

Basic Skills

- solve the heat equation subject to a given set of initial and boundary conditions

Exercises 12.3   *Hints, Suggestions, Solutions, and Examples*

3. Using $u = XT$ and $-\lambda$ as a separation constant we obtain

$$X'' + \lambda X = 0,$$

$$X'(0) = 0,$$

$$X'(L) = 0,$$

and

$$T' + k\lambda T = 0.$$

This leads to

$$X = c_1 \cos \frac{n\pi}{L}x \quad \text{and} \quad T = c_2 e^{-kn^2\pi^2 t/L^2}$$

for $n = 0, 1, 2, \ldots$ ($\lambda = 0$ is an eigenvalue in ths case) so that

$$u = \sum_{n=0}^{\infty} A_n e^{-kn^2\pi^2 t/L^2} \cos \frac{n\pi}{L}x.$$

Imposing

$$u(x, 0) = f(x) = A_0 + \sum_{n=1}^{\infty} A_n \cos \frac{n\pi}{L}x$$

gives

$$u(x, t) = \frac{1}{L}\int_0^L f(x)\,dx + \frac{2}{L}\sum_{n=1}^{\infty}\left(\int_0^L f(x)\cos\frac{n\pi}{L}x\,dx\right) e^{-kn^2\pi^2 t/L^2}\cos\frac{n\pi}{L}x.$$

6. In Problem 5 we instead find that $X(0) = 0$ and $X(L) = 0$ so that

$$X = c_1 \sin \frac{n\pi}{L}x$$

and

$$u = \frac{2e^{-ht}}{L}\sum_{n=1}^{\infty}\left(\int_0^L f(x)\sin\frac{n\pi}{L}x\,dx\right) e^{-kn^2\pi^2 t/L^2}\sin\frac{n\pi}{L}x.$$

Section 12.4   Wave Equation

The terminology and concepts listed below provide an outline of the main ideas encountered in this section. These can be useful when preparing for a quiz or test.

Terminology and Concepts

- solution of the wave equation
- standing waves

The basic skills listed below summarize the more mechanical types of problems encountered in the exercise set for this section.

- solve the wave equation subject to a given set of inital and boundary conditions

3. Using $u = XT$ and $-\lambda$ as a separation constant we obtain

$$X'' + \lambda X = 0,$$

$$X(0) = 0,$$

$$X(L) = 0,$$

and

$$T'' + \lambda a^2 T = 0,$$

$$T'(0) = 0.$$

Solving the DEs we get

$$X = c_1 \sin \frac{n\pi}{L} x + c_2 \cos \frac{n\pi}{L} x \qquad \text{and} \qquad T = c_3 \cos \frac{n\pi a}{L} t + c_4 \sin \frac{n\pi a}{L} t$$

for $n = 1, 2, 3, \ldots$. The boundary and initial conditions give

$$u = \sum_{n=1}^{\infty} A_n \cos \frac{n\pi a}{L} t \, \sin \frac{n\pi}{L} x.$$

Imposing

$$u(x,0) = \sum_{n=1}^{\infty} A_n \sin \frac{n\pi}{L} x$$

gives

$$A_n = \frac{2}{L} \left(\int_0^{L/3} \frac{3}{L} x \sin \frac{n\pi}{L} x \, dx + \int_{L/3}^{2L/3} \sin \frac{n\pi}{L} x \, dx + \int_{2L/3}^{L} \left(3 - \frac{3}{L} x \right) \sin \frac{n\pi}{L} x \, dx \right)$$

so that

$$A_1 = \frac{6\sqrt{3}}{\pi^2},$$

$$A_2 = A_3 = A_4 = 0,$$

$$A_5 = -\frac{6\sqrt{3}}{5^2\pi^2},$$

$$A_6 = 0,$$

$$A_7 = \frac{6\sqrt{3}}{7^2\pi^2}$$

$$\vdots$$

and

$$u(x,t) = \frac{6\sqrt{3}}{\pi^2}\left(\cos\frac{\pi a}{L}t\,\sin\frac{\pi}{L}x - \frac{1}{5^2}\cos\frac{5\pi a}{L}t\,\sin\frac{5\pi}{L}x + \frac{1}{7^2}\cos\frac{7\pi a}{L}t\,\sin\frac{7\pi}{L}x - \cdots\right).$$

6. Using $u = XT$ and $-\lambda$ as a separation constant we obtain

$$X'' + \lambda X = 0,$$

$$X(0) = 0,$$

$$X(1) = 0,$$

and

$$T'' + \lambda a^2 T = 0,$$

$$T'(0) = 0.$$

Solving the DEs we get

$$X = c_1 \sin n\pi x + c_2 \cos n\pi x \qquad \text{and} \qquad T = c_3 \cos n\pi a t + c_4 \sin n\pi a t$$

for $n = 1, 2, 3, \ldots$. The boundary and initial conditions give

$$u = \sum_{n=1}^{\infty} A_n \cos nt\,\sin nx.$$

Imposing

$$u(x,0) = 0.01\sin 3\pi x = \sum_{n=1}^{\infty} A_n \sin n\pi x$$

gives $A_3 = 0.01$, and $A_n = 0$ for $n = 1, 2, 4, 5, 6, \ldots$ so that

$$u(x,t) = 0.01\sin 3\pi x\,\cos 3\pi a t.$$

9. Using $u = XT$ and $-\lambda$ as a separation constant we obtain

$$X'' + \lambda X = 0,$$

$$X(0) = 0,$$

$$X(\pi) = 0,$$

and

$$T'' + 2\beta T' + \lambda T = 0,$$

$$T'(0) = 0.$$

Solving the DEs we get

$$X = c_1 \sin nx + c_2 \cos nx \quad \text{and} \quad T = e^{-\beta t} \left(c_3 \cos \sqrt{n^2 - \beta^2}\, t + c_4 \sin \sqrt{n^2 - \beta^2}\, t \right)$$

The boundary conditions on X imply $c_2 = 0$ so

$$X = c_1 \sin nx \quad \text{and} \quad T = e^{-\beta t} \left(c_3 \cos \sqrt{n^2 - \beta^2}\, t + c_4 \sin \sqrt{n^2 - \beta^2}\, t \right)$$

and

$$u = \sum_{n=1}^{\infty} e^{-\beta t} \left(A_n \cos \sqrt{n^2 - \beta^2}\, t + B_n \sin \sqrt{n^2 - \beta^2}\, t \right) \sin nx.$$

Imposing

$$u(x,0) = f(x) = \sum_{n=1}^{\infty} A_n \sin nx$$

and

$$u_t(x,0) = 0 = \sum_{n=1}^{\infty} \left(B_n \sqrt{n^2 - \beta^2} - \beta A_n \right) \sin nx$$

gives

$$u(x,t) = e^{-\beta t} \sum_{n=1}^{\infty} A_n \left(\cos \sqrt{n^2 - \beta^2}\, t + \frac{\beta}{\sqrt{n^2 - \beta^2}} \sin \sqrt{n^2 - \beta^2}\, t \right) \sin nx,$$

where

$$A_n = \frac{2}{\pi} \int_0^{\pi} f(x) \sin nx \, dx.$$

12. **(a)** Write the DE in X from Problem 11 as $X^{(4)} - \alpha^4 X = 0$ where the eigenvalues are $\lambda = \alpha^2$. Then

$$X = c_1 \cosh \alpha x + c_2 \sinh \alpha x + c_3 \cos \alpha x + c_4 \sin \alpha x$$

and using $X(0) = 0$ and $X'(0) = 0$ we find $c_3 = -c_1$ and $c_4 = -c_2$. The conditions $X(L) = 0$ and $X'(L) = 0$ yield the system of equations

$$c_1(\cosh \alpha L - \cos \alpha L) + c_2(\sinh \alpha L - \sin \alpha L) = 0$$

$$c_1(\alpha \sinh \alpha L + \alpha \sin \alpha L) + c_2(\alpha \cosh \alpha L - \alpha \cos \alpha L) = 0.$$

In order for this system to have nontrivial solutions the determinant of the coefficients must be zero. That is,

$$\alpha(\cosh \alpha L - \cos \alpha L)^2 - \alpha(\sinh^2 \alpha L - \sin^2 \alpha L) = 0.$$

Since $\alpha = 0$ leads to $X = 0$, $\lambda = \alpha^2 = 0^2 = 0$ is not an eigenvalue. Then, dividing the above equation by α, we have

$$(\cosh \alpha L - \cos \alpha L)^2 - (\sinh^2 \alpha L - \sin^2 \alpha L)$$

$$= \cosh^2 \alpha L - 2 \cosh \alpha L \cos \alpha L + \cos^2 \alpha L - \sinh^2 \alpha L + \sin^2 \alpha L$$

$$= -2 \cosh \alpha L \cos \alpha L + 2 = 0$$

or $\cosh \alpha L \cos \alpha L = 1$. Letting $x = \alpha L$ we see that the eigenvalues are $\lambda_n = \alpha_n^2 = x_n^2/L^2$ where x_n, $n = 1, 2, 3, \ldots$, are the positive roots of the equation $\cosh x \cos x = 1$.

(b) The equation $\cosh x \cos x = 1$ is the same as $\cos x = \operatorname{sech} x$. The figure indicates that the equation has an infinite number of roots.

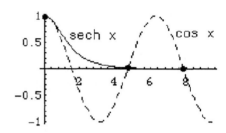

(c) Using a CAS we find the first four positive roots of $\cosh x \cos x = 1$ to be $x_1 = 4.7300$, $x_2 = 7.8532$, $x_3 = 10.9956$, and $x_4 = 14.1372$. Thus the first four eigenvalues are $\lambda_1 = x_1^2/L = 22.3733/L$, $\lambda_2 = x_2^2/L = 61.6728/L$, $\lambda_3 = x_3^2/L = 120.9034/L$, and $\lambda_4 = x_4^2/L = 199.8594/L$.

15. $u(x, t) = \dfrac{1}{2}[\sin(x + at) + \sin(x - at)] + \dfrac{1}{2a}\displaystyle\int_{x-at}^{x+at} ds$

$$= \frac{1}{2}[\sin x \cos at + \cos x \sin at + \sin x \cos at - \cos x \sin at] + \frac{1}{2a}s \Big|_{x-at}^{x+at} = \sin x \cos at + t$$

18. $u(x, t) = \dfrac{1}{2}\left[e^{-(x+at)^2} + e^{-(x-at)^2}\right]$

$$= \frac{1}{2}\left[e^{-(x^2 + 2axt + a^2 t^2)} + e^{-(x^2 - 2axt + a^2 t^2)}\right]$$

$$= e^{-(x^2 + a^2 t^2)}\left[\frac{e^{-2axt} + e^{2axt}}{2}\right] = e^{-(x^2 + a^2 t^2)} \cosh 2axt$$

Section 12.5 Laplace's Equation

The terminology and concepts listed below provide an outline of the main ideas encountered in this section. These can be useful when preparing for a quiz or test.

Terminology and Concepts

- solution of Laplace's equation
- Dirichlet problem

The basic skills listed below summarize the more mechanical types of problems encountered in the exercise set for this section.

Basic Skills

- solve Laplace's equation subject to a given set of boundary conditions

Exercises 12.5   *Hints, Suggestions, Solutions, and Examples*

3. Using $u = XY$ and $-\lambda$ as a separation constant we obtain

$$X'' + \lambda X = 0,$$

$$X(0) = 0,$$

$$X(a) = 0,$$

and

$$Y'' - \lambda Y = 0,$$

$$Y(b) = 0.$$

With $\lambda = \alpha^2 > 0$ the solutions of the DEs are

$$X = c_1 \cos \alpha x + c_2 \sin \alpha x \qquad \text{and} \qquad Y = c_3 \cosh \alpha y + c_4 \sinh \alpha y$$

The boundary and initial conditions imply

$$X = c_2 \sin \frac{n\pi}{a} x \qquad \text{and} \qquad Y = c_2 \cosh \frac{n\pi}{a} y - c_2 \frac{\cosh \frac{n\pi b}{a}}{\sinh \frac{n\pi b}{a}} \sinh \frac{n\pi}{a} y$$

for $n = 1, 2, 3, \ldots$ so that

$$u = \sum_{n=1}^{\infty} A_n \left(\cosh \frac{n\pi}{a} y - \frac{\cosh \frac{n\pi b}{a}}{\sinh \frac{n\pi b}{a}} \sinh \frac{n\pi}{a} y \right) \sin \frac{n\pi}{a} x.$$

Imposing

$$u(x,0) = f(x) = \sum_{n=1}^{\infty} A_n \sin \frac{n\pi}{a} x$$

gives

$$A_n = \frac{2}{a} \int_0^a f(x) \sin \frac{n\pi}{a} x \, dx$$

so that

$$u(x,y) = \frac{2}{a} \sum_{n=1}^{\infty} \left(\int_0^a f(x) \sin \frac{n\pi}{a} x \, dx \right) \left(\cosh \frac{n\pi}{a} y - \frac{\cosh \frac{n\pi b}{a}}{\sinh \frac{n\pi b}{a}} \sinh \frac{n\pi}{a} y \right) \sin \frac{n\pi}{a} x.$$

6. Using $u = XY$ and $-\lambda$ as a separation constant we obtain

$$X'' + \lambda X = 0,$$

$$X'(1) = 0$$

and

$$Y'' - \lambda Y = 0,$$

$$Y'(0) = 0,$$

$$Y'(\pi) = 0.$$

With $\lambda = \alpha^2 < 0$ the solutions of the DEs are

$$X = c_1 \cosh \alpha x + c_2 \sinh \alpha x \qquad \text{and} \qquad Y = c_3 \cos \alpha y + c_4 \sin \alpha y$$

The boundary and initial conditions imply

$$X = c_1 \cosh nx - c_1 \frac{\sinh n}{\cosh n} \sinh nx \qquad \text{and} \qquad Y = c_3 \cos ny$$

for $n = 1, 2, 3, \ldots$. Since $\lambda = 0$ is an eigenvalue for both DEs with corresponding eigenfunctions 1 and 1 we have

$$u = A_0 + \sum_{n=1}^{\infty} A_n \left(\cosh nx - \frac{\sinh n}{\cosh n} \sinh nx \right) \cos ny.$$

Imposing

$$u(0,y) = g(y) = A_0 + \sum_{n=1}^{\infty} A_n \cos ny$$

gives

$$A_0 = \frac{1}{\pi} \int_0^{\pi} g(y) \, dy \qquad \text{and} \qquad A_n = \frac{2}{\pi} \int_0^{\pi} g(y) \cos ny \, dy$$

for $n = 1, 2, 3, \ldots$ so that

$$u(x,y) = \frac{1}{\pi} \int_0^{\pi} g(y) \, dy + \sum_{n=1}^{\infty} \left(\frac{2}{\pi} \int_0^{\pi} g(y) \cos ny \, dy \right) \left(\cosh nx - \frac{\sinh n}{\cosh n} \sinh nx \right) \cos ny.$$

9. This BVP has the form of Problem 1 in this section, with $a = b = 1$, $f(x) = 100$, and $g(x) = 200$. The solution, then, is

$$u(x,y) = \sum_{n=1}^{\infty} \left(A_n \cosh n\pi y + B_n \sinh n\pi y \right) \sin n\pi x,$$

where

$$A_n = 2 \int_0^1 100 \sin n\pi x \, dx = 200 \left(\frac{1 - (-1)^n}{n\pi} \right)$$

and

$$B_n = \frac{1}{\sinh n\pi} \left[2 \int_0^1 200 \sin n\pi x \, dx - A_n \cosh n\pi \right]$$

$$= \frac{1}{\sinh n\pi} \left[400 \left(\frac{1 - (-1)^n}{n\pi} \right) - 200 \left(\frac{1 - (-1)^n}{n\pi} \right) \cosh n\pi \right]$$

$$= 200 \left[\frac{1 - (-1)^n}{n\pi} \right] [2 \operatorname{csch} n\pi - \coth n\pi].$$

12. Using $u = XY$ and $-\lambda$ as a separation constant we obtain

$$X'' + \lambda X = 0,$$

$$X'(0) = 0,$$

$$X'(\pi) = 0,$$

and

$$Y'' - \lambda Y = 0.$$

With $\lambda = \alpha^2 > 0$ the solutions of the DEs are

$$X = c_1 \cos \alpha x + c_2 \sin \alpha x \qquad \text{and} \qquad Y = c_3 e^{\alpha y} + c_4 e^{-\alpha y}$$

The boundary conditions at $x = 0$ and $x = \pi$ imply $c_2 = 0$ so $X = c_1 \cos nx$ for $n = 1, 2, 3, \ldots$. Now the boundedness of u as $y \to \infty$ implies $c_3 = 0$, so $Y = c_4 e^{-ny}$. In this problem $\lambda = 0$ is also an eigenvalue with corresponding eigenfunction 1 so that

$$u = A_0 + \sum_{n=1}^{\infty} A_n e^{-ny} \cos nx.$$

Imposing

$$u(x,0) = f(x) = A_0 + \sum_{n=1}^{\infty} A_n \cos nx$$

gives

$$A_0 = \frac{1}{\pi} \int_0^\pi f(x) \, dx \qquad \text{and} \qquad A_n = \frac{2}{\pi} \int_0^\pi f(x) \cos nx \, dx$$

so that

$$u(x,y) = \frac{1}{\pi}\int_0^\pi f(x)\,dx + \sum_{n=1}^\infty \left(\frac{2}{\pi}\int_0^\pi f(x)\cos nx\,dx\right)e^{-ny}\cos nx.$$

15. Referring to the discussion in this section of the text we identify $a = b = \pi$, $f(x) = 0$, $g(x) = 1$, $F(y) = 1$, and $G(y) = 1$. Then $A_n = 0$ and

$$u_1(x,y) = \sum_{n=1}^\infty B_n \sinh ny \sin nx$$

where

$$B_n = \frac{2}{\pi \sinh n\pi}\int_0^\pi \sin nx\,dx = \frac{2[1-(-1)^n]}{n\pi \sinh n\pi}.$$

Next

$$u_2(x,y) = \sum_{n=1}^\infty \left(A_n \cosh nx + B_n \sinh nx\right)\sin ny$$

where

$$A_n = \frac{2}{\pi}\int_0^\pi \sin ny\,dy = \frac{2[1-(-1)^n]}{n\pi}$$

and

$$B_n = \frac{1}{\sinh n\pi}\left(\frac{2}{\pi}\int_0^\pi \sin ny\,dy - A_n \cosh n\pi\right)$$

$$= \frac{1}{\sinh n\pi}\left(\frac{2[1-(-1)^n]}{n\pi} - \frac{2[1-(-1)^n]}{n\pi}\cosh n\pi\right)$$

$$= \frac{2[1-(-1)^n]}{n\pi \sinh n\pi}(1 - \cosh n\pi).$$

Now

$$A_n \cosh nx + B_n \sinh nx = \frac{2[1-(-1)^n]}{n\pi}\left[\cosh nx + \frac{\sinh nx}{\sinh n\pi}(1 - \cosh n\pi)\right]$$

$$= \frac{2[1-(-1)^n]}{n\pi \sinh n\pi}[\cosh nx \sinh n\pi + \sinh nx - \sinh nx \cosh n\pi]$$

$$= \frac{2[1-(-1)^n]}{n\pi \sinh n\pi}[\sinh nx + \sinh n(\pi - x)]$$

and

$$u(x,y) = u_1 + u_2 = \frac{2}{\pi}\sum_{n=1}^\infty \frac{1-(-1)^n}{n \sinh n\pi}\sinh ny \sin nx$$

$$+ \frac{2}{\pi}\sum_{n=1}^\infty \frac{[1-(-1)^n][\sinh nx + \sinh n(\pi - x)]}{n \sinh n\pi}\sin ny.$$

Section 12.6

Nonhomogeneous Boundary-Value Problems

The terminology and concepts listed below provide an outline of the main ideas encountered in this section. These can be useful when preparing for a quiz or test.

Terminology and Concepts

- nonhomogeneous BVP

- steady-state solution

- transient solution

The basic skills listed below summarize the more mechanical types of problems encountered in the exercise set for this section.

Basic Skills

- solve a nonhomogeneous BVP

Exercises 12.6 *Hints, Suggestions, Solutions, and Examples*

3. If we let $u(x,t) = v(x,t) + \psi(x)$, then we obtain as in Example 1 in the text

$$k\psi'' + r = 0$$

or

$$\psi(x) = -\frac{r}{2k}x^2 + c_1 x + c_2.$$

The boundary conditions become

$$u(0,t) = v(0,t) + \psi(0) = u_0$$

$$u(1,t) = v(1,t) + \psi(1) = u_0.$$

Letting $\psi(0) = \psi(1) = u_0$ we obtain homogeneous boundary conditions in v:

$$v(0,t) = 0 \qquad \text{and} \qquad v(1,t) = 0.$$

Now $\psi(0) = \psi(1) = u_0$ implies $c_2 = u_0$ and $c_1 = r/2k$. Thus

$$\psi(x) = -\frac{r}{2k}x^2 + \frac{r}{2k}x + u_0 = u_0 - \frac{r}{2k}x(x-1).$$

To determine $v(x, t)$ we solve

$$k\frac{\partial^2 v}{\partial x^2} = \frac{\partial v}{dt}, \quad 0 < x < 1, \ t > 0$$

$$v(0, t) = 0, \quad v(1, t) = 0,$$

$$v(x, 0) = \frac{r}{2k}x(x - 1) - u_0.$$

Separating variables, we find

$$v(x, t) = \sum_{n=1}^{\infty} A_n e^{-kn^2\pi^2 t} \sin n\pi x,$$

where

$$A_n = 2\int_0^1 \left[\frac{r}{2k}x(x - 1) - u_0\right] \sin n\pi x \, dx = 2\left[\frac{u_0}{n\pi} + \frac{r}{kn^3\pi^3}\right][(-1)^n - 1]. \tag{1}$$

Hence, a solution of the original problem is

$$u(x, t) = \psi(x) + v(x, t)$$

$$= u_0 - \frac{r}{2k}x(x - 1) + \sum_{n=1}^{\infty} A_n e^{-kn^2\pi^2 t} \sin n\pi x,$$

where A_n is defined in (1).

6. Substituting $u(x, t) = v(x, t) + \psi(x)$ into the PDE gives

$$k\frac{\partial^2 v}{\partial x^2} + k\psi'' - hv - h\psi = \frac{\partial v}{\partial t}.$$

This equation will be homogeneous provided ψ satisfies

$$k\psi'' - h\psi = 0.$$

Since k and h are positive, the general solution of this latter linear second-order equation is

$$\psi(x) = c_1 \cosh \sqrt{\frac{h}{k}}\, x + c_2 \sinh \sqrt{\frac{h}{k}}\, x.$$

From $\psi(0) = 0$ and $\psi(\pi) = u_0$ we find $c_1 = 0$ and $c_2 = u_0/\sinh\sqrt{h/k}\,\pi$. Hence

$$\psi(x) = u_0 \frac{\sinh\sqrt{h/k}\, x}{\sinh\sqrt{h/k}\,\pi}.$$

Now the new problem is

$$k\frac{\partial^2 v}{\partial x^2} - hv = \frac{\partial v}{\partial t}, \quad 0 < x < \pi, \ t > 0$$

$$v(0, t) = 0, \quad v(\pi, t) = 0, \quad t > 0$$

$$v(x, 0) = -\psi(x), \quad 0 < x < \pi.$$

Section 12.6 Nonhomogeneous Boundary-Value Problems

If we let $v = XT$ then

$$\frac{X''}{X} = \frac{T' + hT}{kT} = -\lambda.$$

With $\lambda = \alpha^2 > 0$, the separated differential equations

$$X'' + \alpha^2 X = 0 \quad \text{and} \quad T' + \left(h + k\alpha^2\right) T = 0.$$

have the respective solutions

$$X(x) = c_3 \cos \alpha x + c_4 \sin \alpha x$$

$$T(t) = c_5 e^{-(h+k\alpha^2)t}.$$

From $X(0) = 0$ we get $c_3 = 0$ and from $X(\pi) = 0$ we find $\alpha = n$ for $n = 1, 2, 3, \ldots$. Consequently, it follows that

$$v(x,t) = \sum_{n=1}^{\infty} A_n e^{-(h+kn^2)t} \sin nx$$

where

$$A_n = -\frac{2}{\pi} \int_0^{\pi} \psi(x) \sin nx \, dx.$$

Hence a solution of the original problem is

$$u(x,t) = u_0 \frac{\sinh \sqrt{h/k}\, x}{\sinh \sqrt{h/k}\, \pi} + e^{-ht} \sum_{n=1}^{\infty} A_n e^{-kn^2 t} \sin nx$$

where

$$A_n = -\frac{2}{\pi} \int_0^{\pi} u_0 \frac{\sinh \sqrt{h/k}\, x}{\sinh \sqrt{h/k}\, \pi} \sin nx \, dx.$$

Using the exponential definition of the hyperbolic sine and integration by parts we find

$$A_n = \frac{2u_0 nk(-1)^n}{\pi \left(h + kn^2\right)}.$$

9. Substituting $u(x,t) = v(x,t) + \psi(x)$ into the PDE gives

$$a^2 \frac{\partial^2 v}{\partial x^2} + a^2 \psi'' + Ax = \frac{\partial^2 v}{\partial t^2}.$$

This equation will be homogeneous provided ψ satisfies

$$a^2 \psi'' + Ax = 0.$$

The solution of this DE is

$$\psi(x) = -\frac{A}{6a^2} x^3 + c_1 x + c_2.$$

From $\psi(0) = 0$ we obtain $c_2 = 0$, and from $\psi(1) = 0$ we obtain $c_1 = A/6a^2$. Hence

$$\psi(x) = \frac{A}{6a^2}(x - x^3).$$

Now the new problem is

$$a^2 \frac{\partial^2 v}{\partial x^2} = \frac{\partial^2 v}{\partial t^2}$$

$$v(0,t) = 0, \quad v(1,t) = 0, \quad t > 0,$$

$$v(x,0) = -\psi(x), \quad v_t(x,0) = 0, \quad 0 < x < 1.$$

Identifying this as the wave equation solved in Section 12.4 in the text with $L = 1$, $f(x) = -\psi(x)$, and $g(x) = 0$ we obtain

$$v(x,t) = \sum_{n=1}^{\infty} A_n \cos n\pi a t \sin n\pi x$$

where

$$A_n = 2\int_0^1 [-\psi(x)] \sin n\pi x \, dx = \frac{A}{3a^2} \int_0^1 (x^3 - x) \sin n\pi x \, dx = \frac{2A(-1)^n}{a^2 \pi^3 n^3}.$$

Thus

$$u(x,t) = \frac{A}{6a^2}(x - x^3) + \frac{2A}{a^2 \pi^3} \sum_{n=1}^{\infty} \frac{(-1)^n}{n^3} \cos n\pi a t \sin n\pi x.$$

12. Substituting $u(x,y) = v(x,y) + \psi(x)$ into Poisson's equation we obtain

$$\frac{\partial^2 v}{\partial x^2} + \psi''(x) + h + \frac{\partial^2 v}{\partial y^2} = 0.$$

The equation will be homogeneous provided ψ satisfies $\psi''(x) + h = 0$ or $\psi(x) = -\frac{h}{2}x^2 + c_1 x + c_2$. From $\psi(0) = 0$ we obtain $c_2 = 0$. From $\psi(\pi) = 1$ we obtain

$$c_1 = \frac{1}{\pi} + \frac{h\pi}{2}.$$

Then

$$\psi(x) = \left(\frac{1}{\pi} + \frac{h\pi}{2}\right) x - \frac{h}{2}x^2.$$

The new boundary-value problem is

$$\frac{\partial^2 v}{\partial x^2} + \frac{\partial^2 v}{\partial y^2} = 0$$

$$v(0,y) = 0, \quad v(\pi,y) = 0,$$

$$v(x,0) = -\psi(x), \quad 0 < x < \pi.$$

This is Problem 11 in Section 12.5. The solution is

$$v(x,y) = \sum_{n=1}^{\infty} A_n e^{-ny} \sin nx$$

where

$$A_n = \frac{2}{\pi} \int_0^\pi [-\psi(x) \sin nx] \, dx$$

$$= \frac{2(-1)^n}{m} \left(\frac{1}{\pi} + \frac{h\pi}{2} \right) - h(-1)^n \left(\frac{\pi}{n} + \frac{2}{n^2} \right).$$

Thus

$$u(x, y) = v(x, y) + \psi(x) = \left(\frac{1}{\pi} + \frac{h\pi}{2} \right) x - \frac{h}{2} x^2 + \sum_{n=1}^\infty A_n e^{-ny} \sin nx.$$

15. With $k = 1$ and $L = 1$ in Method 2 the eigenfunctions of $X'' + \lambda X = 0$, $X(0) = 0$, $X(1) = 0$ are $\sin n\pi x$, $n = 1, 2, 3, \ldots$. Assuming that $u(x, t) = \sum_{n=1}^\infty u_n(t) \sin n\pi x$, the formal partial derivatives of u are

$$\frac{\partial^2 u}{\partial x^2} = \sum_{n=1}^\infty u_n(t)(-n^2\pi^2) \sin n\pi x \qquad \text{and} \qquad \frac{\partial u}{\partial t} = \sum_{n=1}^\infty u'_n(t) \sin n\pi x.$$

Assuming that $-1 + x - x \cos t = \sum_{n=1}^\infty F_n(t) \sin n\pi x$ we have

$$F_n(t) = \frac{2}{1} \int_0^1 (-1 + x - x \cos t) \sin n\pi x \, dx = \frac{2[(-1)^n - 1]}{n\pi} \cos t.$$

Then

$$-1 + x - x \cos t = \frac{2}{\pi} \sum_{n=1}^\infty \frac{[(-1)^n - 1]}{n} \cos t \sin n\pi x$$

and

$$u_t - u_{xx} = \sum_{n=1}^\infty [u'_n(t) + n^2\pi^2 u_n(t)] \sin n\pi x$$

$$= -1 + x - x \cos t = \frac{2}{\pi} \sum_{n=1}^\infty \frac{-1 + (-1)^n \cos t}{n} \sin n\pi x.$$

Equating coefficients we obtain

$$u'_n(t) + n^2\pi^2 u_n(t) = \frac{2[-1 + (-1)^n \cos t]}{n\pi}.$$

This is a linear first-order DE whose solution is

$$u_n(t) = \frac{2}{n\pi} \left[-\frac{1}{n^2\pi^2} + (-1)^n \frac{n^2\pi^2 \cos t + \sin t}{n^4\pi^4 + 1} \right] + C_n e^{-n^2\pi^2 t}.$$

Thus

$$u(x, t) = \sum_{n=1}^\infty \frac{2}{n\pi} \left[-\frac{1}{n^2\pi^2} + (-1)^n \frac{n^2\pi^2 \cos t + \sin t}{n^4\pi^4 + 1} \right] \sin n\pi x + \sum_{n=1}^\infty C_n e^{-n^2\pi^2 t} \sin n\pi x$$

and $u(x, 0) = x(1 - x)$ implies

$$\sum_{n=1}^\infty \frac{2}{n\pi} \left[-\frac{1}{n^2\pi^2} + (-1)^n \frac{n^2\pi^2}{n^4\pi^4 + 1} + C_n \right] \sin n\pi x = x(1 - x).$$

Hence

$$\frac{2}{n\pi}\left[-\frac{1}{n^2\pi^2}+(-1)^n\frac{n^2\pi^2}{n^4\pi^4+1}+C_n\right]=\frac{2}{1}\int_0^1 x(1-x)\sin n\pi x\,dx=2\left[\frac{1-(-1)^n}{n^3\pi^3}\right]$$

and

$$C_n=\frac{4-2(-1)^n}{n^3\pi^3}-(-1)^n\frac{2n\pi}{n^4\pi^4+1}.$$

Therefore

$$u(x,t)=\sum_{n=1}^{\infty}\frac{2}{n\pi}\left[-\frac{1}{n^2\pi^2}+(-1)^n\frac{n^2\pi^2\cos t+\sin t}{n^4\pi^4+1}\right]\sin n\pi x$$

$$+\sum_{n=1}^{\infty}\left[\frac{4-2(-1)^n}{n^3\pi^3}-(-1)^n\frac{2n\pi}{n^4\pi^4+1}\right]e^{-n^2\pi^2 t}\sin n\pi x.$$

Section 12.7 Orthogonal Series Expansions

The terminology and concepts listed below provide an outline of the main ideas encountered in this section. These can be useful when preparing for a quiz or test.

Terminology and Concepts

- orthogonal series expansion

The basic skills listed below summarize the more mechanical types of problems encountered in the exercise set for this section.

Basic Skills

- solve a BVP involving orthogonal series expansions or generalized Fourier series

Exercises 12.7 *Hints, Suggestions, Solutions, and Examples*

3. Separating variables in Laplace's equation gives

$$X''+\alpha^2 X=0$$

$$Y''-\alpha^2 Y=0$$

and

$$X(x)=c_1\cos\alpha x+c_2\sin\alpha x$$

$$Y(y)=c_3\cosh\alpha y+c_4\sinh\alpha y.$$

259

From $u(0, y) = 0$ we obtain $X(0) = 0$ and $c_1 = 0$. From $u_x(a, y) = -hu(a, y)$ we obtain $X'(a) = -hX(a)$ and

$$\alpha \cos \alpha a = -h \sin \alpha a \qquad \text{or} \qquad \tan \alpha a = -\frac{\alpha}{h}.$$

Let α_n, where $n = 1, 2, 3, \ldots$, be the consecutive positive roots of this equation. From $u(x, 0) = 0$ we obtain $Y(0) = 0$ and $c_3 = 0$. Thus

$$u(x, y) = \sum_{n=1}^{\infty} A_n \sinh \alpha_n y \sin \alpha_n x.$$

Now

$$f(x) = \sum_{n=1}^{\infty} A_n \sinh \alpha_n b \sin \alpha_n x$$

and

$$A_n \sinh \alpha_n b = \frac{\int_0^a f(x) \sin \alpha_n x \, dx}{\int_0^a \sin^2 \alpha_n x \, dx}.$$

Since

$$\int_0^a \sin^2 \alpha_n x \, dx = \frac{1}{2}\left[a - \frac{1}{2\alpha_n}\sin 2\alpha_n a\right] = \frac{1}{2}\left[a - \frac{1}{\alpha_n}\sin \alpha_n a \cos \alpha_n a\right]$$

$$= \frac{1}{2}\left[a - \frac{1}{h\alpha_n}(h\sin \alpha_n a)\cos \alpha_n a\right]$$

$$= \frac{1}{2}\left[a - \frac{1}{h\alpha_n}(-\alpha_n \cos \alpha_n a)\cos \alpha_n a\right] = \frac{1}{2h}\left[ah + \cos^2 \alpha_n a\right],$$

we have

$$A_n = \frac{2h}{\sinh \alpha_n b[ah + \cos^2 \alpha_n a]}\int_0^a f(x)\sin \alpha_n x \, dx.$$

6. Substituting $u(x, t) = v(x, t) + \psi(x)$ into the PDE gives

$$a^2 \frac{\partial^2 v}{\partial x^2} + \psi''(x) = \frac{\partial^2 v}{\partial t^2}.$$

This equation will be homogeneous if $\psi''(x) = 0$ or $\psi(x) = c_1 x + c_2$. The boundary condition $u(0, t) = 0$ implies $\psi(0) = 0$ which implies $c_2 = 0$. Thus $\psi(x) = c_1 x$. Using the second boundary condition, we obtain

$$E\left(\frac{\partial v}{\partial x} + \psi'\right)\bigg|_{x=L} = F_0,$$

which will be homogeneous when

$$E\psi'(L) = F_0.$$

Since $\psi'(x) = c_1$ we conclude that $c_1 = F_0/E$ and

$$\psi(x) = \frac{F_0}{E}x.$$

The new BVP is

$$a^2 \frac{\partial^2 v}{\partial x^2} = \frac{\partial^2 v}{\partial t^2}, \quad 0 < x < L, \quad t > 0$$

$$v(0,t) = 0, \quad \frac{\partial v}{\partial x}\bigg|_{x=L} = 0, \quad t > 0,$$

$$v(x,0) = -\frac{F_0}{E}x, \quad \frac{\partial v}{\partial t}\bigg|_{t=0} = 0, \quad 0 < x < L.$$

Referring to Example 2 in the text we see that

$$v(x,t) = \sum_{n=1}^{\infty} A_n \cos a \left(\frac{2n-1}{2L}\right) \pi t \sin \left(\frac{2n-1}{2L}\right) \pi x$$

where

$$-\frac{F_0}{E}x = \sum_{n=1}^{\infty} A_n \sin \left(\frac{2n-1}{2L}\right) \pi x$$

and

$$A_n = \frac{-F_0 \int_0^L x \sin \left(\frac{2n-1}{2L}\right) \pi x \, dx}{E \int_0^L \sin^2 \left(\frac{2n-1}{2L}\right) \pi x \, dx} = \frac{8 F_0 L (-1)^n}{E \pi^2 (2n-1)^2}.$$

Thus

$$u(x,t) = v(x,t) + \psi(x)$$

$$= \frac{F_0}{E}x + \frac{8 F_0 L}{E \pi^2} \sum_{n=1}^{\infty} \frac{(-1)^n}{(2n-1)^2} \cos a \left(\frac{2n-1}{2L}\right) \pi t \sin \left(\frac{2n-1}{2L}\right) \pi x.$$

9. The eigenfunctions of the associated homogeneous BVP are $\sin \alpha_n x$, $n = 1, 2, 3, \ldots$, where the α_n are the consecutive positive roots of $\tan \alpha = -\alpha$. We assume that

$$u(x,t) = \sum_{n=1}^{\infty} u_n(t) \sin \alpha_n x \quad \text{and} \quad x e^{-2t} = \sum_{n=1}^{\infty} F_n(t) \sin \alpha_n x.$$

Then

$$F_n(t) = \frac{e^{-2t} \int_0^1 x \sin \alpha_n x \, dx}{\int_0^1 \sin^2 \alpha_n x \, dx}.$$

Since $\alpha_n \cos \alpha_n = -\sin \alpha_n$ and

$$\int_0^1 \sin^2 \alpha_n x \, dx = \frac{1}{2} \left[1 - \frac{1}{2\alpha_n} \sin 2\alpha_n\right],$$

we have

$$e^{-2t} \int_0^1 x \sin \alpha_n x \, dx = e^{-2t} \left(\frac{\sin \alpha_n - \alpha_n \cos \alpha_n}{\alpha_n^2}\right) = \frac{2 \sin \alpha_n}{\alpha_n^2} e^{-2t}$$

$$\int_0^1 \sin^2 \alpha_n x \, dx = \frac{1}{2}[1 + \cos^2 \alpha_n]$$

and so

$$F_n(t) = \frac{4\sin\alpha_n}{\alpha_n^2(1+\cos^2\alpha_n)}e^{-2t}.$$

Substituting the assumptions into $u_t - ku_{xx} = xe^{-2t}$ and equating coefficients leads to the linear first-order DE

$$u_n'(t) + k\alpha_n^2 u(t) = \frac{4\sin\alpha_n}{\alpha_n^2(1+\cos^2\alpha_n)}e^{-2t}$$

whose solution is

$$u_n(t) = \frac{4\sin\alpha_n}{\alpha_n^2(1+\cos^2\alpha_n)(k\alpha_n^2 - 2)}e^{-2t} + C_n e^{-k\alpha_n^2 t}.$$

From

$$u(x,t) = \sum_{n=1}^{\infty}\left[\frac{4\sin\alpha_n}{\alpha_n^2(1+\cos^2\alpha_n)(k\alpha_n^2 - 2)}e^{-2t} + C_n e^{-k\alpha_n^2 t}\right]\sin\alpha_n x$$

and the initial condition $u(x,0) = 0$ we see

$$C_n = -\frac{4\sin\alpha_n}{\alpha_n^2(1+\cos^2\alpha_n)(k\alpha_n^2 - 2)}.$$

The formal solution of the original problem is then

$$u(x,t) = \sum_{n=1}^{\infty}\frac{4\sin\alpha_n}{\alpha_n^2(1+\cos^2\alpha_n)(k\alpha_n^2 - 2)}(e^{-2t} - e^{-k\alpha_n^2 t})\sin\alpha_n x.$$

10. (*Hint*) Use the separation constant $-\lambda = \alpha^4$.

Section 12.8   Higher-Dimensional Problems

The terminology and concepts listed below provide an outline of the main ideas encountered in this section. These can be useful when preparing for a quiz or test.

Terminology and Concepts

- two-dimensional heat equation
- two-dimensional wave equation
- double sine series
- double cosine series

The basic skills listed below summarize the more mechanical types of problems encountered in the exercise set for this section.

Basic Skills

- solve a two-dimensional heat equation with specified boundary and initial conditions
- solve a two-dimensional wave equation with specified boundary and initial conditions

Exercises 12.8   *Hints, Suggestions, Solutions, and Examples*

3. The conditions $X(0) = 0$ and $Y(0) = 0$ give $c_1 = 0$ and $c_3 = 0$. The conditions $X(\pi) = 0$ and $Y(\pi) = 0$ yield two sets of eigenvalues:

$$\alpha = m, \ m = 1, 2, 3, \dots \qquad \text{and} \qquad \beta = n, \ n = 1, 2, 3, \dots .$$

A product solution of the PDE that satisfies the boundary conditions is

$$u_{mn}(x, y, t) = \left(A_{mn} \cos a\sqrt{m^2 + n^2}\, t + B_{mn} \sin a\sqrt{m^2 + n^2}\, t \right) \sin mx \sin ny.$$

To satisfy the initial conditions we use the Superposition Principle:

$$u(x, y, t) = \sum_{m=1}^{\infty} \sum_{n=1}^{\infty} \left(A_{mn} \cos a\sqrt{m^2 + n^2}\, t + B_{mn} \sin a\sqrt{m^2 + n^2}\, t \right) \sin mx \sin ny.$$

The initial condition $u_t(x, y, 0) = 0$ implies $B_{mn} = 0$ and

$$u(x, y, t) = \sum_{m=1}^{\infty} \sum_{n=1}^{\infty} A_{mn} \cos a\sqrt{m^2 + n^2}\, t \sin mx \sin ny.$$

At $t = 0$ we have

$$xy(x - \pi)(y - \pi) = \sum_{m=1}^{\infty} \sum_{n=1}^{\infty} A_{mn} \sin mx \sin ny.$$

Using (11) and (12) in the text, it follows that

$$A_{mn} = \frac{4}{\pi^2} \int_0^{\pi} \int_0^{\pi} xy(x - \pi)(y - \pi) \sin mx \sin ny \, dx \, dy$$

$$= \frac{4}{\pi^2} \int_0^{\pi} x(x - \pi) \sin mx \, dx \int_0^{\pi} y(y - \pi) \sin ny \, dy$$

$$= \frac{16}{m^3 n^3 \pi^2} [(-1)^m - 1][(-1)^n - 1].$$

6. The boundary and initial conditions are

$$u(0, y, z) = 0, \qquad u(a, y, z) = 0,$$

$$u(x, 0, z) = 0, \qquad u(x, b, z) = 0,$$

$$u(x, y, 0) = f(x, y), \qquad u(x, y, c) = 0.$$

The conditions $X(0) = Y(0) = 0$ give $c_1 = c_3 = 0$. The conditions $X(a) = Y(b) = 0$ yield two sets of eigenvalues:

$$\alpha = \frac{m\pi}{a}, \ m = 1, 2, 3, \ldots \qquad \text{and} \qquad \beta = \frac{n\pi}{b}, \ n = 1, 2, 3, \ldots.$$

Let

$$\omega_{mn}^2 = \frac{m^2\pi^2}{a^2} + \frac{n^2\pi^2}{b^2}.$$

Then the boundary condition $Z(c) = 0$ gives

$$c_5 \cosh c\omega_{mn} + c_6 \sinh c\omega_{mn} = 0$$

from which we obtain

$$Z(z) = c_5 \left(\cosh w_{mn}z - \frac{\cosh c\omega_{mn}}{\sinh c\omega_{mn}} \sinh \omega z \right)$$

$$= \frac{c_5}{\sinh c\omega_{mn}} (\sinh c\omega_{mn} \cosh \omega_{mn}z - \cosh c\omega_{mn} \sinh \omega_{mn}z) = c_{mn} \sinh \omega_{mn}(c - z).$$

By the Superposition Principle

$$u(x, y, t) = \sum_{m=1}^{\infty} \sum_{n=1}^{\infty} A_{mn} \sinh \omega_{mn}(c - z) \sin \frac{m\pi}{a}x \sin \frac{n\pi}{b}y$$

where

$$A_{mn} = \frac{4}{ab \sinh c\omega_{mn}} \int_0^b \int_0^a f(x, y) \sin \frac{m\pi}{a}x \sin \frac{n\pi}{b}y \, dx \, dy.$$

Chapter 12 in Review
Hints, Suggestions, Solutions, and Examples

3. Substituting $u(x, t) = v(x, t) + \psi(x)$ into the PDE we obtain

$$k \frac{\partial^2 v}{\partial x^2} + k\psi''(x) = \frac{\partial v}{\partial t}.$$

This equation will be homogeneous provided ψ satisfies

$$k\psi'' = 0 \qquad \text{or} \qquad \psi = c_1 x + c_2.$$

Considering

$$u(0, t) = v(0, t) + \psi(0) = u_0$$

we set $\psi(0) = u_0$ so that $\psi(x) = c_1 x + u_0$. Now

$$-\frac{\partial u}{\partial x}\bigg|_{x=\pi} = -\frac{\partial v}{\partial x}\bigg|_{x=\pi} - \psi'(x) = v(\pi, t) + \psi(\pi) - u_1$$

is equivalent to

$$\left.\frac{\partial v}{\partial x}\right|_{x=\pi} + v(\pi, t) = u_1 - \psi'(x) - \psi(\pi) = u_1 - c_1 - (c_1\pi + u_0),$$

which will be homogeneous when

$$u_1 - c_1 - c_1\pi - u_0 = 0 \qquad \text{or} \qquad c_1 = \frac{u_1 - u_0}{1 + \pi}.$$

The steady-state solution is

$$\psi(x) = \left(\frac{u_1 - u_0}{1 + \pi}\right) x + u_0.$$

6. The BVP is

$$\frac{\partial^2 u}{\partial x^2} + x^2 = \frac{\partial^2 u}{\partial t^2}, \qquad 0 < x < 1, \quad t > 0,$$

$$u(0, t) = 1, \quad u(1, t) = 0, \quad t > 0,$$

$$u(x, 0) = f(x), \quad u_t(x, 0) = 0, \quad 0 < x < 1.$$

Substituting $u(x, t) = v(x, t) + \psi(x)$ into the PDE gives

$$\frac{\partial^2 v}{\partial x^2} + \psi''(x) + x^2 = \frac{\partial^2 v}{\partial t^2}.$$

This equation will be homogeneous provided $\psi''(x) + x^2 = 0$ or

$$\psi(x) = -\frac{1}{12}x^4 + c_1 x + c_2.$$

From $\psi(0) = 1$ and $\psi(1) = 0$ we obtain $c_1 = -11/12$ and $c_2 = 1$. The new problem is

$$\frac{\partial^2 v}{\partial x^2} = \frac{\partial^2 v}{\partial t^2}, \qquad 0 < x < 1, \quad t > 0,$$

$$v(0, t) = 0, \quad v(1, t) = 0, \quad t > 0,$$

$$v(x, 0) = f(x) - \psi(x), \quad v_t(x, 0) = 0, \quad 0 < x < 1.$$

From Section 12.4 in the text we see that $B_n = 0$,

$$A_n = 2\int_0^1 [f(x) - \psi(x)]\sin n\pi x\, dx = 2\int_0^1 \left[f(x) + \frac{1}{12}x^4 + \frac{11}{12}x - 1\right]\sin n\pi x\, dx,$$

and

$$v(x, t) = \sum_{n=1}^{\infty} A_n \cos n\pi t \sin n\pi x.$$

Thus

$$u(x, t) = v(x, t) + \psi(x) = -\frac{1}{12}x^4 - \frac{11}{12}x + 1 + \sum_{n=1}^{\infty} A_n \cos n\pi t \sin n\pi x.$$

Chapter 12 in Review

9. Using $u = XY$ and $-\lambda$ as a separation constant leads to

$$X'' - \lambda X = 0,$$

and

$$Y'' + \lambda Y = 0,$$
$$Y(0) = 0,$$
$$Y(\pi) = 0.$$

Then

$$X = c_1 e^{nx} + c_2 e^{-nx} \qquad \text{and} \qquad Y = c_3 \cos ny + c_4 \sin ny$$

for $n = 1, 2, 3, \dots$. Since u must be bounded as $x \to \infty$) we define $c_1 = 0$. Also $Y(0) = 0$ implies $c_3 = 0$ so

$$u = \sum_{n=1}^{\infty} A_n e^{-nx} \sin ny.$$

Imposing

$$u(0, y) = 50 = \sum_{n=1}^{\infty} A_n \sin ny$$

gives

$$A_n = \frac{2}{\pi} \int_0^{\pi} 50 \sin ny\, dy = \frac{100}{n\pi}[1 - (-1)^n]$$

so that

$$u(x, y) = \sum_{n=1}^{\infty} \frac{100}{n\pi}[1 - (-1)^n] e^{-nx} \sin ny.$$

12. Substituting $u(x, t) = v(x, t) + \psi(x)$ into the PDE results in $\psi'' = -\sin x$ and $\psi(x) = c_1 x + c_2 + \sin x$. The boundary conditions $\psi(0) = 400$ and $\psi(\pi) = 200$ imply $c_1 = -200/\pi$ and $c_2 = 400$ so

$$\psi(x) = -\frac{200}{\pi}x + 400 + \sin x.$$

Solving

$$\frac{\partial^2 v}{\partial x^2} = \frac{\partial v}{\partial t}, \qquad 0 < x < \pi, \quad t > 0$$

$$v(0, t) = 0, \qquad v(\pi, t) = 0, \quad t > 0$$

$$u(x, 0) = 400 + \sin x - \left(-\frac{200}{\pi}x + 400 + \sin x\right) = \frac{200}{\pi}x, \qquad 0 < x < \pi$$

using separation of variables with separation constant $-\lambda$, where $\lambda = \alpha^2$, gives

$$X'' + \alpha^2 X = 0 \qquad \text{and} \qquad T' + \alpha^2 T = 0.$$

Using $X(0) = 0$ and $X(\pi) = 0$ we determine $\alpha^2 = n^2$, $X(x) = c_2 \sin nx$, and $T(t) = c_3 e^{-n^2 t}$. Then

$$v(x,t) = \sum_{n=1}^{\infty} A_n e^{-n^2 t} \sin nx$$

and

$$v(x,0) = \frac{200}{\pi} x = \sum_{n=1}^{\infty} A_n \sin nx$$

so

$$A_n = \frac{400}{\pi^2} \int_0^{\pi} x \sin nx \, dx = \frac{400}{n\pi} (-1)^{n+1}.$$

Thus

$$u(x,t) = -\frac{200}{\pi} x + 400 + \sin x + \frac{400}{\pi} \sum_{n=1}^{\infty} \frac{(-1)^{n+1}}{n} e^{-n^2 t} \sin nx.$$

13

Boundary-Value Problems in Other Coordinate Systems

The terminology and concepts listed below provide an outline of the main ideas encountered in this section. These can be useful when preparing for a quiz or test.

Terminology and Concepts

- Laplacian in polar coordinates
- double eigenvalue

The basic skills listed below summarize the more mechanical types of problems encountered in the exercise set for this section.

Basic Skills

- find the steady-state of a plate that is described in terms of polar coordinates

Exercises 13.1   *Hints, Suggestions, Solutions, and Examples*

3. We have

$$A_0 = \frac{1}{2\pi} \int_0^{2\pi} \left(2\pi\theta - \theta^2\right) d\theta = \frac{2\pi^2}{3}$$

$$A_n = \frac{1}{\pi} \int_0^{2\pi} \left(2\pi\theta - \theta^2\right) \cos n\theta \, d\theta = -\frac{4}{n^2}$$

$$B_n = \frac{1}{\pi} \int_0^{2\pi} \left(2\pi\theta - \theta^2\right) \sin n\theta \, d\theta = 0$$

and so

$$u(r, \theta) = \frac{2\pi^2}{3} - 4 \sum_{n=1}^{\infty} \frac{r^n}{n^2} \cos n\theta.$$

268

6. We solve

$$\frac{\partial^2 u}{\partial r^2} + \frac{1}{r}\frac{\partial u}{\partial r} + \frac{1}{r^2}\frac{\partial^2 u}{\partial \theta^2} = 0, \quad 0 < \theta < \frac{\pi}{2}, \quad 0 < r < c,$$

$$u(c, \theta) = f(\theta), \quad 0 < \theta < \frac{\pi}{2},$$

$$u(r, 0) = 0, \quad u(r, \pi/2) = 0, \quad 0 < r < c.$$

Proceeding as in Example 1 in the text we obtain the separated DEs

$$r^2 R'' + r R' - \lambda R = 0$$

$$\Theta'' + \lambda \Theta = 0.$$

Taking $\lambda = \alpha^2$ the solutions are

$$\Theta(\theta) = c_1 \cos \alpha\theta + c_2 \sin \alpha\theta$$

$$R(r) = c_3 r^\alpha + c_4 r^{-\alpha}.$$

Since we want $R(r)$ to be bounded as $r \to 0$ we require $c_4 = 0$. Applying the boundary conditions $\Theta(0) = 0$ and $\Theta(\pi/2) = 0$ we find that $c_1 = 0$ and $\alpha = 2n$ for $n = 1, 2, 3, \ldots$. Therefore

$$u(r, \theta) = \sum_{n=1}^{\infty} A_n r^{2n} \sin 2n\theta.$$

From

$$u(c, \theta) = f(\theta) = \sum_{n=1}^{\infty} A_n c^n \sin 2n\theta$$

we find

$$A_n = \frac{4}{\pi c^{2n}} \int_0^{\pi/2} f(\theta) \sin 2n\theta \, d\theta.$$

9. Proceeding as in Example 1 in the text and again using the periodicity of $u(r, \theta)$, we have

$$\Theta(\theta) = c_1 \cos \alpha\theta + c_2 \sin \alpha\theta$$

where $\alpha = n$ for $n = 0, 1, 2, \ldots$. Then

$$R(r) = c_3 r^n + c_4 r^{-n}.$$

[We do not have $c_4 = 0$ in this case since $0 < a \le r$.] Since $u(b, \theta) = 0$ we have

$$u(r, \theta) = A_0 \ln \frac{r}{b} + \sum_{n=1}^{\infty} \left[\left(\frac{b}{r}\right)^n - \left(\frac{r}{b}\right)^n \right] [A_n \cos n\theta + B_n \sin n\theta].$$

From

$$u(a, \theta) = f(\theta) = A_0 \ln \frac{a}{b} + \sum_{n=1}^{\infty} \left[\left(\frac{b}{a}\right)^n - \left(\frac{a}{b}\right)^n \right] [A_n \cos n\theta + B_n \sin n\theta]$$

269

we find

$$A_0 \ln \frac{a}{b} = \frac{1}{2\pi} \int_0^{2\pi} f(\theta)\, d\theta,$$

$$\left[\left(\frac{b}{a}\right)^n - \left(\frac{a}{b}\right)^n \right] A_n = \frac{1}{\pi} \int_0^{2\pi} f(\theta) \cos n\theta\, d\theta,$$

and

$$\left[\left(\frac{b}{a}\right)^n - \left(\frac{a}{b}\right)^n \right] B_n = \frac{1}{\pi} \int_0^{2\pi} f(\theta) \sin n\theta\, d\theta.$$

12. Letting $u(r,\theta) = v(r,\theta) + \psi(\theta)$ we obtain $\psi''(\theta) = 0$ and so $\psi(\theta) = c_1\theta + c_2$. From $\psi(0) = 0$ and $\psi(\pi) = u_0$ we find, in turn, $c_2 = 0$ and $c_1 = u_0/\pi$. Therefore $\psi(\theta) = \dfrac{u_0}{\pi}\theta$. Now $u(1,\theta) = v(1,\theta) + \psi(\theta)$ so that $v(1,\theta) = u_0 - \dfrac{u_0}{\pi}\theta$. From

$$v(r,\theta) = \sum_{n=1}^{\infty} A_n r^n \sin n\theta \qquad \text{and} \qquad v(1,\theta) = \sum_{n=1}^{\infty} A_n \sin n\theta$$

we obtain

$$A_n = \frac{2}{\pi} \int_0^{\pi} \left(u_0 - \frac{u_0}{\pi}\theta \right) \sin n\theta\, d\theta = \frac{2u_0}{\pi n}.$$

Thus

$$u(r,\theta) = \frac{u_0}{\pi}\theta + \frac{2u_0}{\pi} \sum_{n=1}^{\infty} \frac{r^n}{n} \sin n\theta.$$

Section 13.2   Polar and Cylindrical Coordinates

The terminology and concepts listed below provide an outline of the main ideas encountered in this section. These can be useful when preparing for a quiz or test.

Terminology and Concepts

- radial symmetry
- radial vibrations
- standing waves
- nodal lines
- Laplacian in cylindrical coordinates

The basic skills listed below summarize the more mechanical types of problems encountered in the exercise set for this section.

Basic Skills

- find the displacement of a circular membrane that is clamped along its circumference and given an initial velocity possessing radial symmetry
- find the steady-state temperature in a cylinder subject to given boundary conditions

Exercises 13.2   *Hints, Suggestions, Solutions, and Examples*

3. Referring to Example 2 in the text we have

$$R(r) = c_1 J_0(\alpha r) + c_2 Y_0(\alpha r)$$

$$Z(z) = c_3 \cosh \alpha z + c_4 \sinh \alpha z$$

where $c_2 = 0$ and $J_0(2\alpha) = 0$ defines the positive eigenvalues $\lambda_n = \alpha_n^2$. From $Z(4) = 0$ we obtain

$$c_3 \cosh 4\alpha_n + c_4 \sinh 4\alpha_n = 0 \qquad \text{or} \qquad c_4 = -c_3 \frac{\cosh 4\alpha_n}{\sinh 4\alpha_n}.$$

Then

$$Z(z) = c_3 \left[\cosh \alpha_n z - \frac{\cosh 4\alpha_n}{\sinh 4\alpha_n} \sinh \alpha_n z \right] = c_3 \frac{\sinh 4\alpha_n \cosh \alpha_n z - \cosh 4\alpha_n \sinh \alpha_n z}{\sinh 4\alpha_n}$$

$$= c_3 \frac{\sinh \alpha_n (4 - z)}{\sinh 4\alpha_n}$$

and

$$u(r, z) = \sum_{n=1}^{\infty} A_n \frac{\sinh \alpha_n (4 - z)}{\sinh 4\alpha_n} J_0(\alpha_n r).$$

From

$$u(r, 0) = u_0 = \sum_{n=1}^{\infty} A_n J_0(\alpha_n r)$$

we obtain

$$A_n = \frac{2u_0}{4 J_1^2(2\alpha_n)} \int_0^2 r J_0(\alpha_n r) \, dr = \frac{u_0}{\alpha_n J_1(2\alpha_n)}.$$

Thus the temperature in the cylinder is

$$u(r, z) = u_0 \sum_{n=1}^{\infty} \frac{\sinh \alpha_n (4 - z) J_0(\alpha_n r)}{\alpha_n \sinh 4\alpha_n J_1(2\alpha_n)}.$$

6. If the edge $r = c$ is insulated we have the boundary condition $u_r(c, t) = 0$. Referring to the solution of Problem 5 above we have

$$R'(c) = \alpha c_1 J_0'(\alpha c) = 0$$

which defines an eigenvalue $\lambda = \alpha^2 = 0$ and positive eigenvalues $\lambda_n = \alpha_n^2$. Thus

$$u(r,t) = A_0 + \sum_{n=1}^{\infty} A_n J_0(\alpha_n r) e^{-\alpha_n^2 kt}.$$

From

$$u(r,0) = f(r) = A_0 + \sum_{n=1}^{\infty} A_n J_0(\alpha_n r)$$

we find

$$A_0 = \frac{2}{c^2} \int_0^c r f(r)\, dr$$

$$A_n = \frac{2}{c^2 J_0^2(\alpha_n c)} \int_0^c r J_0(\alpha_n r) f(r)\, dr.$$

9. Substituting $u(r,t) = v(r,t) + \psi(r)$ into the partial DE gives

$$\frac{\partial^2 v}{\partial r^2} + \frac{1}{r}\frac{\partial v}{\partial r} + \psi'' + \frac{1}{r}\psi' = \frac{\partial v}{\partial t}.$$

This equation will be homogeneous provided $\psi'' + \frac{1}{r}\psi' = 0$ or $\psi(r) = c_1 \ln r + c_2$. Since $\ln r$ is unbounded as $r \to 0$ we take $c_1 = 0$. Then $\psi(r) = c_2$ and using $u(2,t) = v(2,t) + \psi(2) = 100$ we set $c_2 = \psi(2) = 100$. Therefore $\psi(r) = 100$. Referring to Problem 5 above, the solution of the BVP

$$\frac{\partial^2 v}{\partial r^2} + \frac{1}{r}\frac{\partial v}{\partial r} = \frac{\partial v}{\partial t}, \quad 0 < r < 2, \quad t > 0,$$

$$v(2,t) = 0, \quad t > 0,$$

$$v(r,0) = u(r,0) - \psi(r)$$

is

$$v(r,t) = \sum_{n=1}^{\infty} A_n J_0(\alpha_n r) e^{-\alpha_n^2 t}$$

where

$$A_n = \frac{2}{2^2 J_1^2(2\alpha_n)} \int_0^2 r J_0(\alpha_n r)[u(r,0) - \psi(r)]\, dr$$

$$= \frac{1}{2 J_1^2(2\alpha_n)} \left[\int_0^1 r J_0(\alpha_n r)[200 - 100]\, dr + \int_1^2 r J_0(\alpha_n r)[100 - 100]\, dr \right]$$

$$= \frac{50}{J_1^2(2\alpha_n)} \int_0^1 r J_0(\alpha_n r)\, dr \qquad \boxed{x = \alpha_n r, \ dx = \alpha_n\, dr}$$

$$= \frac{50}{J_1^2(2\alpha_n)} \int_0^{\alpha_n} \frac{1}{\alpha_n^2} x J_0(x)\, dx$$

$$= \frac{50}{\alpha_n^2 J_1^2(2\alpha_n)} \int_0^{\alpha_n} \frac{d}{dx}[x J_1(x)] \, dx \qquad \boxed{\text{see (5) of Section 11.5 in text}}$$

$$= \frac{50}{\alpha_n^2 J_1^2(2\alpha_n)} (x J_1(x)) \Big|_0^{\alpha_n} = \frac{50 J_1(\alpha_n)}{\alpha_n J_1^2(2\alpha_n)}.$$

Thus

$$u(r,t) = v(r,t) + \psi(r) = 100 + 50 \sum_{n=1}^{\infty} \frac{J_1(\alpha_n) J_0(\alpha_n r)}{\alpha_n J_1^2(2\alpha_n)} e^{-\alpha_n^2 t}.$$

12. (a) First we see that

$$\frac{R''\Theta + \frac{1}{r} R'\Theta + \frac{1}{r^2} R\Theta''}{R\Theta} = \frac{T''}{a^2 T} = -\lambda.$$

This gives $T'' + a^2 \lambda T = 0$ and from

$$\frac{R'' + \frac{1}{r} R' + \lambda R}{-R/r^2} = \frac{\Theta''}{\Theta} = -\nu$$

we get $\Theta'' + \nu\Theta = 0$ and $r^2 R'' + r R' + (\lambda r^2 - \nu)R = 0$.

(b) With $\lambda = \alpha^2$ and $\nu = \beta^2$ the general solutions of the differential equations in part (a) are

$$T = c_1 \cos a\alpha t + c_2 \sin a\alpha t$$

$$\Theta = c_3 \cos \beta\theta + c_4 \cos \beta\theta$$

$$R = c_5 J_\beta(\alpha r) + c_6 Y_\beta(\alpha r).$$

(c) Implicitly we expect $u(r,\theta,t) = u(r,\theta+2\pi,t)$ and so Θ must be 2π-periodic. Therefore $\beta = n$, $n = 0, 1, 2, \ldots$. The corresponding eigenfunctions are 1, $\cos\theta$, $\cos 2\theta$, $\ldots$, $\sin\theta$, $\sin 2\theta$, $\ldots$. Arguing that $u(r,\theta,t)$ is bounded as $r \to 0$ we then define $c_6 = 0$ and so $R = c_3 J_n(\alpha r)$. But $R(c) = 0$ gives $J_n(\alpha c) = 0$; this equation defines the eigenvalues $\lambda_n = \alpha_n^2$. For each n, $\alpha_{ni} = x_{ni}/c$, $i = 1, 2, 3, \ldots$.

(d) $u(r,\theta,t) = \displaystyle\sum_{i=1}^{n} (A_{0i} \cos a\alpha_{0i} t + B_{0i} \sin a\alpha_{0i} t) J_0(\alpha_{0i} r)$

$$+ \sum_{n=1}^{\infty} \sum_{i=1}^{\infty} \Big[(A_{ni} \cos a\alpha_{ni} t + B_{ni} \sin a\alpha_{ni} t) \cos n\theta$$

$$+ (C_{ni} \cos a\alpha_{ni} t + D_{ni} \sin a\alpha_{ni} t) \sin n\theta \Big] J_n(\alpha_{ni} r)$$

Section 13.3

Spherical Coordinates

The terminology and concepts listed below provide an outline of the main ideas encountered in this section. These can be useful when preparing for a quiz or test.

Terminology and Concepts

- Laplacian in spherical coordinates

The basic skills listed below summarize the more mechanical types of problems encountered in the exercise set for this section.

Basic Skills

- find the steady-state temperature in a sphere or a portion of a sphere subject to given boundary conditions

Exercises 13.3   *Hints, Suggestions, Solutions, and Examples*

3. The coefficients are given by

$$A_n = \frac{2n+1}{2c^n} \int_0^\pi \cos\theta \, P_n(\cos\theta) \sin\theta \, d\theta = \frac{2n+1}{2c^n} \int_0^\pi P_1(\cos\theta) P_n(\cos\theta) \sin\theta \, d\theta$$

$$\boxed{x = \cos\theta, \ dx = -\sin\theta \, d\theta}$$

$$= \frac{2n+1}{2c^n} \int_{-1}^1 P_1(x) P_n(x) \, dx.$$

Since $P_n(x)$ and $P_m(x)$ are orthogonal for $m \neq n$, $A_n = 0$ for $n \neq 1$ and

$$A_1 = \frac{2(1)+1}{2c^1} \int_{-1}^1 P_1(x) P_1(x) \, dx = \frac{3}{2c} \int_{-1}^1 x^2 dx = \frac{1}{c}.$$

Thus

$$u(r,\theta) = \frac{r}{c} P_1(\cos\theta) = \frac{r}{c} \cos\theta.$$

6. Referring to Example 1 in the text we have

$$R(r) = c_1 r^n \qquad \text{and} \qquad \Theta(\theta) = P_n(\cos\theta).$$

Now $\Theta(\pi/2) = 0$ implies that n is odd, so

$$u(r,\theta) = \sum_{n=0}^\infty A_{2n+1} r^{2n+1} P_{2n+1}(\cos\theta).$$

274

From

$$u(c, \theta) = f(\theta) = \sum_{n=0}^{\infty} A_{2n+1} c^{2n+1} P_{2n+1}(\cos \theta)$$

we see that

$$A_{2n+1} c^{2n+1} = (4n + 3) \int_0^{\pi/2} f(\theta) \sin \theta \, P_{2n+1}(\cos \theta) \, d\theta.$$

Thus

$$u(r, \theta) = \sum_{n=0}^{\infty} A_{2n+1} r^{2n+1} P_{2n+1}(\cos \theta)$$

where

$$A_{2n+1} = \frac{4n + 3}{c^{2n+1}} \int_0^{\pi/2} f(\theta) \sin \theta \, P_{2n+1}(\cos \theta) \, d\theta.$$

9. Checking the hint, we find

$$\frac{1}{r} \frac{\partial^2}{\partial r^2}(ru) = \frac{1}{r} \frac{\partial}{\partial r} \left[r \frac{\partial u}{\partial r} + u \right] = \frac{1}{r} \left[r \frac{\partial^2 u}{\partial r^2} + \frac{\partial u}{\partial r} + \frac{\partial u}{\partial r} \right] = \frac{\partial^2 u}{\partial r^2} + \frac{2}{r} \frac{\partial u}{\partial r}.$$

The partial differential equation then becomes

$$\frac{\partial^2}{\partial r^2}(ru) = r \frac{\partial u}{\partial t}.$$

Now, letting $ru(r, t) = v(r, t) + \psi(r)$, since the boundary condition is nonhomogeneous, we obtain

$$\frac{\partial^2}{\partial r^2}[v(r, t) + \psi(r)] = r \frac{\partial}{\partial t}\left[\frac{1}{r} v(r, t) + \psi(r) \right]$$

or

$$\frac{\partial^2 v}{\partial r^2} + \psi''(r) = \frac{\partial v}{\partial t}.$$

This differential equation will be homogeneous if $\psi''(r) = 0$ or $\psi(r) = c_1 r + c_2$. Now

$$u(r, t) = \frac{1}{r} v(r, t) + \frac{1}{r} \psi(r) \qquad \text{and} \qquad \frac{1}{r} \psi(r) = c_1 + \frac{c_2}{r}.$$

Since we want $u(r, t)$ to be bounded as r approaches 0, we require $c_2 = 0$. Then $\psi(r) = c_1 r$. When $r = 1$

$$u(1, t) = v(1, t) + \psi(1) = v(1, t) + c_1 = 100,$$

and we will have the homogeneous boundary condition $v(1, t) = 0$ when $c_1 = 100$. Consequently, $\psi(r) = 100r$. The initial condition

$$u(r, 0) = \frac{1}{r} v(r, 0) + \frac{1}{r} \psi(r) = \frac{1}{r} v(r, 0) + 100 = 0$$

implies $v(r, 0) = -100r$. We are thus led to solve the new boundary-value problem

$$\frac{\partial^2 v}{\partial r^2} = \frac{\partial v}{\partial t}, \quad 0 < r < 1, \quad t > 0,$$

$$v(1, t) = 0, \quad \lim_{r \to 0} \frac{1}{r} v(r, t) < \infty,$$

$$v(r, 0) = -100r.$$

Letting $v(r, t) = R(r)T(t)$ and using the separation constant $-\lambda$ we obtain

$$R'' + \lambda R = 0 \quad \text{and} \quad T' + \lambda T = 0.$$

Using $\lambda = \alpha^2 > 0$ we then have

$$R(r) = c_3 \cos \alpha r + c_4 \sin \alpha r \quad \text{and} \quad T(t) = c_5 e^{-\alpha^2 t}.$$

The boundary conditions are equivalent to $R(1) = 0$ and $\lim_{r \to 0} R(r)/r < \infty$. Since

$$\lim_{r \to 0} \frac{\cos \alpha r}{r}$$

does not exist we must have $c_3 = 0$. Then $R(r) = c_4 \sin \alpha r$, and $R(1) = 0$ implies $\alpha = n\pi$ for $n = 1$, 2, 3, ... Thus

$$v_n(r, t) = A_n e^{-n^2 \pi^2 t} \sin n\pi r$$

for $n = 1$, 2, 3, ... Using the condition $\lim_{r \to 0} R(r)/r < \infty$ it is easily shown that there are no eigenvalues for $\lambda = 0$, nor does setting the common constant to $-\lambda = \alpha^2$ when separating variables lead to any solutions. Now, by the Superposition Principle,

$$v(r, t) = \sum_{n=1}^{\infty} A_n e^{-n^2 \pi^2 t} \sin n\pi r.$$

The initial condition $v(r, 0) = -100r$ implies

$$-100r = \sum_{n=1}^{\infty} A_n \sin n\pi r.$$

This is a Fourier sine series and so

$$A_n = 2 \int_0^1 (-100r \sin n\pi r)\, dr = -200 \left[-\frac{r}{n\pi} \cos n\pi r \Big|_0^1 + \int_0^1 \frac{1}{n\pi} \cos n\pi r\, dr \right]$$

$$= -200 \left[-\frac{\cos n\pi}{n\pi} + \frac{1}{n^2 \pi^2} \sin n\pi r \Big|_0^1 \right] = -200 \left[-\frac{(-1)^n}{n\pi} \right] = \frac{(-1)^n 200}{n\pi}.$$

A solution of the problem is thus

$$u(r, t) = \frac{1}{r} v(r, t) + \frac{1}{r} \psi(r) = \frac{1}{r} \sum_{n=1}^{\infty} (-1)^n \frac{20}{n\pi} e^{-n^2 \pi^2 t} \sin n\pi r + \frac{1}{r} (100r)$$

$$= \frac{200}{\pi r} \sum_{n=1}^{\infty} \frac{(-1)^n}{n} e^{-n^2 \pi^2 t} \sin n\pi r + 100.$$

12. Proceeding as in Example 1 we obtain

$$\Theta(\theta) = P_n(\cos\theta) \qquad \text{and} \qquad R(r) = c_1 r^n + c_2 r^{-(n+1)}$$

so that

$$u(r, \theta) = \sum_{n=0}^{\infty} (A_n r^n + B_n r^{-(n+1)}) P_n(\cos\theta).$$

To satisfy $\lim_{r\to\infty} u(r,\theta) = -Er\cos\theta$ we must have $A_n = 0$ for $n = 2, 3, 4, \ldots$. Then

$$\lim_{r\to\infty} u(r,\theta) = -Er\cos\theta = A_0 \cdot 1 + A_1 r \cos\theta,$$

so $A_0 = 0$ and $A_1 = -E$. Thus

$$u(r, \theta) = -Er\cos\theta + \sum_{n=0}^{\infty} B_n r^{-(n+1)} P_n(\cos\theta).$$

Now

$$u(c, \theta) = 0 = -Ec\cos\theta + \sum_{n=0}^{\infty} B_n c^{-(n+1)} P_n(\cos\theta)$$

so

$$\sum_{n=0}^{\infty} B_n c^{-(n+1)} P_n(\cos\theta) = Ec\cos\theta$$

and

$$B_n c^{-(n+1)} = \frac{2n+1}{2} \int_0^{\pi} Ec\cos\theta\, P_n(\cos\theta)\sin\theta\, d\theta.$$

Now $\cos\theta = P_1(\cos\theta)$ so, for $n \neq 1$,

$$\int_0^{\pi} \cos\theta\, P_n(\cos\theta)\sin\theta\, d\theta = 0$$

by orthogonality. Thus $B_n = 0$ for $n \neq 1$ and

$$B_1 = \frac{3}{2} Ec^3 \int_0^{\pi} \cos^2\theta \sin\theta\, d\theta = Ec^3.$$

Therefore,

$$u(r, \theta) = -Er\cos\theta + Ec^3 r^{-2} \cos\theta.$$

3. The conditions $\Theta(0) = 0$ and $\Theta(\pi) = 0$ applied to $\Theta = c_1 \cos \alpha\theta + c_2 \sin \alpha\theta$ give $c_1 = 0$ and $\alpha = n$, $n = 1, 2, 3, \ldots$, respectively. Thus we have the Fourier sine-series coefficients

$$A_n = \frac{2}{\pi} \int_0^\pi u_0(\pi\theta - \theta^2) \sin n\theta \, d\theta = \frac{4u_0}{n^3\pi}[1 - (-1)^n].$$

Thus

$$u(r, \theta) = \frac{4u_0}{\pi} \sum_{n=1}^\infty \frac{1 - (-1)^n}{n^3} r^n \sin n\theta.$$

6. We solve

$$\frac{\partial^2 u}{\partial r^2} + \frac{1}{r}\frac{\partial u}{\partial r} + \frac{1}{r^2}\frac{\partial^2 u}{\partial \theta^2} = 0, \quad r > 1, \quad 0 < \theta < \pi,$$

$$u(r, 0) = 0, \quad u(r, \pi) = 0, \quad r > 1,$$

$$u(1, \theta) = f(\theta), \quad 0 < \theta < \pi.$$

Separating variables we obtain

$$\Theta(\theta) = c_1 \cos \alpha\theta + c_2 \sin \alpha\theta$$

$$R(r) = c_3 r^\alpha + c_4 r^{-\alpha}.$$

Applying the boundary conditions $\Theta(0) = 0$, and $\Theta(\pi) = 0$ gives $c_1 = 0$ and $\alpha = n$ for $n = 1, 2, 3, \ldots$. Assuming $f(\theta)$ to be bounded, we expect the solution $u(r, \theta)$ to also be bounded as $r \to \infty$. This requires that $c_3 = 0$. Therefore

$$u(r, \theta) = \sum_{n=1}^\infty A_n r^{-n} \sin n\theta.$$

From

$$u(1, \theta) = f(\theta) = \sum_{n=1}^\infty A_n \sin n\theta$$

we obtain

$$A_n = \frac{2}{\pi} \int_0^\pi f(\theta) \sin n\theta \, d\theta.$$

9. Letting the separation constant be $\lambda = \alpha^2$ and referring to Example 2 in Section 13.2 in the text we have

$$R(r) = c_1 J_0(\alpha r) + c_2 Y_0(\alpha r)$$

$$Z(z) = c_3 \cosh \alpha z + c_4 \sinh \alpha z$$

where $c_2 = 0$ and the positive eigenvalues λ_n are determined by $J_0(2\alpha) = 0$. From $Z'(0) = 0$ we obtain $c_4 = 0$. Then

$$u(r, z) = \sum_{n=1}^{\infty} A_n \cosh \alpha_n z J_0(\alpha_n r).$$

From

$$u(r, 4) = 50 = \sum_{n=1}^{\infty} A_n \cosh 4\alpha_n J_0(\alpha_n r)$$

we obtain (as in Example 1 of Section 13.1)

$$A_n \cosh 4\alpha_n = \frac{2(50)}{4J_1^2(2\alpha_n)} \int_0^2 r J_0(\alpha_n r)\, dr = \frac{50}{\alpha_n J_1(2\alpha_n)}.$$

Thus the temperature in the cylinder is

$$u(r, z) = 50 \sum_{n=1}^{\infty} \frac{\cosh \alpha_n z J_0(\alpha_n r)}{\alpha_n \cosh 4\alpha_n J_1(2\alpha_n)}.$$

12. Since

$$\frac{1}{r} \frac{\partial^2}{\partial r^2}(ru) = \frac{1}{r} \frac{\partial}{\partial r}\left[r \frac{\partial u}{\partial r} + u\right] = \frac{1}{r}\left[r \frac{\partial^2 u}{\partial r^2} + \frac{\partial u}{\partial r} + \frac{\partial u}{\partial r}\right] = \frac{\partial^2 u}{\partial r^2} + \frac{2}{r}\frac{\partial u}{r}$$

the DE becomes

$$\frac{1}{r} \frac{\partial^2}{\partial r^2}(ru) = \frac{\partial^2 u}{\partial t^2} \quad \text{or} \quad \frac{\partial^2}{\partial r^2}(ru) = r \frac{\partial^2 u}{\partial t^2}.$$

Letting $v(r, t) = ru(r, t)$ we obtain the boundary-value problem

$$\frac{\partial^2 v}{\partial r^2} = \frac{\partial^2 v}{\partial t^2}, \quad 0 < r < 1, \ t > 0$$

$$\left.\frac{\partial v}{\partial r}\right|_{r=1} - v(1, t) = 0, \quad t > 0$$

$$v(r, 0) = rf(r), \quad \left.\frac{\partial v}{\partial t}\right|_{t=0} = rg(r), \quad 0 < r < 1.$$

If we separate variables using $v(r, t) = R(r)T(t)$ and separation constant $-\lambda$ then we obtain

$$\frac{R''}{R} = \frac{T''}{T} = -\lambda$$

so that

$$R'' + \lambda R = 0$$

$$T'' + \lambda T = 0.$$

Letting $\lambda = \alpha^2 > 0$ and solving the DEs we get

$$R(r) = c_1 \cos \alpha r + c_2 \sin \alpha r$$

$$T(t) = c_3 \cos \alpha t + c_4 \sin \alpha t.$$

Since $u(r, t) = v(r, t)/r$, in order to insure boundedness at $r = 0$ we define $c_1 = 0$. Then $R(r) = c_2 \sin \alpha r$. Now the boundary condition $R'(1) - R(1) = 0$ implies $\alpha \cos \alpha - \sin \alpha = 0$. Thus, the eigenvalues λ_n are determined by the positive solutions of $\tan \alpha = \alpha$. We now have

$$v_n(r, t) = (A_n \cos \alpha_n t + B_n \sin \alpha_n t) \sin \alpha_n r.$$

For the eigenvalue $\lambda = 0$,

$$R(r) = c_1 r + c_2 \qquad \text{and} \qquad T(t) = c_3 t + c_4,$$

and boundedness at $r = 0$ implies $c_2 = 0$. We then take

$$v_0(r, t) = A_0 tr + B_0 r$$

so that

$$v(r, t) = A_0 tr + B_0 r + \sum_{n=1}^{\infty} (a_n \cos \alpha_n t + B_n \sin \alpha_n t) \sin \alpha_n r.$$

Now

$$v(r, 0) = rf(r) = B_0 r + \sum_{n=1}^{\infty} A_n \sin \alpha_n r.$$

Since $\{r, \sin \alpha_n r\}$ is an orthogonal set on $[0, 1]$,

$$\int_0^1 r \sin \alpha_n r \, dr = 0 \qquad \text{and} \qquad \int_0^1 \sin \alpha_n r \sin \alpha_n r \, dr = 0$$

for $m \neq n$. Therefore

$$\int_0^1 r^2 f(r) \, dr = B_0 \int_0^1 r^2 \, dr = \frac{1}{3} B_0$$

and

$$B_0 = 3 \int_0^1 r^2 f(r) \, dr.$$

Also

$$\int_0^1 rf(r) \sin \alpha_n r \, dr = A_n \int_0^1 \sin^2 \alpha_n r \, dr$$

and

$$A_n = \frac{\int_0^1 rf(r) \sin \alpha_n r \, dr}{\int_0^1 \sin^2 \alpha_n r \, dr}.$$

Now

$$\int_0^1 \sin^2 \alpha_n r \, dr = \frac{1}{2} \int_0^1 (1 - \cos 2\alpha_n r) \, dr = \frac{1}{2} \left[1 - \frac{\sin 2\alpha_n}{2\alpha_n} \right] = \frac{1}{2} [1 - \cos^2 \alpha_n].$$

Since $\tan \alpha_n = \alpha_n$,

$$1 + \alpha_n^2 = 1 + \tan^2 \alpha_n = \sec^2 \alpha_n = \frac{1}{\cos^2 \alpha_n}$$

and

$$\cos^2 \alpha_n = \frac{1}{1 + \alpha_n^2}.$$

Then

$$\int_0^1 \sin^2 \alpha_n r \, dr = \frac{1}{2}\left[1 - \frac{1}{1+\alpha_n^2}\right] = \frac{\alpha_n^2}{2(1+\alpha_n^2)}$$

and

$$A_n = \frac{2(1+\alpha_n^2)}{\alpha_n^2}\int_0^1 r f(r) \sin \alpha_n r \, dr.$$

Similarly, setting

$$\frac{\partial v}{\partial t}\bigg|_{t=0} = rg(r) = A_0 r + \sum_{n=1}^{\infty} B_n \alpha_n \sin \alpha_n r$$

we obtain

$$A_0 = 3\int_0^1 r^2 g(r)\, dr$$

and

$$B_n = \frac{2(1+\alpha_n^2)}{\alpha_n^3}\int_0^1 rg(r) \sin \alpha_n r \, dr.$$

Therefore, since $v(r,t) = ru(r,t)$ we have

$$u(r,t) = A_0 t + B_0 + \sum_{n=1}^{\infty}(A_n \cos \alpha_n t + B_n \sin \alpha_n t)\frac{\sin \alpha_n r}{r},$$

where the α_n are solutions of $\tan \alpha = \alpha$ and

$$A_0 = 3\int_0^1 r^2 g(r)\, dr$$

$$B_0 = 3\int_0^1 r^2 f(r)\, dr$$

$$A_n = \frac{2(1+\alpha_n^2)}{\alpha_n^2}\int_0^1 r f(r) \sin \alpha_n r \, dr$$

$$B_n = \frac{2(1+\alpha_n^2)}{\alpha_n^3}\int_0^1 rg(r) \sin \alpha_n r \, dr$$

for $n = 1, 2, 3, \ldots$.

15. We use the Superposition Principle for Laplace's equation discussed in Section 12.5 and shown schematically in Figure 12.15 in the text. That is,

Solution u = Solution u_1 of Problem 1 + Solution u_2 of Problem 2,

where in Problem 1 the boundary condition on the top and bottom of the cylinder is $u = 0$, while on the lateral surface $r = c$ it is $u = h(z)$, and in Problem 2 the boundary condition on the top of the cylinder $z = L$ is $u = f(r)$, on the bottom $z = 0$ it is $u = g(r)$, and on the lateral surface $r = c$ it is $u = 0$.

Chapter 13 in Review

Using λ as a separation constant we have

$$\frac{R'' + \frac{1}{r}R'}{R} = -\frac{Z''}{Z} = \lambda,$$

so

$$rR'' + R' - \lambda r R = 0 \qquad \text{and} \qquad Z'' + \lambda Z = 0.$$

The DE in Z, together with the boundary conditions $Z(0) = 0$ and $Z(L) = 0$ is a Sturm-Liouville problem. Letting $\lambda = \alpha^2 > 0$ we note that the above DE in R is a modified parametric Bessel equation which is discussed in Section 6.3 in the text. Also, we have $Z(z) = c_1 \cos \alpha z + c_2 \sin \alpha z$. The boundary conditions imply $c_1 = 0$ and $\sin \alpha L = 0$. Thus, $\alpha_n = n\pi/L$, $n = 1, 2, 3, \ldots$, so $\lambda_n = n^2 \pi^2 / L^2$ and

$$R(r) = c_3 I_0 \left(\frac{n\pi}{L} r \right) + c_4 K_0 \left(\frac{n\pi}{L} r \right).$$

Now boundedness at $r = 0$ implies $c_4 = 0$, so $R(r) = c_3 I_0(n\pi r/L)$ and

$$u_1(r,z) = \sum_{n=1}^{\infty} A_n I_0 \left(\frac{n\pi}{L} r \right) \sin \left(\frac{n\pi}{L} z \right).$$

At $r = c$ for $0 < z < L$ we have

$$h(z) = u_1(c,z) = \sum_{n=1}^{\infty} A_n I_0 \left(\frac{n\pi}{L} c \right) \sin \left(\frac{n\pi}{L} z \right)$$

which gives

$$A_n = \frac{2}{L I_0(n\pi c/L)} \int_0^L h(z) \sin \left(\frac{n\pi}{L} z \right) dz.$$

In this case we use $-\lambda$ as a separation constant which leads to

$$\frac{R'' + \frac{1}{r}R'}{R} = -\frac{Z''}{Z} = -\lambda,$$

so

$$rR'' + R' + \lambda r R = 0 \qquad \text{and} \qquad Z'' - \lambda Z = 0.$$

The DE in R is a parametric Bessel equation. Using $\lambda = \alpha^2$ we find $R(r) = c_1 J_0(\alpha r) + c_2 Y_0(\alpha r)$. Boundedness at $r = 0$ implies $c_2 = 0$ so $R(r) = c_3 J_0(\alpha r)$. The boundary condition $R(c) = 0$ then gives the defining equation for the eigenvalues: $J_0(\alpha c) = 0$. Let $\lambda_n = \alpha_n^2$ where $\alpha_n c = x_n$ are the roots. The solution of the DE in Z is $Z(z) = c_4 \cosh \alpha_n z + c_5 \sinh \alpha_n z$, so

$$u_2(r,z) = \sum_{n=1}^{\infty} (B_n \cosh \alpha_n z + C_n \sinh \alpha_n z) J_0(\alpha_n r).$$

At $z = 0$, for $0 < r < c$, we have

$$f(r) = u_2(r,0) = \sum_{n=1}^{\infty} B_n J_0(\alpha_n r),$$

so

$$B_n = \frac{2}{c^2 J_1^2(\alpha_n c)} \int_0^c r f(r) J_0(\alpha_n r)\, dr.$$

At $z = L$, for $0 < r < c$, we have

$$g(r) = u_2(r, L) = \sum_{n=1}^{\infty} (B_n \cosh \alpha_n L + C_n \sinh \alpha_n L) J_0(\alpha_n r),$$

so

$$B_n \cosh \alpha_n L + C_n \sinh \alpha_n L = \frac{2}{c^2 J_1^2(\alpha_n c)} \int_0^c r g(r) J_0(\alpha_n r)\, dr$$

and

$$C_n = -B_n \frac{\cosh \alpha_n L}{\sinh \alpha_n L} + \frac{2}{c^2 (\sinh \alpha_n L) J_1^2(\alpha_n c)} \int_0^c r g(r) J_0(\alpha_n r)\, dr.$$

By the Superposition Principle the solution of the original problem is

$$u(r, z) = u_1(r, z) + u_2(r, z).$$

14

Integral Transform Method

Error Function

The terminology and concepts listed below provide an outline of the main ideas encountered in this section. These can be useful when preparing for a quiz or test.

Terminology and Concepts

- error function
- complementary error function

The basic skills listed below summarize the more mechanical types of problems encountered in the exercise set for this section.

Basic Skills

- find the Laplace transforms of the error function and complementary error function

Exercises 14.1   *Hints, Suggestions, Solutions, and Examples*

3. By the first translation theorem,

$$\mathscr{L}\left\{e^t\operatorname{erf}(\sqrt{t})\right\} = \mathscr{L}\left\{\operatorname{erf}(\sqrt{t})\right\}\bigg|_{s\to s-1} = \frac{1}{s\sqrt{s+1}}\bigg|_{s\to s-1} = \frac{1}{\sqrt{s}\,(s-1)}.$$

6. We first compute

$$\frac{\sinh a\sqrt{s}}{s\sinh\sqrt{s}} = \frac{e^{a\sqrt{s}} - e^{-a\sqrt{s}}}{s(e^{\sqrt{s}} - e^{-\sqrt{s}})} = \frac{e^{(a-1)\sqrt{s}} - e^{-(a+1)\sqrt{s}}}{s(1 - e^{-2\sqrt{s}})}$$

$$= \frac{e^{(a-1)\sqrt{s}}}{s}\left[1 + e^{-2\sqrt{s}} + e^{-4\sqrt{s}} + \cdots\right] - \frac{e^{-(a+1)\sqrt{s}}}{s}\left[1 + e^{-2\sqrt{s}} + e^{-4\sqrt{s}} + \cdots\right]$$

$$= \left[\frac{e^{-(1-a)\sqrt{s}}}{s} + \frac{e^{-(3-a)\sqrt{s}}}{s} + \frac{e^{-(5-a)\sqrt{s}}}{s} + \cdots\right]$$

$$- \left[\frac{e^{-(1+a)\sqrt{s}}}{s} + \frac{e^{-(3+a)\sqrt{s}}}{s} + \frac{e^{-(5+a)\sqrt{s}}}{s} + \cdots\right]$$

$$= \sum_{n=0}^{\infty}\left[\frac{e^{-(2n+1-a)\sqrt{s}}}{s} - \frac{e^{-(2n+1+a)\sqrt{s}}}{s}\right].$$

Then

$$\mathscr{L}^{-1}\left\{\frac{\sinh a\sqrt{s}}{s\sinh\sqrt{s}}\right\} = \sum_{n=0}^{\infty}\left[\mathscr{L}^{-1}\left\{\frac{e^{-(2n+1-a)\sqrt{s}}}{s}\right\} - \mathscr{L}^{-1}\left\{-\frac{e^{-(2n+1+a)\sqrt{s}}}{s}\right\}\right]$$

$$= \sum_{n=0}^{\infty}\left[\text{erfc}\left(\frac{2n+1-a}{2\sqrt{t}}\right) - \text{erfc}\left(\frac{2n+1+a}{2\sqrt{t}}\right)\right]$$

$$= \sum_{n=0}^{\infty}\left(\left[1 - \text{erf}\left(\frac{2n+1-a}{2\sqrt{t}}\right)\right] - \left[1 - \text{erf}\left(\frac{2n+1+a}{2\sqrt{t}}\right)\right]\right)$$

$$= \sum_{n=0}^{\infty}\left[\text{erf}\left(\frac{2n+1+a}{2\sqrt{t}}\right) - \text{erf}\left(\frac{2n+1-a}{2\sqrt{t}}\right)\right].$$

9. $\displaystyle\int_a^b e^{-u^2}\,du = \int_a^0 e^{-u^2}\,du + \int_0^b e^{-u^2}\,du = \int_0^b e^{-u^2}\,du - \int_0^a e^{-u^2}\,du$

$$= \frac{\sqrt{\pi}}{2}\text{erf}(b) - \frac{\sqrt{\pi}}{2}\text{erf}(a) = \frac{\sqrt{\pi}}{2}[\text{erf}(b) - \text{erf}(a)]$$

Section 14.2 Applications of the Laplace Transform

The terminology and concepts listed below provide an outline of the main ideas encountered in this section. These can be useful when preparing for a quiz or test.

Terminology and Concepts

- Laplace transform of a function of two variables
- Laplace transform of partial derivatives

The basic skills listed below summarize the more mechanical types of problems encountered in the exercise set for this section.

Basic Skills

- use the Laplace transform to solve a BVP

Exercises 14.2 *Hints, Suggestions, Solutions, and Examples*

3. The solution of

$$a^2 \frac{d^2 U}{dx^2} - s^2 U = 0$$

is in this case

$$U(x, s) = c_1 e^{-(x/a)s} + c_2 e^{(x/a)s}.$$

Since $\lim_{x \to \infty} u(x, t) = 0$ we have $\lim_{x \to \infty} U(x, s) = 0$. Thus $c_2 = 0$ and

$$U(x, s) = c_1 e^{-(x/a)s}.$$

If $\mathscr{L}\{u(0, t)\} = \mathscr{L}\{f(t)\} = F(s)$ then $U(0, s) = F(s)$. From this we have $c_1 = F(s)$ and

$$U(x, s) = F(s) e^{-(x/a)s}.$$

Hence, by the second translation theorem,

$$u(x, t) = f\left(t - \frac{x}{a}\right) \mathscr{U}\left(t - \frac{x}{a}\right).$$

6. Transforming the partial DE gives

$$\frac{d^2 U}{dx^2} - s^2 U = -\frac{\omega}{s^2 + \omega^2} \sin \pi x.$$

Using undetermined coefficients we obtain

$$U(x, s) = c_1 \cosh sx + c_2 \sinh sx + \frac{\omega}{(s^2 + \pi^2)(s^2 + \omega^2)} \sin \pi x.$$

The transformed boundary conditions $U(0, s) = 0$ and $U(1, s) = 0$ give, in turn, $c_1 = 0$ and $c_2 = 0$. Therefore

$$U(x, s) = \frac{\omega}{(s^2 + \pi^2)(s^2 + \omega^2)} \sin \pi x$$

and

$$u(x, t) = \omega \sin \pi x \, \mathscr{L}^{-1}\left\{\frac{1}{(s^2 + \pi^2)(s^2 + \omega^2)}\right\}$$

$$= \frac{\omega}{\omega^2 - \pi^2} \sin \pi x \, \mathscr{L}^{-1}\left\{\frac{1}{\pi} \frac{\pi}{s^2 + \pi^2} - \frac{1}{\omega} \frac{\omega}{s^2 + \omega^2}\right\}$$

$$= \frac{\omega}{\pi(\omega^2 - \pi^2)} \sin \pi t \sin \pi x - \frac{1}{\omega^2 - \pi^2} \sin \omega t \sin \pi x.$$

9. Transforming the partial DE gives

$$\frac{d^2U}{dx^2} - s^2U = -sxe^{-x}.$$

Using undetermined coefficients we obtain

$$U(x,s) = c_1e^{-sx} + c_2e^{sx} - \frac{2s}{(s^2-1)^2}e^{-x} + \frac{s}{s^2-1}xe^{-x}.$$

The transformed boundary conditions $\lim_{x\to\infty} U(x,s) = 0$ and $U(0,s) = 0$ give, in turn, $c_2 = 0$ and $c_1 = 2s/(s^2-1)^2$. Therefore

$$U(x,s) = \frac{2s}{(s^2-1)^2}e^{-sx} - \frac{2s}{(s^2-1)^2}e^{-x} + \frac{s}{s^2-1}xe^{-x}.$$

From entries (13) and (26) in the Table of Laplace transforms we obtain

$$u(x,t) = \mathscr{L}^{-1}\left\{\frac{2s}{(s^2-1)^2}e^{-sx} - \frac{2s}{(s^2-1)^2}e^{-x} + \frac{s}{s^2-1}xe^{-x}\right\}$$

$$= 2(t-x)\sinh(t-x)\,\mathscr{U}(t-x) - te^{-x}\sinh t + xe^{-x}\cosh t.$$

12. We use

$$U(x,s) = c_1e^{-\sqrt{s}\,x} + c_2e^{\sqrt{s}\,x} + \frac{u_1x}{s}.$$

The condition $\lim_{x\to\infty} u(x,t)/x = u_1$ implies $\lim_{x\to\infty} U(x,s)/x = u_1/s$, so we define $c_2 = 0$. Then

$$U(x,s) = c_1e^{-\sqrt{s}\,x} + \frac{u_1x}{s}.$$

From $U(0,s) = u_0/s$ we obtain $c_1 = u_0/s$. Hence

$$U(x,s) = u_0\frac{e^{-\sqrt{s}\,x}}{s} + \frac{u_1x}{s}$$

and

$$u(x,t) = u_0\mathscr{L}^{-1}\left\{\frac{e^{-x\sqrt{s}}}{s}\right\} + u_1x\,\mathscr{L}^{-1}\left\{\frac{1}{s}\right\} = u_0\,\text{erfc}\left(\frac{x}{2\sqrt{t}}\right) + u_1x.$$

15. We use

$$U(x,s) = c_1e^{-\sqrt{s}\,x} + c_2e^{\sqrt{s}\,x}.$$

The condition $\lim_{x\to\infty} u(x,t) = 0$ implies $\lim_{x\to\infty} U(x,s) = 0$, so we define $c_2 = 0$. Hence

$$U(x,s) = c_1e^{-\sqrt{s}\,x}.$$

The transform of $u(0,t) = f(t)$ is $U(0,s) = F(s)$. Therefore

$$U(x,s) = F(s)e^{-\sqrt{s}\,x}$$

and

$$u(x,t) = \mathscr{L}^{-1}\left\{F(s)e^{-x\sqrt{s}}\right\} = \frac{x}{2\sqrt{\pi}}\int_0^t \frac{f(t-\tau)e^{-x^2/4\tau}}{\tau^{3/2}}\,d\tau.$$

18. The solution of the transformed equation

$$\frac{d^2U}{dx^2} - sU = -100$$

by undetermined coefficients is

$$U(x,s) = c_1 e^{\sqrt{s}\,x} + c_2 e^{-\sqrt{s}\,x} + \frac{100}{s}.$$

From the fact that $\lim_{x\to\infty} U(x,s) = 100/s$ we see that $c_1 = 0$. Thus

$$U(x,s) = c_2 e^{-\sqrt{s}\,x} + \frac{100}{s}. \tag{1}$$

Now the transform of the boundary condition at $x = 0$ is

$$U(0,s) = 20\left[\frac{1}{s} - \frac{1}{s}e^{-s}\right].$$

It follows from (1) that

$$\frac{20}{s} - \frac{20}{s}e^{-s} = c_2 + \frac{100}{s} \qquad \text{or} \qquad c_2 = -\frac{80}{s} - \frac{20}{s}e^{-s}$$

and so

$$U(x,s) = \left(-\frac{80}{s} - \frac{20}{s}e^{-s}\right)e^{-\sqrt{s}\,x} + \frac{100}{s}$$

$$= \frac{100}{s} - \frac{80}{s}e^{-\sqrt{s}\,x} - \frac{20}{s}e^{-\sqrt{s}\,x}e^{-s}.$$

Thus

$$u(x,t) = 100\,\mathscr{L}^{-1}\left\{\frac{1}{s}\right\} - 80\,\mathscr{L}^{-1}\left\{\frac{e^{-\sqrt{s}\,x}}{s}\right\} - 20\,\mathscr{L}^{-1}\left\{\frac{e^{-\sqrt{s}\,x}}{s}e^{-s}\right\}$$

$$= 100 - 80\,\mathrm{erfc}\left(x/2\sqrt{t}\right) - 20\,\mathrm{erfc}\left(x/2\sqrt{t-1}\right)\mathscr{U}(t-1).$$

21. The solution of

$$\frac{d^2U}{dx^2} - sU = -u_0 - u_0\sin\frac{\pi}{L}x$$

is

$$U(x,s) = c_1\cosh(\sqrt{s}\,x) + c_2\sinh(\sqrt{s}\,x) + \frac{u_0}{s} + \frac{u_0}{s+\pi^2/L^2}\sin\frac{\pi}{L}x.$$

The transformed boundary conditions $U(0,s) = u_0/s$ and $U(L,s) = u_0/s$ give, in turn, $c_1 = 0$ and $c_2 = 0$. Therefore

$$U(x,s) = \frac{u_0}{s} + \frac{u_0}{s+\pi^2/L^2}\sin\frac{\pi}{L}x$$

and

$$u(x,t) = u_0\,\mathscr{L}^{-1}\left\{\frac{1}{s}\right\} + u_0\,\mathscr{L}^{-1}\left\{\frac{1}{s+\pi^2/L^2}\right\}\sin\frac{\pi}{L}x = u_0 + u_0 e^{-\pi^2 t/L^2}\sin\frac{\pi}{L}x.$$

24. We use

$$c(x, s) = c_1 \cosh \sqrt{\frac{s}{D}}\, x + c_2 \sinh \sqrt{\frac{s}{D}}\, x.$$

The transform of the two boundary conditions are $c(0, s) = c_0/s$ and $c(1, s) = c_0/s$. From these conditions we obtain $c_1 = c_0/s$ and

$$c_2 = c_0(1 - \cosh \sqrt{s/D}\,)/s \sinh \sqrt{s/D}\,.$$

Therefore

$$c(x, s) = c_0 \left[\frac{\cosh \sqrt{s/D}\, x}{s} + \frac{(1 - \cosh \sqrt{s/D}\,)}{s \sinh \sqrt{s/D}} \sinh \sqrt{s/D}\, x \right]$$

$$= c_0 \left[\frac{\sinh \sqrt{s/D}\,(1 - x)}{s \sinh \sqrt{s/D}} + \frac{\sin \sqrt{s/D}\, x}{s \sinh \sqrt{s/D}} \right]$$

$$= c_0 \left[\frac{e^{\sqrt{s/D}\,(1-x)} - e^{-\sqrt{s/D}\,(1-x)}}{s(e^{\sqrt{s/D}} - e^{-\sqrt{s/D}}\,)} + \frac{e^{\sqrt{s/D}\, x} - e^{-\sqrt{s/D}\, x}}{s(e^{\sqrt{s/D}} - e^{-\sqrt{s/D}}\,)} \right]$$

$$= c_0 \left[\frac{e^{-\sqrt{s/D}\, x} - e^{-\sqrt{s/D}\,(2-x)}}{s(1 - e^{-2\sqrt{s/D}}\,)} + \frac{e^{\sqrt{s/D}\,(x-1)} - e^{-\sqrt{s/D}\,(x+1)}}{s(1 - e^{-2\sqrt{s/D}}\,)} \right]$$

$$= c_0 \frac{(e^{-\sqrt{s/D}\, x} - e^{-\sqrt{s/D}\,(2-x)})}{s} \left(1 + e^{-2\sqrt{s/D}} + e^{-4\sqrt{s/D}} + \cdots \right)$$

$$+ c_0 \frac{(e^{\sqrt{s/D}\,(x-1)} - e^{-\sqrt{s/D}\,(x+1)})}{s} \left(1 + e^{-2\sqrt{s/D}} + e^{-4\sqrt{s/D}} + \cdots \right)$$

$$= c_0 \sum_{n=0}^{\infty} \left[\frac{e^{-(2n+x)\sqrt{s/D}}}{s} - \frac{e^{-(2n+2-x)\sqrt{s/D}}}{s} \right]$$

$$+ c_0 \sum_{n=0}^{\infty} \left[\frac{e^{-(2n+1-x)\sqrt{s/D}}}{s} - \frac{e^{-(2n+1+x)\sqrt{s/D}}}{s} \right]$$

and so

$$c(x, t) = c_0 \sum_{n=0}^{\infty} \left[\mathscr{L}^{-1}\left\{ \frac{e^{-\frac{(2n+x)}{\sqrt{D}}\sqrt{s}}}{s} \right\} - \mathscr{L}^{-1}\left\{ \frac{e^{-\frac{(2n+2-x)}{\sqrt{D}}\sqrt{s}}}{s} \right\} \right]$$

$$+ c_0 \sum_{n=0}^{\infty} \left[\mathscr{L}^{-1}\left\{ \frac{e^{-\frac{(2n+1-x)}{\sqrt{D}}\sqrt{s}}}{s} \right\} - \mathscr{L}^{-1}\left\{ \frac{e^{-\frac{(2n+1+x)}{\sqrt{D}}\sqrt{s}}}{s} \right\} \right]$$

$$= c_0 \sum_{n=0}^{\infty} \left[\text{erfc}\left(\frac{2n+x}{2\sqrt{Dt}} \right) - \text{erfc}\left(\frac{2n+2-x}{2\sqrt{Dt}} \right) \right]$$

$$+ c_0 \sum_{n=0}^{\infty} \left[\text{erfc}\left(\frac{2n+1-x}{2\sqrt{Dt}} \right) - \text{erfc}\left(\frac{2n+1+x}{2\sqrt{Dt}} \right) \right].$$

Now using $\text{erfc}(x) = 1 - \text{erf}(x)$ we get

$$c(x,t) = c_0 \sum_{n=0}^{\infty} \left[\text{erf}\left(\frac{2n+2-x}{2\sqrt{Dt}} \right) - \text{erf}\left(\frac{2n+x}{2\sqrt{Dt}} \right) \right]$$

$$+ c_0 \sum_{n=0}^{\infty} \left[\text{erf}\left(\frac{2n+1+x}{2\sqrt{Dt}} \right) - \text{erf}\left(\frac{2n+1-x}{2\sqrt{Dt}} \right) \right].$$

27. (a) We use

$$U(x,s) = c_1 e^{-(s/a)x} + c_2 e^{(s/a)x} + \frac{v_0^2 F_0}{(a^2 - v_0^2)s^2} e^{-(s/v_0)x}.$$

The condition $\lim_{x \to \infty} u(x,t) = 0$ implies $\lim_{x \to \infty} U(x,s) = 0$, so we must define $c_2 = 0$. Consequently

$$U(x,s) = c_1 e^{-(s/a)x} + \frac{v_0^2 F_0}{(a^2 - v_0^2)s^2} e^{-(s/v_0)x}.$$

The remaining boundary condition transforms into $U(0,s) = 0$. From this we find

$$c_1 = -v_0^2 F_0 / (a^2 - v_0^2)s^2.$$

Therefore, by the second translation theorem

$$U(x,s) = -\frac{v_0^2 F_0}{(a^2 - v_0^2)s^2} e^{-(s/a)x} + \frac{v_0^2 F_0}{(a^2 - v_0^2)s^2} e^{-(s/v_0)x}$$

and

$$u(x,t) = \frac{v_0^2 F_0}{a^2 - v_0^2} \left[\mathscr{L}^{-1}\left\{ \frac{e^{-(x/v_0)s}}{s^2} \right\} - \mathscr{L}^{-1}\left\{ \frac{e^{-(x/a)s}}{s^2} \right\} \right]$$

$$= \frac{v_0^2 F_0}{a^2 - v_0^2} \left[\left(t - \frac{x}{v_0} \right) \mathscr{U}\left(t - \frac{x}{v_0} \right) - \left(t - \frac{x}{a} \right) \mathscr{U}\left(t - \frac{x}{a} \right) \right].$$

(b) In the case when $v_0 = a$ the solution of the transformed equation is

$$U(x,s) = c_1 e^{-(s/a)x} + c_2 e^{(s/a)x} - \frac{F_0}{2as} x e^{-(s/a)x}.$$

The usual analysis then leads to $c_1 = 0$ and $c_2 = 0$. Therefore

$$U(x,s) = -\frac{F_0}{2as} x e^{-(s/a)x}$$

and

$$u(x,t) = -\frac{xF_0}{2a}\,\mathscr{L}^{-1}\left\{\frac{e^{-(x/a)s}}{s}\right\} = -\frac{xF_0}{2a}\,\mathscr{U}\left(t - \frac{x}{a}\right).$$

Section 14.3   Fourier Integral

The terminology and concepts listed below provide an outline of the main ideas encountered in this section. These can be useful when preparing for a quiz or test.

Terminology and Concepts

- Fourier integral representation of a function
- convergence conditions for a Fourier integral
- cosine integral representation of an even function
- sine integral representation of an odd function
- complex or exponential form of a Fourier integral

The basic skills listed below summarize the more mechanical types of problems encountered in the exercise set for this section.

Basic Skills

- find the Fourier integral representation of a function defined on $(-\infty, \infty)$
- find the cosine integral representation of an even function
- find the sine integral representation of an odd function
- find both the cosine and sine integral representations of a function defined on $(0, \infty)$

Exercises 14.3   *Hints, Suggestions, Solutions, and Examples*

3. From formulas (5) and (6) in the text,

$$A(\alpha) = \int_0^3 x\cos\alpha x\,dx = \frac{x\sin\alpha x}{\alpha}\bigg|_0^3 - \frac{1}{\alpha}\int_0^3 \sin\alpha x\,dx$$

$$= \frac{3\sin 3\alpha}{\alpha} + \frac{\cos\alpha x}{\alpha^2}\bigg|_0^3 = \frac{3\alpha\sin 3\alpha + \cos 3\alpha - 1}{\alpha^2}$$

291

and

$$B(\alpha) = \int_0^3 x \sin \alpha x \, dx = -\frac{x \cos \alpha x}{\alpha} \Big|_0^3 + \frac{1}{\alpha} \int_0^3 \cos \alpha x \, dx$$

$$= -\frac{3 \cos 3\alpha}{\alpha} + \frac{\sin \alpha x}{\alpha^2} \Big|_0^3 = \frac{\sin 3\alpha - 3\alpha \cos 3\alpha}{\alpha^2}.$$

Hence

$$f(x) = \frac{1}{\pi} \int_0^\infty \frac{(3\alpha \sin 3\alpha + \cos 3\alpha - 1) \cos \alpha x + (\sin 3\alpha - 3\alpha \cos 3\alpha) \sin \alpha x}{\alpha^2} \, d\alpha$$

$$= \frac{1}{\pi} \int_0^\infty \frac{3\alpha(\sin 3\alpha \cos \alpha x - \cos 3\alpha \sin \alpha x) + \cos 3\alpha \cos \alpha x + \sin 3\alpha \sin \alpha x - \cos \alpha x}{\alpha^2} \, d\alpha$$

$$= \frac{1}{\pi} \int_0^\infty \frac{3\alpha \sin \alpha(3 - x) + \cos \alpha(3 - x) - \cos \alpha x}{\alpha^2} \, d\alpha.$$

6. From formulas (5) and (6) in the text,

$$A(\alpha) = \int_{-1}^1 e^x \cos \alpha x \, dx$$

$$= \frac{e(\cos \alpha + \alpha \sin \alpha) - e^{-1}(\cos \alpha - \alpha \sin \alpha)}{1 + \alpha^2}$$

$$= \frac{2(\sinh 1) \cos \alpha - 2\alpha(\cosh 1) \sin \alpha}{1 + \alpha^2}$$

and

$$B(\alpha) = \int_{-1}^1 e^x \sin \alpha x \, dx$$

$$= \frac{e(\sin \alpha - \alpha \cos \alpha) - e^{-1}(-\sin \alpha - \alpha \cos \alpha)}{1 + \alpha^2}$$

$$= \frac{2(\cosh 1) \sin \alpha - 2\alpha(\sinh 1) \cos \alpha}{1 + \alpha^2}.$$

Hence

$$f(x) = \frac{1}{\pi} \int_0^\infty [A(\alpha) \cos \alpha x + B(\alpha) \sin \alpha x] \, d\alpha.$$

9. The function is even. Thus from formula (9) in the text

$$A(\alpha) = \int_0^\pi x \cos \alpha x \, dx = \frac{x \sin \alpha x}{\alpha} \Big|_0^\pi - \frac{1}{\alpha} \int_0^\pi \sin \alpha x \, dx$$

$$= \frac{\pi \alpha \sin \pi \alpha}{\alpha} + \frac{1}{\alpha^2} \cos \alpha x \Big|_0^\pi = \frac{\pi \alpha \sin \pi \alpha + \cos \pi \alpha - 1}{\alpha^2}.$$

Hence from formula (8) in the text

$$f(x) = \frac{2}{\pi} \int_0^\infty \frac{(\pi \alpha \sin \pi \alpha + \cos \pi \alpha - 1) \cos \alpha x}{\alpha^2} \, d\alpha.$$

12. The function is odd. Thus from formula (11) in the text

$$B(\alpha) = \int_0^\infty xe^{-x} \sin \alpha x \, dx.$$

Now recall

$$\mathscr{L}\{t \sin kt\} = -\frac{d}{ds} \mathscr{L}\{\sin kt\} = 2ks/(s^2 + k^2)^2.$$

If we set $s = 1$ and $k = \alpha$ we obtain

$$B(\alpha) = \frac{2\alpha}{(1 + \alpha^2)^2}.$$

Hence from formula (10) in the text

$$f(x) = \frac{4}{\pi} \int_0^\infty \frac{\alpha \sin \alpha x}{(1 + \alpha^2)^2} \, d\alpha.$$

15. For the cosine integral,

$$A(\alpha) = \int_0^\infty xe^{-2x} \cos \alpha x \, dx.$$

But we know

$$\mathscr{L}\{t \cos kt\} = -\frac{d}{ds} \frac{s}{(s^2 + k^2)} = \frac{(s^2 - k^2)}{(s^2 + k^2)^2}.$$

If we set $s = 2$ and $k = \alpha$ we obtain

$$A(\alpha) = \frac{4 - \alpha^2}{(4 + \alpha^2)^2}.$$

Hence

$$f(x) = \frac{2}{\pi} \int_0^\infty \frac{(4 - \alpha^2) \cos \alpha x}{(4 + \alpha^2)^2} \, d\alpha.$$

For the sine integral,

$$B(\alpha) = \int_0^\infty xe^{-2x} \sin \alpha x \, dx.$$

From Problem 12, we know

$$\mathscr{L}\{t \sin kt\} = \frac{2ks}{(s^2 + k^2)^2}.$$

If we set $s = 2$ and $k = \alpha$ we obtain

$$B(\alpha) = \frac{4\alpha}{(4 + \alpha^2)^2}.$$

Hence

$$f(x) = \frac{8}{\pi} \int_0^\infty \frac{\alpha \sin \alpha x}{(4 + \alpha^2)^2} \, d\alpha.$$

293

18. From the formula for sine integral of $f(x)$ we have

$$f(x) = \frac{2}{\pi} \int_0^\infty \left(\int_0^\infty f(x) \sin \alpha x \, dx \right) \sin \alpha x \, dx$$

$$= \frac{2}{\pi} \left[\int_0^1 1 \cdot \sin \alpha x \, d\alpha + \int_1^\infty 0 \cdot \sin \alpha x \, d\alpha \right]$$

$$= \frac{2}{\pi} \frac{(- \cos \alpha x)}{x} \bigg|_0^1 = \frac{2}{\pi} \frac{1 - \cos x}{x}.$$

Section 14.4   Fourier Transforms

The terminology and concepts listed below provide an outline of the main ideas encountered in this section. These can be useful when preparing for a quiz or test.

Terminology and Concepts

- integral transform of a function
- inverse transform of a function
- transform pairs
- kernel of a transform
- Fourier transform
- inverse Fourier transform
- Fourier sine transform
- inverse Fourier sine transform
- Fourier cosine transform
- inverse Fourier cosine transform

The basic skills listed below summarize the more mechanical types of problems encountered in the exercise set for this section.

Basic Skills

- use Fourier integral transforms to solve BVPs

For the BVPs in this section it is sometimes useful to note that the identities

$$e^{i\alpha} = \cos \alpha + i \sin \alpha \quad \text{and} \quad e^{-i\alpha} = \cos \alpha - i \sin \alpha$$

imply

$$e^{i\alpha} + e^{-i\alpha} = 2 \cos \alpha \quad \text{and} \quad e^{i\alpha} - e^{-i\alpha} = 2i \sin \alpha.$$

3. Using the Fourier sine transform, the partial DE becomes

$$\frac{dU}{dt} + k\alpha^2 U = k\alpha u_0.$$

The general solution of this linear equation is

$$U(\alpha, t) = ce^{-k\alpha^2 t} + \frac{u_0}{\alpha}.$$

But $U(\alpha, 0) = 0$ implies $c = -u_0/\alpha$ and so

$$U(\alpha, t) = u_0 \frac{1 - e^{-k\alpha^2 t}}{\alpha}$$

and

$$u(x, t) = \frac{2u_0}{\pi} \int_0^\infty \frac{1 - e^{-k\alpha^2 t}}{\alpha} \sin \alpha x \, d\alpha.$$

6. Since the domain of x is $(0, \infty)$ and the condition at $x = 0$ involves $\partial u/\partial x$ we use the Fourier cosine transform:

$$-k\alpha^2 U(\alpha, t) - ku_x(0, t) = \frac{dU}{dt}$$

$$\frac{dU}{dt} + k\alpha^2 U = kA$$

$$U(\alpha, t) = ce^{-k\alpha^2 t} + \frac{A}{\alpha^2}.$$

Since

$$\mathscr{F}\{u(x, 0)\} = U(\alpha, 0) = 0$$

we find $c = -A/\alpha^2$, so that

$$U(\alpha, t) = A \frac{1 - e^{-k\alpha^2 t}}{\alpha^2}.$$

Applying the inverse Fourier cosine transform we obtain

$$u(x, t) = \mathscr{F}_C^{-1}\{U(\alpha, t)\} = \frac{2A}{\pi} \int_0^\infty \frac{1 - e^{-k\alpha^2 t}}{\alpha^2} \cos \alpha x \, d\alpha.$$

9. (a) Using the Fourier transform we obtain

$$U(\alpha, t) = c_1 \cos \alpha at + c_2 \sin \alpha at.$$

If we write

$$\mathscr{F}\{u(x,0)\} = \mathscr{F}\{f(x)\} = F(\alpha)$$

and

$$\mathscr{F}\{u_t(x,0)\} = \mathscr{F}\{g(x)\} = G(\alpha)$$

we first obtain $c_1 = F(\alpha)$ from $U(\alpha,0) = F(\alpha)$ and then $c_2 = G(\alpha)/\alpha a$ from $dU/dt \,|_{t=0} = G(\alpha)$. Thus

$$U(\alpha,t) = F(\alpha)\cos\alpha at + \frac{G(\alpha)}{\alpha a}\sin\alpha at$$

and

$$u(x,t) = \frac{1}{2\pi}\int_{-\infty}^{\infty}\left(F(\alpha)\cos\alpha at + \frac{G(\alpha)}{\alpha a}\sin\alpha at\right)e^{-i\alpha x}d\alpha.$$

(b) If $g(x) = 0$ then $c_2 = 0$ and

$$u(x,t) = \frac{1}{2\pi}\int_{-\infty}^{\infty} F(\alpha)\cos\alpha at\, e^{-i\alpha x}d\alpha$$

$$= \frac{1}{2\pi}\int_{-\infty}^{\infty} F(\alpha)\left(\frac{e^{\alpha ati}+e^{-\alpha ati}}{2}\right)e^{-i\alpha x}d\alpha$$

$$= \frac{1}{2}\left[\frac{1}{2\pi}\int_{-\infty}^{\infty} F(\alpha)e^{-i(x-at)\alpha}d\alpha + \frac{1}{2\pi}\int_{-\infty}^{\infty} F(\alpha)e^{-i(x+at)\alpha}d\alpha\right]$$

$$= \frac{1}{2}\left[f(x-at) + f(x+at)\right].$$

12. Since the boundary condition at $y = 0$ now involves $u(x,0)$ rather than $u'(x,0)$, we use the Fourier sine transform. The transform of the partial differential equation is then

$$\frac{d^2U}{dx^2} - \alpha^2 U + \alpha u(x,0) = 0 \qquad \text{or} \qquad \frac{d^2U}{dx^2} - \alpha^2 U = -\alpha.$$

The solution of this differential equation is

$$U(x,\alpha) = c_1\cosh\alpha x + c_2\sinh\alpha x + \frac{1}{\alpha}.$$

The transforms of the boundary conditions at $x = 0$ and $x = \pi$ in turn imply that $c_1 = 1/\alpha$ and

$$c_2 = \frac{\cosh\alpha\pi}{\alpha\sinh\alpha\pi} - \frac{1}{\alpha\sinh\alpha\pi} + \frac{\alpha}{(1+\alpha^2)\sinh\alpha\pi}.$$

Hence

$$U(x,\alpha) = \frac{1}{\alpha} - \frac{\cosh\alpha x}{\alpha} + \frac{\cosh\alpha\pi}{\alpha\sinh\alpha\pi}\sinh\alpha x - \frac{\sinh\alpha x}{\alpha\sinh\alpha\pi} + \frac{\alpha\sinh\alpha x}{(1+\alpha^2)\sinh\alpha\pi}$$

$$= \frac{1}{\alpha} - \frac{\sinh\alpha(\pi-x)}{\alpha\sinh\alpha\pi} - \frac{\sinh\alpha x}{\alpha(1+\alpha^2)\sinh\alpha\pi}.$$

296

Taking the inverse transform it follows that

$$u(x,y) = \frac{2}{\pi} \int_0^\infty \left(\frac{1}{\alpha} - \frac{\sinh \alpha(\pi - x)}{\alpha \sinh \alpha\pi} - \frac{\sinh \alpha x}{\alpha(1 + \alpha^2) \sinh \alpha\pi} \right) \sin \alpha y \, d\alpha.$$

15. We use the Fourier sine transform with respect to x to obtain

$$U(\alpha, y) = c_1 \cosh \alpha y + c_2 \sinh \alpha y.$$

The transforms of $u(x,0) = f(x)$ and $u(x,2) = 0$ give, in turn, $U(\alpha,0) = F(\alpha)$ and $U(\alpha,2) = 0$. The first condition gives $c_1 = F(\alpha)$ and the second condition then yields

$$c_2 = -\frac{F(\alpha) \cosh 2\alpha}{\sinh 2\alpha}.$$

Hence

$$U(\alpha, y) = F(\alpha) \cosh \alpha y - \frac{F(\alpha) \cosh 2\alpha \sinh \alpha y}{\sinh 2\alpha}$$

$$= F(\alpha) \frac{\sinh 2\alpha \cosh \alpha y - \cosh 2\alpha \sinh \alpha y}{\sinh 2\alpha}$$

$$= F(\alpha) \frac{\sinh \alpha(2 - y)}{\sinh 2\alpha}$$

and

$$u(x,y) = \frac{2}{\pi} \int_0^\infty F(\alpha) \frac{\sinh \alpha(2 - y)}{\sinh 2\alpha} \sin \alpha x \, d\alpha.$$

18. We solve the three boundary-value problems:

Using separation of variables we find the solution of the first problem is

$$u_1(x,y) = \sum_{n=1}^\infty A_n e^{-ny} \sin nx \quad \text{where} \quad A_n = \frac{2}{\pi} \int_0^\pi f(x) \sin nx \, dx.$$

Using the Fourier sine transform with respect to y gives the solution of the second problem:

$$u_2(x,y) = \frac{200}{\pi} \int_0^\infty \frac{(1 - \cos \alpha) \sinh \alpha(\pi - x)}{\alpha \sinh \alpha\pi} \sin \alpha y \, d\alpha.$$

Also, the Fourier sine transform with respect to y gives the solution of the third problem:

$$u_3(x,y) = \frac{2}{\pi} \int_0^\infty \frac{\alpha \sinh \alpha x}{(1 + \alpha^2) \sinh \alpha\pi} \sin \alpha y \, d\alpha.$$

The solution of the original problem is

$$u(x,y) = u_1(x,y) + u_2(x,y) + u_3(x,y).$$

297

21. Using the Fourier transform with respect to x gives

$$U(\alpha, y) = c_1 \cosh \alpha y + c_2 \sinh \alpha y.$$

The transform of the boundary condition $\partial u/\partial y\, |_{y=0} = 0$ is $dU/dy\, |_{y=0} = 0$. This condition gives $c_2 = 0$. Hence

$$U(\alpha, y) = c_1 \cosh \alpha y.$$

Now by the given information the transform of the boundary condition $u(x, 1) = e^{-x^2}$ is $U(\alpha, 1) = \sqrt{\pi}\, e^{-\alpha^2/4}$. This condition then gives $c_1 = \sqrt{\pi}\, e^{-\alpha^2/4} \cosh \alpha$. Therefore

$$U(\alpha, y) = \sqrt{\pi}\, \frac{e^{-\alpha^2/4} \cosh \alpha y}{\cosh \alpha}$$

and

$$U(x, y) = \frac{1}{2\sqrt{\pi}} \int_{-\infty}^{\infty} \frac{e^{-\alpha^2/4} \cosh \alpha y}{\cosh \alpha} e^{-i\alpha x} d\alpha$$

$$= \frac{1}{2\sqrt{\pi}} \int_{-\infty}^{\infty} \frac{e^{-\alpha^2/4} \cosh \alpha y}{\cosh \alpha} \cos \alpha x\, d\alpha$$

$$= \frac{1}{\sqrt{\pi}} \int_{0}^{\infty} \frac{e^{-\alpha^2/4} \cosh \alpha y}{\cosh \alpha} \cos \alpha x\, d\alpha.$$

Chapter 14 in Review

Hints, Suggestions, Solutions, and Examples

3. The Laplace transform gives

$$U(x, s) = c_1 e^{-\sqrt{s+h}\, x} + c_2 e^{\sqrt{s+h}\, x} + \frac{u_0}{s+h}.$$

The condition $\lim_{x \to \infty} \partial u/\partial x = 0$ implies $\lim_{x \to \infty} dU/dx = 0$ and so we define $c_2 = 0$. Thus

$$U(x, s) = c_1 e^{-\sqrt{s+h}\, x} + \frac{u_0}{s+h}.$$

The condition $U(0, s) = 0$ then gives $c_1 = -u_0/(s+h)$ and so

$$U(x, s) = \frac{u_0}{s+h} - u_0 \frac{e^{-\sqrt{s+h}\, x}}{s+h}.$$

With the help of the first translation theorem we then obtain

$$u(x,t) = u_0 \mathscr{L}^{-1}\left\{\frac{1}{s+h}\right\} - u_0 \mathscr{L}^{-1}\left\{\frac{e^{-\sqrt{s+h}\,x}}{s+h}\right\} = u_0 e^{-ht} - u_0 e^{-ht}\,\text{erfc}\left(\frac{x}{2\sqrt{t}}\right)$$

$$= u_0 e^{-ht}\left[1 - \text{erfc}\left(\frac{x}{2\sqrt{t}}\right)\right] = u_0 e^{-ht}\,\text{erf}\left(\frac{x}{2\sqrt{t}}\right).$$

6. The Laplace transform and undetermined coefficients give

$$U(x,s) = c_1 \cosh sx + c_2 \sinh sx + \frac{s-1}{s^2+\pi^2}\sin \pi x.$$

The conditions $U(0,s) = 0$ and $U(1,s) = 0$ give, in turn, $c_1 = 0$ and $c_2 = 0$. Thus

$$U(x,s) = \frac{s-1}{s^2+\pi^2}\sin \pi x$$

and

$$u(x,t) = \sin \pi x \,\mathscr{L}^{-1}\left\{\frac{s}{s^2+\pi^2}\right\} - \frac{1}{\pi}\sin \pi x \,\mathscr{L}^{-1}\left\{\frac{\pi}{s^2+\pi^2}\right\}$$

$$= (\sin \pi x)\cos \pi t - \frac{1}{\pi}(\sin \pi x)\sin \pi t.$$

9. We solve the two problems

$$\frac{\partial^2 u_1}{\partial x^2} + \frac{\partial^2 u_1}{\partial y^2} = 0, \quad x > 0, \quad y > 0,$$

$$u_1(0,y) = 0, \quad y > 0,$$

$$u_1(x,0) = \begin{cases} 100, & 0 < x < 1 \\ 0, & x > 1 \end{cases}$$

and

$$\frac{\partial^2 u_2}{\partial x^2} + \frac{\partial^2 u_2}{\partial y^2} = 0, \quad x > 0, \quad y > 0,$$

$$u_2(0,y) = \begin{cases} 50, & 0 < y < 1 \\ 0, & y > 1 \end{cases}$$

$$u_2(x,0) = 0.$$

Using the Fourier sine transform with respect to x we find

$$u_1(x,y) = \frac{200}{\pi}\int_0^\infty \left(\frac{1-\cos\alpha}{\alpha}\right)e^{-\alpha y}\sin \alpha x \, d\alpha.$$

Using the Fourier sine transform with respect to y we find

$$u_2(x,y) = \frac{100}{\pi}\int_0^\infty \left(\frac{1-\cos\alpha}{\alpha}\right)e^{-\alpha x}\sin \alpha y \, d\alpha.$$

Chapter 14 in Review

The solution of the problem is then

$$u(x, y) = u_1(x, y) + u_2(x, y).$$

12. Using the Laplace transform gives

$$U(x, s) = c_1 \cosh \sqrt{s}\, x + c_2 \sinh \sqrt{s}\, x.$$

The condition $u(0, t) = u_0$ transforms into $U(0, s) = u_0/s$. This gives $c_1 = u_0/s$. The condition $u(1, t) = u_0$ transforms into $U(1, s) = u_0/s$. This implies that $c_2 = u_0(1 - \cosh \sqrt{s}\,)/s \sinh \sqrt{s}$. Hence

$$U(x, s) = \frac{u_0}{s} \cosh \sqrt{s}\, x + u_0 \left[\frac{1 - \cosh \sqrt{s}}{s \sinh \sqrt{s}} \right] \sinh \sqrt{s}\, x$$

$$= u_0 \left[\frac{\sinh \sqrt{s}\, \cosh \sqrt{s}\, x - \cosh \sqrt{s}\, \sinh \sqrt{s}\, x + \sinh \sqrt{s}\, x}{s \sinh \sqrt{s}} \right]$$

$$= u_0 \left[\frac{\sinh \sqrt{s}\, (1 - x) + \sinh \sqrt{s}\, x}{s \sinh \sqrt{s}} \right]$$

$$= u_0 \left[\frac{\sinh \sqrt{s}\, (1 - x)}{s \sinh \sqrt{s}} + \frac{\sinh \sqrt{s}\, x}{s \sinh \sqrt{s}} \right]$$

and

$$u(x, t) = u_0 \left[\mathscr{L}^{-1} \left\{ \frac{\sinh \sqrt{s}\, (1 - x)}{s \sinh \sqrt{s}} \right\} + \mathscr{L}^{-1} \left\{ \frac{\sinh \sqrt{s}\, x}{s \sinh \sqrt{s}} \right\} \right]$$

$$= u_0 \sum_{n=0}^{\infty} \left[\operatorname{erf} \left(\frac{2n + 2 - x}{2\sqrt{t}} \right) - \operatorname{erf} \left(\frac{2n + x}{2\sqrt{t}} \right) \right]$$

$$+ u_0 \sum_{n=0}^{\infty} \left[\operatorname{erf} \left(\frac{2n + 1 + x}{2\sqrt{t}} \right) - \operatorname{erf} \left(\frac{2n + 1 - x}{2\sqrt{t}} \right) \right].$$

15. Using the Fourier transform with respect to x we obtain

$$\frac{d^2 U}{dy^2} - \alpha^2 U = 0.$$

Since $0 < y < 1$ is a finite interval we use the general solution

$$U(\alpha, y) = c_1 \cosh \alpha y + c_2 \sinh \alpha y.$$

The boundary condition at $y = 0$ transforms into $U'(\alpha, 0) = 0$, so $c_2 = 0$ and $U(\alpha, y) = c_1 \cosh \alpha y$. Now denote the Fourier transform of f as $F(\alpha)$. Then $U(\alpha, 1) = F(\alpha)$ so $F(\alpha) = c_1 \cosh \alpha$ and

$$U(\alpha, y) = F(\alpha) \frac{\cosh \alpha y}{\cosh \alpha}.$$

Taking the inverse Fourier transform we obtain

$$u(x, y) = \frac{1}{2\pi} \int_{-\infty}^{\infty} F(\alpha) \frac{\cosh \alpha y}{\cosh \alpha} e^{-i\alpha x}\, d\alpha.$$

But

$$F(\alpha) = \int_{-\infty}^{\infty} f(t)e^{i\alpha t}\, dt,$$

and so

$$u(x,y) = \frac{1}{2\pi}\int_{-\infty}^{\infty}\left(\int_{-\infty}^{\infty} f(t)e^{i\alpha t}\, dt\right)\frac{\cosh\alpha y}{\cosh\alpha}e^{-i\alpha x}\, d\alpha$$

$$= \frac{1}{2\pi}\int_{-\infty}^{\infty}\int_{-\infty}^{\infty} f(t)e^{i\alpha(t-x)}\frac{\cosh\alpha y}{\cosh\alpha}\, dt\, d\alpha$$

$$= \frac{1}{2\pi}\int_{-\infty}^{\infty}\int_{-\infty}^{\infty} f(t)(\cos\alpha(t-x)+i\sin\alpha(t-x))\frac{\cosh\alpha y}{\cosh\alpha}\, dt\, d\alpha$$

$$= \frac{1}{2\pi}\int_{-\infty}^{\infty}\int_{-\infty}^{\infty} f(t)\cos\alpha(t-x)\frac{\cosh\alpha y}{\cosh\alpha}\, dt\, d\alpha$$

$$= \frac{1}{\pi}\int_{0}^{\infty}\int_{-\infty}^{\infty} f(t)\cos\alpha(t-x)\frac{\cosh\alpha y}{\cosh\alpha}\, dt\, d\alpha,$$

since the imaginary part of the integrand is an odd function of α followed by the fact that the remaining integrand is an even function of α.

15

Numerical Solutions of Partial Differential Equations

Section 15.1   Laplace's Equation

The terminology and concepts listed below provide an outline of the main ideas encountered in this section. These can be useful when preparing for a quiz or test.

Terminology and Concepts

- difference equation replacement

- mesh size

- interior points

- boundary points

- sparse matrix

- banded matrix

- Gauss-Seidel iteration

The basic skills listed below summarize the more mechanical types of problems encountered in the exercise set for this section.

Basic Skills

- use difference equation replacement to approximate the solution of Laplace's equation with a given mesh size subject to given boundary conditions

- use Gauss-Seidel iteration to approximate the solution of Laplace's equation on the interior of a unit square subject to given boundary conditions

Exercises 15.1 *Hints, Suggestions, Solutions, and Examples*

3. The figure shows the values of $u(x, y)$ along the boundary. We need to determine u_{11}, u_{21}, u_{12}, and u_{22}. By symmetry $u_{11} = u_{21}$ and $u_{12} = u_{22}$. The system is

$$u_{21} + u_{12} + 0 + 0 - 4u_{11} = 0$$

$$0 + u_{22} + u_{11} + 0 - 4u_{21} = 0$$

$$u_{22} + \sqrt{3}/2 + 0 + u_{11} - 4u_{12} = 0$$

$$0 + \sqrt{3}/2 + u_{12} + u_{21} - 4u_{22} = 0$$

or

$$3u_{11} + u_{12} = 0$$

$$u_{11} - 3u_{12} = -\frac{\sqrt{3}}{2}.$$

Solving we obtain $u_{11} = u_{21} = \sqrt{3}/16$ and $u_{12} = u_{22} = 3\sqrt{3}/16$.

6. The coefficients of the unknowns are the same as shown above in Problem 5. The constant terms are 7.5, 5, 20, 10, 0, 15, 17.5, 5, 27.5. We use 32.5 as the initial guess for each variable. Then $u_{11} = 21.92$, $u_{21} = 28.30$, $u_{31} = 38.17$, $u_{12} = 29.38$, $u_{22} = 33.13$, $u_{32} = 44.38$, $u_{13} = 22.46$, $u_{23} = 30.45$, and $u_{33} = 46.21$.

Section 15.2   Heat Equation

The terminology and concepts listed below provide an outline of the main ideas encountered in this section. These can be useful when preparing for a quiz or test.

Terminology and Concepts

- explicit finite difference method
- unstable numerical method
- Crank-Nicholson method
- tridiagonal matrix

The basic skills listed below summarize the more mechanical types of problems encountered in the exercise set for this section.

Basic Skills

- solve the heat equation subject to given boundary and initial conditions using the finite difference method
- solve the heat equation subject to given boundary and initial conditions using the Crank-Nicholson method

Exercises 15.2   *Hints, Suggestions, Solutions, and Examples*

3. We identify $c = 1$, $a = 2$, $T = 1$, $n = 8$, and $m = 40$. Then $h = 2/8 = 0.25$, $k = 1/40 = 0.025$, and $\lambda = 2/5 = 0.4$.

TIME	X=0.25	X=0.50	X=0.75	X=1.00	X=1.25	X=1.50	X=1.75
0.000	1.0000	1.0000	1.0000	1.0000	0.0000	0.0000	0.0000
0.025	0.7074	0.9520	0.9566	0.7444	0.2545	0.0371	0.0053
0.050	0.5606	0.8499	0.8685	0.6633	0.3303	0.1034	0.0223
0.075	0.4684	0.7473	0.7836	0.6191	0.3614	0.1529	0.0462
0.100	0.4015	0.6577	0.7084	0.5837	0.3753	0.1871	0.0684
0.125	0.3492	0.5821	0.6428	0.5510	0.3797	0.2101	0.0861
0.150	0.3069	0.5187	0.5857	0.5199	0.3778	0.2247	0.0990
0.175	0.2721	0.4652	0.5359	0.4901	0.3716	0.2329	0.1078
0.200	0.2430	0.4198	0.4921	0.4617	0.3622	0.2362	0.1132
0.225	0.2186	0.3809	0.4533	0.4348	0.3507	0.2358	0.1160
0.250	0.1977	0.3473	0.4189	0.4093	0.3378	0.2327	0.1166
0.275	0.1798	0.3181	0.3881	0.3853	0.3240	0.2275	0.1157
0.300	0.1643	0.2924	0.3604	0.3626	0.3097	0.2208	0.1136
0.325	0.1507	0.2697	0.3353	0.3412	0.2953	0.2131	0.1107
0.350	0.1387	0.2495	0.3125	0.3211	0.2808	0.2047	0.1071
0.375	0.1281	0.2313	0.2916	0.3021	0.2666	0.1960	0.1032
0.400	0.1187	0.2150	0.2725	0.2843	0.2528	0.1871	0.0989
0.425	0.1102	0.2002	0.2549	0.2675	0.2393	0.1781	0.0946
0.450	0.1025	0.1867	0.2387	0.2517	0.2263	0.1692	0.0902
0.475	0.0955	0.1743	0.2236	0.2368	0.2139	0.1606	0.0858
0.500	0.0891	0.1630	0.2097	0.2228	0.2020	0.1521	0.0814
0.525	0.0833	0.1525	0.1967	0.2096	0.1906	0.1439	0.0772
0.550	0.0779	0.1429	0.1846	0.1973	0.1798	0.1361	0.0731
0.575	0.0729	0.1339	0.1734	0.1856	0.1696	0.1285	0.0691
0.600	0.0683	0.1256	0.1628	0.1746	0.1598	0.1214	0.0653
0.625	0.0641	0.1179	0.1530	0.1643	0.1506	0.1145	0.0617
0.650	0.0601	0.1106	0.1438	0.1546	0.1419	0.1080	0.0582
0.675	0.0564	0.1039	0.1351	0.1455	0.1336	0.1018	0.0549
0.700	0.0530	0.0976	0.1270	0.1369	0.1259	0.0959	0.0518
0.725	0.0497	0.0917	0.1194	0.1288	0.1185	0.0904	0.0488
0.750	0.0467	0.0862	0.1123	0.1212	0.1116	0.0852	0.0460
0.775	0.0439	0.0810	0.1056	0.1140	0.1050	0.0802	0.0433
0.800	0.0413	0.0762	0.0993	0.1073	0.0989	0.0755	0.0408
0.825	0.0388	0.0716	0.0934	0.1009	0.0931	0.0711	0.0384
0.850	0.0365	0.0674	0.0879	0.0950	0.0876	0.0669	0.0362
0.875	0.0343	0.0633	0.0827	0.0894	0.0824	0.0630	0.0341
0.900	0.0323	0.0596	0.0778	0.0841	0.0776	0.0593	0.0321
0.925	0.0303	0.0560	0.0732	0.0791	0.0730	0.0558	0.0302
0.950	0.0285	0.0527	0.0688	0.0744	0.0687	0.0526	0.0284
0.975	0.0268	0.0496	0.0647	0.0700	0.0647	0.0495	0.0268
1.000	0.0253	0.0466	0.0609	0.0659	0.0608	0.0465	0.0252

(x,y)	exact	approx	abs error
(0.25,0.1)	0.3794	0.4015	0.0221
(1,0.5)	0.1854	0.2228	0.0374
(1.5,0.8)	0.0623	0.0755	0.0132

6. (a) We identify $c = 15/88 \approx 0.1705$, $a = 20$, $T = 10$, $n = 10$, and $m = 10$. Then $h = 2$, $k = 1$, and $\lambda = 15/352 \approx 0.0426$.

TIME	X=2	X=4	X=6	X=8	X=10	X=12	X=14	X=16	X=18
0	30.0000	30.0000	30.0000	30.0000	30.0000	30.0000	30.0000	30.0000	30.0000
1	28.7216	30.0000	30.0000	30.0000	30.0000	30.0000	30.0000	30.0000	28.7216
2	27.5521	29.9455	30.0000	30.0000	30.0000	30.0000	30.0000	29.9455	27.5521
3	26.4800	29.8459	29.9977	30.0000	30.0000	30.0000	29.9977	29.8459	26.4800
4	25.4951	29.7089	29.9913	29.9999	30.0000	29.9999	29.9913	29.7089	25.4951
5	24.5882	29.5414	29.9796	29.9995	30.0000	29.9995	29.9796	29.5414	24.5882
6	23.7515	29.3490	29.9618	29.9987	30.0000	29.9987	29.9618	29.3490	23.7515
7	22.9779	29.1365	29.9373	29.9972	29.9998	29.9972	29.9373	29.1365	22.9779
8	22.2611	28.9082	29.9057	29.9948	29.9996	29.9948	29.9057	28.9082	22.2611
9	21.5958	28.6675	29.8670	29.9912	29.9992	29.9912	29.8670	28.6675	21.5958
10	20.9768	28.4172	29.8212	29.9862	29.9985	29.9862	29.8212	28.4172	20.9768

(b) We identify $c = 15/88 \approx 0.1705$, $a = 50$, $T = 10$, $n = 10$, and $m = 10$. Then $h = 5$, $k = 1$, and $\lambda = 3/440 \approx 0.0068$.

TIME	X=5	X=10	X=15	X=20	X=25	X=30	X=35	X=40	X=45
0	30.0000	30.0000	30.0000	30.0000	30.0000	30.0000	30.0000	30.0000	30.0000
1	29.7955	30.0000	30.0000	30.0000	30.0000	30.0000	30.0000	30.0000	29.7955
2	29.5937	29.9986	30.0000	30.0000	30.0000	30.0000	30.0000	29.9986	29.5937
3	29.3947	29.9959	30.0000	30.0000	30.0000	30.0000	30.0000	29.9959	29.3947
4	29.1984	29.9918	30.0000	30.0000	30.0000	30.0000	30.0000	29.9918	29.1984
5	29.0047	29.9864	29.9999	30.0000	30.0000	30.0000	29.9999	29.9864	29.0047
6	28.8136	29.9798	29.9998	30.0000	30.0000	30.0000	29.9998	29.9798	28.8136
7	28.6251	29.9720	29.9997	30.0000	30.0000	30.0000	29.9997	29.9720	28.6251
8	28.4391	29.9630	29.9995	30.0000	30.0000	30.0000	29.9995	29.9630	28.4391
9	28.2556	29.9529	29.9992	30.0000	30.0000	30.0000	29.9992	29.9529	28.2556
10	28.0745	29.9416	29.9989	30.0000	30.0000	30.0000	29.9989	29.9416	28.0745

(c) We identify $c = 50/27 \approx 1.8519$, $a = 20$, $T = 10$, $n = 10$, and $m = 10$. Then $h = 2$, $k = 1$, and $\lambda = 25/54 \approx 0.4630$.

TIME	X=2	X=4	X=6	X=8	X=10	X=12	X=14	X=16	X=18
0	18.0000	32.0000	42.0000	48.0000	50.0000	48.0000	42.0000	32.0000	18.0000
1	16.1481	30.1481	40.1481	46.1481	48.1481	46.1481	40.1481	30.1481	16.1481
2	15.1536	28.2963	38.2963	44.2963	46.2963	44.2963	38.2963	28.2963	15.1536
3	14.2226	26.8414	36.4444	42.4444	44.4444	42.4444	36.4444	26.8414	14.2226
4	13.4801	25.4452	34.7764	40.5926	42.5926	40.5926	34.7764	25.4452	13.4801
5	12.7787	24.2258	33.1491	38.8258	40.7407	38.8258	33.1491	24.2258	12.7787
6	12.1622	23.0574	31.6460	37.0842	38.9677	37.0842	31.6460	23.0574	12.1622
7	11.5756	21.9895	30.1875	35.4385	37.2238	35.4385	30.1875	21.9895	11.5756
8	11.0378	20.9636	28.8232	33.8340	35.5707	33.8340	28.8232	20.9636	11.0378
9	10.5230	20.0070	27.5043	32.3182	33.9626	32.3182	27.5043	20.0070	10.5230
10	10.0420	19.0872	26.2620	30.8509	32.4400	30.8509	26.2620	19.0872	10.0420

(d) We identify $c = 260/159 \approx 1.6352$, $a = 100$, $T = 10$, $n = 10$, and $m = 10$. Then $h = 10$, $k = 1$, and $\lambda = 13/795 \approx 00164$.

TIME	X=10	X=20	X=30	X=40	X=50	X=60	X=70	X=80	X=90
0	8.0000	16.0000	24.0000	32.0000	40.0000	32.0000	24.0000	16.0000	8.0000
1	8.0000	16.0000	23.6075	31.3459	39.2151	31.6075	23.7384	15.8692	8.0000
2	8.0000	15.9936	23.2279	30.7068	38.4452	31.2151	23.4789	15.7384	7.9979
3	7.9999	15.9812	22.8606	30.0824	37.6900	30.8229	23.2214	15.6076	7.9937
4	7.9996	15.9631	22.5050	29.4724	36.9492	30.4312	22.9660	15.4769	7.9874
5	7.9990	15.9399	22.1606	28.8765	36.2228	30.0401	22.7125	15.3463	7.9793
6	7.9981	15.9118	21.8270	28.2945	35.5103	29.6500	22.4610	15.2158	7.9693
7	7.9967	15.8791	21.5037	27.7261	34.8117	29.2610	22.2112	15.0854	7.9575
8	7.9948	15.8422	21.1902	27.1709	34.1266	28.8733	21.9633	14.9553	7.9439
9	7.9924	15.8013	20.8861	26.6288	33.4548	28.4870	21.7172	14.8253	7.9287
10	7.9894	15.7568	20.5911	26.0995	32.7961	28.1024	21.4727	14.6956	7.9118

9. (a) We identify $c = 15/88 \approx 0.1705$, $a = 20$, $T = 10$, $n = 10$, and $m = 10$. Then $h = 2$, $k = 1$, and $\lambda = 15/352 \approx 0.0426$.

TIME	X=2.00	X=4.00	X=6.00	X=8.00	X=10.00	X=12.00	X=14.00	X=16.00	X=18.00
0.00	30.0000	30.0000	30.0000	30.0000	30.0000	30.0000	30.0000	30.0000	30.0000
1.00	28.7733	29.9749	29.9995	30.0000	30.0000	30.0000	29.9998	29.9916	29.5911
2.00	27.6450	29.9037	29.9970	29.9999	30.0000	30.0000	29.9990	29.9679	29.2150
3.00	26.6051	29.7938	29.9911	29.9997	30.0000	29.9999	29.9970	29.9313	28.8684
4.00	25.6452	29.6517	29.9805	29.9991	30.0000	29.9997	29.9935	29.8839	28.5484
5.00	24.7573	29.4829	29.9643	29.9981	29.9999	29.9994	29.9881	29.8276	28.2524
6.00	23.9347	29.2922	29.9421	29.9963	29.9997	29.9988	29.9807	29.7641	27.9782
7.00	23.1711	29.0836	29.9134	29.9936	29.9995	29.9979	29.9711	29.6945	27.7237
8.00	22.4612	28.8606	29.8782	29.9899	29.9991	29.9966	29.9594	29.6202	27.4870
9.00	21.7999	28.6263	29.8362	29.9848	29.9985	29.9949	29.9454	29.5421	27.2666
10.00	21.1829	28.3831	29.7878	29.9783	29.9976	29.9927	29.9293	29.4610	27.0610

(b) We identify $c = 15/88 \approx 0.1705$, $a = 50$, $T = 10$, $n = 10$, and $m = 10$. Then $h = 5$, $k = 1$, and $\lambda = 3/440 \approx 0.0068$.

TIME	X=5.00	X=10.00	X=15.00	X=20.00	X=25.00	X=30.00	X=35.00	X=40.00	X=45.00
0.00	30.0000	30.0000	30.0000	30.0000	30.0000	30.0000	30.0000	30.0000	30.0000
1.00	29.7968	29.9993	30.0000	30.0000	30.0000	30.0000	30.0000	29.9998	29.9323
2.00	29.5964	29.9973	30.0000	30.0000	30.0000	30.0000	30.0000	29.9991	29.8655
3.00	29.3987	29.9939	30.0000	30.0000	30.0000	30.0000	30.0000	29.9980	29.7996
4.00	29.2036	29.9893	29.9999	30.0000	30.0000	30.0000	30.0000	29.9964	29.7345
5.00	29.0112	29.9834	29.9998	30.0000	30.0000	30.0000	29.9999	29.9945	29.6704
6.00	28.8212	29.9762	29.9997	30.0000	30.0000	30.0000	29.9999	29.9921	29.6071
7.00	28.6339	29.9679	29.9995	30.0000	30.0000	30.0000	29.9998	29.9893	29.5446
8.00	28.4490	29.9585	29.9992	30.0000	30.0000	30.0000	29.9997	29.9862	29.4830
9.00	28.2665	29.9479	29.9989	30.0000	30.0000	30.0000	29.9996	29.9827	29.4222
10.00	28.0864	29.9363	29.9986	30.0000	30.0000	30.0000	29.9995	29.9788	29.3621

(c) We identify $c = 50/27 \approx 1.8519$, $a = 20$, $T = 10$, $n = 10$, and $m = 10$. Then $h = 2$, $k = 1$, and $\lambda = 25/54 \approx 0.4630$.

TIME	X=2.00	X=4.00	X=6.00	X=8.00	X=10.00	X=12.00	X=14.00	X=16.00	X=18.00
0.00	18.0000	32.0000	42.0000	48.0000	50.0000	48.0000	42.0000	32.0000	18.0000
1.00	16.4489	30.1970	40.1562	46.1502	48.1531	46.1773	40.3274	31.2520	22.9449
2.00	15.3312	28.5350	38.3477	44.3130	46.3327	44.4671	39.0872	31.5755	24.6930
3.00	14.4219	27.0429	36.6090	42.5113	44.5759	42.9362	38.1976	31.7478	25.4131
4.00	13.6381	25.6913	34.9606	40.7728	42.9127	41.5716	37.4340	31.7086	25.6986
5.00	12.9409	24.4545	33.4091	39.1182	41.3519	40.3240	36.7033	31.5136	25.7663
6.00	12.3088	23.3146	31.9546	37.5566	39.8880	39.1565	35.9745	31.2134	25.7128
7.00	11.7294	22.2589	30.5939	36.0884	38.5109	38.0470	35.2407	30.8434	25.5871
8.00	11.1946	21.2785	29.3217	34.7092	37.2109	36.9834	34.5032	30.4279	25.4167
9.00	10.6987	20.3660	28.1318	33.4130	35.9801	35.9591	33.7660	29.9836	25.2181
10.00	10.2377	19.5150	27.0178	32.1929	34.8117	34.9710	33.0338	29.5224	25.0019

(d) We identify $c = 260/159 \approx 1.6352$, $a = 100$, $T = 10$, $n = 10$, and $m = 10$. Then $h = 10$, $k = 1$, and $\lambda = 13/795 \approx 00164$.

TIME	X=10.00	X=20.00	X=30.00	X=40.00	X=50.00	X=60.00	X=70.00	X=80.00	X=90.00
0.00	8.0000	16.0000	24.0000	32.0000	40.0000	32.0000	24.0000	16.0000	8.0000
1.00	8.0000	16.0000	24.0000	31.9979	39.7425	31.9979	24.0000	16.0026	8.3218
2.00	8.0000	16.0000	23.9999	31.9918	39.4932	31.9918	24.0000	16.0102	8.6333
3.00	8.0000	16.0000	23.9997	31.9820	39.2517	31.9820	24.0001	16.0225	8.9350
4.00	8.0000	16.0000	23.9993	31.9687	39.0176	31.9687	24.0002	16.0392	9.2272
5.00	8.0000	16.0000	23.9987	31.9520	38.7905	31.9521	24.0003	16.0599	9.5103
6.00	8.0000	15.9999	23.9978	31.9323	38.5701	31.9324	24.0005	16.0845	9.7846
7.00	8.0000	15.9999	23.9966	31.9097	38.3561	31.9098	24.0008	16.1126	10.0506
8.00	8.0000	15.9998	23.9951	31.8844	38.1483	31.8846	24.0012	16.1441	10.3084
9.00	8.0000	15.9997	23.9931	31.8566	37.9463	31.8569	24.0017	16.1786	10.5585
10.00	8.0000	15.9996	23.9908	31.8265	37.7499	31.8270	24.0023	16.2160	10.8012

12. We identify $c = 1$, $a = 1$, $T = 1$, $n = 5$, and $m = 20$. Then $h = 0.2$, $k = 0.04$, and $\lambda = 1$. The values below were obtained using *Excel*, which carries more than 12 significant digits. In order to see evidence of instability use $0 \le t \le 2$.

TIME	X=0.2	X=0.4	X=0.6	X=0.8
0.00	0.5878	0.9511	0.9511	0.5878
0.04	0.3633	0.5878	0.5878	0.3633
0.08	0.2245	0.3633	0.3633	0.2245
0.12	0.1388	0.2245	0.2245	0.1388
0.16	0.0858	0.1388	0.1388	0.0858
0.20	0.0530	0.0858	0.0858	0.0530
0.24	0.0328	0.0530	0.0530	0.0328
0.28	0.0202	0.0328	0.0328	0.0202
0.32	0.0125	0.0202	0.0202	0.0125
0.36	0.0077	0.0125	0.0125	0.0077
0.40	0.0048	0.0077	0.0077	0.0048
0.44	0.0030	0.0048	0.0048	0.0030
0.48	0.0018	0.0030	0.0030	0.0018
0.52	0.0011	0.0018	0.0018	0.0011
0.56	0.0007	0.0011	0.0011	0.0007
0.60	0.0004	0.0007	0.0007	0.0004
0.64	0.0003	0.0004	0.0004	0.0003
0.68	0.0002	0.0003	0.0003	0.0002
0.72	0.0001	0.0002	0.0002	0.0001
0.76	0.0001	0.0001	0.0001	0.0001
0.80	0.0000	0.0001	0.0001	0.0000
0.84	0.0000	0.0000	0.0000	0.0000
0.88	0.0000	0.0000	0.0000	0.0000
0.92	0.0000	0.0000	0.0000	0.0000
0.96	0.0000	0.0000	0.0000	0.0000
1.00	0.0000	0.0000	0.0000	0.0000

TIME	X=0.2	X=0.4	X=0.6	X=0.8
1.04	0.0000	0.0000	0.0000	0.0000
1.08	0.0000	0.0000	0.0000	0.0000
1.12	0.0000	0.0000	0.0000	0.0000
1.16	0.0000	0.0000	0.0000	0.0000
1.20	-0.0001	0.0001	-0.0001	0.0001
1.24	0.0001	-0.0002	0.0002	-0.0001
1.28	-0.0004	0.0006	-0.0006	0.0004
1.32	0.0010	-0.0015	0.0015	-0.0010
1.36	-0.0025	0.0040	-0.0040	0.0025
1.40	0.0065	-0.0106	0.0106	-0.0065
1.44	-0.0171	0.0277	-0.0277	0.0171
1.48	0.0448	-0.0724	0.0724	-0.0448
1.52	-0.1172	0.1897	-0.1897	0.1172
1.56	0.3069	-0.4965	0.4965	-0.3069
1.60	-0.8034	1.2999	-1.2999	0.8034
1.64	2.1033	-3.4032	3.4032	-2.1033
1.68	-5.5064	8.9096	-8.9096	5.5064
1.72	14.416	-23.326	23.326	-14.416
1.76	-37.742	61.067	-61.067	37.742
1.80	98.809	-159.88	159.88	-98.809
1.84	-258.68	418.56	-418.56	258.685
1.88	677.24	-1095.8	1095.8	-677.245
1.92	-1773.1	2868.9	-2868.9	1773.1
1.96	4641.9	-7510.8	7510.8	-4641.9
2.00	-12153	19663	-19663	12153

Section 15.3 Wave Equation

The basic skills listed below summarize the more mechanical types of problems encountered in the exercise set for this section.

Basic Skills

- solve the wave equation subject to given boundary and initial conditions using the finite difference method

Exercises 15.3   *Hints, Suggestions, Solutions, and Examples*

3. (a) Identifying $h = 1/5$ and $k = 0.5/10 = 0.05$ we see that $\lambda = 0.25$.

TIME	X=0.2	X=0.4	X=0.6	X=0.8
0.00	0.5878	0.9511	0.9511	0.5878
0.05	0.5808	0.9397	0.9397	0.5808
0.10	0.5599	0.9059	0.9059	0.5599
0.15	0.5256	0.8505	0.8505	0.5256
0.20	0.4788	0.7748	0.7748	0.4788
0.25	0.4206	0.6806	0.6806	0.4206
0.30	0.3524	0.5701	0.5701	0.3524
0.35	0.2757	0.4460	0.4460	0.2757
0.40	0.1924	0.3113	0.3113	0.1924
0.45	0.1046	0.1692	0.1692	0.1046
0.50	0.0142	0.0230	0.0230	0.0142

(b) Identifying $h = 1/5$ and $k = 0.5/20 = 0.025$ we see that $\lambda = 0.125$.

TIME	X=0.2	X=0.4	X=0.6	X=0.8
0.00	0.5878	0.9511	0.9511	0.5878
0.03	0.5860	0.9482	0.9482	0.5860
0.05	0.5808	0.9397	0.9397	0.5808
0.08	0.5721	0.9256	0.9256	0.5721
0.10	0.5599	0.9060	0.9060	0.5599
0.13	0.5445	0.8809	0.8809	0.5445
0.15	0.5257	0.8507	0.8507	0.5257
0.18	0.5039	0.8153	0.8153	0.5039
0.20	0.4790	0.7750	0.7750	0.4790
0.23	0.4513	0.7302	0.7302	0.4513
0.25	0.4209	0.6810	0.6810	0.4209
0.28	0.3879	0.6277	0.6277	0.3879
0.30	0.3527	0.5706	0.5706	0.3527
0.33	0.3153	0.5102	0.5102	0.3153
0.35	0.2761	0.4467	0.4467	0.2761
0.38	0.2352	0.3806	0.3806	0.2352
0.40	0.1929	0.3122	0.3122	0.1929
0.43	0.1495	0.2419	0.2419	0.1495
0.45	0.1052	0.1701	0.1701	0.1052
0.48	0.0602	0.0974	0.0974	0.0602
0.50	0.0149	0.0241	0.0241	0.0149

Section 15.3   Wave Equation

6. We identify $c = 24944.4$, $k = 0.00010022$ seconds $= 0.10022$ milliseconds, and $\lambda = 0.25$. Time in the table is expressed in milliseconds.

TIME	X=10	X=20	X=30	X=40	X=50
0.00000	0.2000	0.2667	0.2000	0.1333	0.0667
0.10022	0.1958	0.2625	0.2000	0.1333	0.0667
0.20045	0.1836	0.2503	0.1997	0.1333	0.0667
0.30067	0.1640	0.2307	0.1985	0.1333	0.0667
0.40089	0.1384	0.2050	0.1952	0.1332	0.0667
0.50111	0.1083	0.1744	0.1886	0.1328	0.0667
0.60134	0.0755	0.1407	0.1777	0.1318	0.0666
0.70156	0.0421	0.1052	0.1615	0.1295	0.0665
0.80178	0.0100	0.0692	0.1399	0.1253	0.0661
0.90201	-0.0190	0.0340	0.1129	0.1184	0.0654
1.00223	-0.0435	0.0004	0.0813	0.1077	0.0638
1.10245	-0.0626	-0.0309	0.0464	0.0927	0.0610
1.20268	-0.0758	-0.0593	0.0095	0.0728	0.0564
1.30290	-0.0832	-0.0845	-0.0278	0.0479	0.0493
1.40312	-0.0855	-0.1060	-0.0639	0.0184	0.0390
1.50334	-0.0837	-0.1237	-0.0974	-0.0150	0.0250
1.60357	-0.0792	-0.1371	-0.1275	-0.0511	0.0069
1.70379	-0.0734	-0.1464	-0.1533	-0.0882	-0.0152
1.80401	-0.0675	-0.1515	-0.1747	-0.1249	-0.0410
1.90424	-0.0627	-0.1528	-0.1915	-0.1595	-0.0694
2.00446	-0.0596	-0.1509	-0.2039	-0.1904	-0.0991
2.10468	-0.0585	-0.1467	-0.2122	-0.2165	-0.1283
2.20491	-0.0592	-0.1410	-0.2166	-0.2368	-0.1551
2.30513	-0.0614	-0.1349	-0.2175	-0.2507	-0.1772
2.40535	-0.0643	-0.1294	-0.2154	-0.2579	-0.1929
2.50557	-0.0672	-0.1251	-0.2105	-0.2585	-0.2005
2.60580	-0.0696	-0.1227	-0.2033	-0.2524	-0.1993
2.70602	-0.0709	-0.1219	-0.1942	-0.2399	-0.1889
2.80624	-0.0710	-0.1225	-0.1833	-0.2214	-0.1699
2.90647	-0.0699	-0.1236	-0.1711	-0.1972	-0.1435
3.00669	-0.0678	-0.1244	-0.1575	-0.1681	-0.1115
3.10691	-0.0649	-0.1237	-0.1425	-0.1348	-0.0761
3.20713	-0.0617	-0.1205	-0.1258	-0.0983	-0.0395
3.30736	-0.0583	-0.1139	-0.1071	-0.0598	-0.0042
3.40758	-0.0547	-0.1035	-0.0859	-0.0209	0.0279
3.50780	-0.0508	-0.0889	-0.0617	0.0171	0.0552
3.60803	-0.0460	-0.0702	-0.0343	0.0525	0.0767
3.70825	-0.0399	-0.0478	-0.0037	0.0840	0.0919
3.80847	-0.0318	-0.0221	0.0297	0.1106	0.1008
3.90870	-0.0211	0.0062	0.0648	0.1314	0.1041
4.00892	-0.0074	0.0365	0.1005	0.1464	0.1025
4.10914	0.0095	0.0680	0.1350	0.1558	0.0973
4.20936	0.0295	0.1000	0.1666	0.1602	0.0897
4.30959	0.0521	0.1318	0.1937	0.1606	0.0808
4.40981	0.0764	0.1625	0.2148	0.1581	0.0719
4.51003	0.1013	0.1911	0.2291	0.1538	0.0639
4.61026	0.1254	0.2164	0.2364	0.1485	0.0575
4.71048	0.1475	0.2373	0.2369	0.1431	0.0532
4.81070	0.1659	0.2526	0.2315	0.1379	0.0512
4.91093	0.1794	0.2611	0.2217	0.1331	0.0514
5.01115	0.1867	0.2620	0.2087	0.1288	0.0535

310

3. (a)

TIME	X=0.0	X=0.2	X=0.4	X=0.6	X=0.8	X=1.0
0.00	0.0000	0.2000	0.4000	0.6000	0.8000	0.0000
0.01	0.0000	0.2000	0.4000	0.6000	0.5500	0.0000
0.02	0.0000	0.2000	0.4000	0.5375	0.4250	0.0000
0.03	0.0000	0.2000	0.3844	0.4750	0.3469	0.0000
0.04	0.0000	0.1961	0.3609	0.4203	0.2922	0.0000
0.05	0.0000	0.1883	0.3346	0.3734	0.2512	0.0000

(b)

TIME	X=0.0	X=0.2	X=0.4	X=0.6	X=0.8	X=1.0
0.00	0.0000	0.2000	0.4000	0.6000	0.8000	0.0000
0.01	0.0000	0.2000	0.4000	0.6000	0.8000	0.0000
0.02	0.0000	0.2000	0.4000	0.6000	0.5500	0.0000
0.03	0.0000	0.2000	0.4000	0.5375	0.4250	0.0000
0.04	0.0000	0.2000	0.3844	0.4750	0.3469	0.0000
0.05	0.0000	0.1961	0.3609	0.4203	0.2922	0.0000

(c) The table in part (b) is the same as the table in part (a) shifted downward one row.

Appendices

Appendix I

Gamma Function

Solutions for Appendix I

3. If $t = x^3$, then $dt = 3x^2\,dx$ and $x^4\,dx = \frac{1}{3}t^{2/3}\,dt$. Now

$$\int_0^\infty x^4 e^{-x^3}\,dx = \int_0^\infty \frac{1}{3}t^{2/3}e^{-t}\,dt = \frac{1}{3}\int_0^\infty t^{2/3}e^{-t}\,dt$$

$$= \frac{1}{3}\Gamma\left(\frac{5}{3}\right) = \frac{1}{3}(0.89) \approx 0.297.$$

6. For $x > 0$ we integrate by parts:

$$\Gamma(x+1) = \int_0^\infty t^x e^{-t}\,dt$$

$u = t^x$	$dv = e^{-t}\,dt$
$du = xt^{x-1}\,dt$	$v = -e^{-t}$

$$= -t^x e^{-t}\,\Big|_0^\infty - \int_0^\infty xt^{x-1}(-e^{-t})\,dt$$

$$= x\int_0^\infty t^{x-1}e^{-t}\,dt$$

$$= x\Gamma(x).$$

Appendix II **Introduction to Matrices**

Solutions for Appendix II

3. (a) $\mathbf{AB} = \begin{pmatrix} -2-9 & 12-6 \\ 5+12 & -30+8 \end{pmatrix} = \begin{pmatrix} -11 & 6 \\ 17 & -22 \end{pmatrix}$

(b) $\mathbf{BA} = \begin{pmatrix} -2-30 & 3+24 \\ 6-10 & -9+8 \end{pmatrix} = \begin{pmatrix} -32 & 27 \\ -4 & -1 \end{pmatrix}$

(c) $\mathbf{A}^2 = \begin{pmatrix} 4+15 & -6-12 \\ -10-20 & 15+16 \end{pmatrix} = \begin{pmatrix} 19 & -18 \\ -30 & 31 \end{pmatrix}$

(d) $\mathbf{B}^2 = \begin{pmatrix} 1+18 & -6+12 \\ -3+6 & 18+4 \end{pmatrix} = \begin{pmatrix} 19 & 6 \\ 3 & 22 \end{pmatrix}$

6. (a) $\mathbf{AB} = \begin{pmatrix} 5 & -6 & 7 \end{pmatrix} \begin{pmatrix} 3 \\ 4 \\ -1 \end{pmatrix} = (-16)$

(b) $\mathbf{BA} = \begin{pmatrix} 3 \\ 4 \\ -1 \end{pmatrix} \begin{pmatrix} 5 & -6 & 7 \end{pmatrix} = \begin{pmatrix} 15 & -18 & 21 \\ 20 & -24 & 28 \\ -5 & 6 & -7 \end{pmatrix}$

(c) $(\mathbf{BA})\mathbf{C} = \begin{pmatrix} 15 & -18 & 21 \\ 20 & -24 & 28 \\ -5 & 6 & -7 \end{pmatrix} \begin{pmatrix} 1 & 2 & 4 \\ 0 & 1 & -1 \\ 3 & 2 & 1 \end{pmatrix} = \begin{pmatrix} 78 & 54 & 99 \\ 104 & 72 & 132 \\ -26 & -18 & -33 \end{pmatrix}$

(d) Since $\mathbf{AB}$ is 1×1 and $\mathbf{C}$ is 3×3 the product $(\mathbf{AB})\mathbf{C}$ is not defined.

9. (a) $(\mathbf{AB})^T = \begin{pmatrix} 7 & 10 \\ 38 & 75 \end{pmatrix}^T = \begin{pmatrix} 7 & 38 \\ 10 & 75 \end{pmatrix}$

(b) $\mathbf{B}^T\mathbf{A}^T = \begin{pmatrix} 5 & -2 \\ 10 & -5 \end{pmatrix} \begin{pmatrix} 3 & 8 \\ 4 & 1 \end{pmatrix} = \begin{pmatrix} 7 & 38 \\ 10 & 75 \end{pmatrix}$

12. $\begin{pmatrix} 6t \\ 3t^2 \\ -3t \end{pmatrix} + \begin{pmatrix} -t+1 \\ -t^2+t \\ 3t-3 \end{pmatrix} - \begin{pmatrix} 6t \\ 8 \\ -10t \end{pmatrix} = \begin{pmatrix} -t+1 \\ 2t^2+t-8 \\ 10t-3 \end{pmatrix}$

15. Since $\det \mathbf{A} = 0$, $\mathbf{A}$ is singular.

18. Since $\det \mathbf{A} = -6$, $\mathbf{A}$ is nonsingular.

$$\mathbf{A}^{-1} = -\frac{1}{6}\begin{pmatrix} 2 & -10 \\ -2 & 7 \end{pmatrix}$$

21. Since $\det \mathbf{A} = -9$, $\mathbf{A}$ is nonsingular. The cofactors are

$$\begin{array}{lll} A_{11} = -2 & A_{12} = -13 & A_{13} = 8 \\ A_{21} = -2 & A_{22} = 5 & A_{23} = -1 \\ A_{31} = -1 & A_{32} = 7 & A_{33} = -5. \end{array}$$

Then

$$\mathbf{A}^{-1} = -\frac{1}{9}\begin{pmatrix} -2 & -13 & 8 \\ -2 & 5 & -1 \\ -1 & 7 & -5 \end{pmatrix}^T = -\frac{1}{9}\begin{pmatrix} -2 & -2 & -1 \\ -13 & 5 & 7 \\ 8 & -1 & -5 \end{pmatrix}.$$

24. Since $\det \mathbf{A}(t) = 2e^{2t} \neq 0$, $\mathbf{A}$ is nonsingular.

$$\mathbf{A}^{-1} = \frac{1}{2}e^{-2t}\begin{pmatrix} e^t \sin t & 2e^t \cos t \\ -e^t \cos t & 2e^t \sin t \end{pmatrix}$$

27. $\mathbf{X} = \begin{pmatrix} 2e^{2t} + 8e^{-3t} \\ -2e^{2t} + 4e^{-3t} \end{pmatrix}$ so that $\dfrac{d\mathbf{X}}{dt} = \begin{pmatrix} 4e^{2t} - 24e^{-3t} \\ -4e^{2t} - 12e^{-3t} \end{pmatrix}$.

30. (a) $\dfrac{d\mathbf{A}}{dt} = \begin{pmatrix} -2t/(t^2+1)^2 & 3 \\ 2t & 1 \end{pmatrix}$

(b) $\dfrac{d\mathbf{B}}{dt} = \begin{pmatrix} 6 & 0 \\ -1/t^2 & 4 \end{pmatrix}$

(c) $\displaystyle\int_0^1 \mathbf{A}(t)\,dt = \begin{pmatrix} \tan^{-1} t & \frac{3}{2}t^2 \\ \frac{1}{3}t^3 & \frac{1}{2}t^2 \end{pmatrix}\Big|_{t=0}^{t=1} = \begin{pmatrix} \frac{\pi}{4} & \frac{3}{2} \\ \frac{1}{3} & \frac{1}{2} \end{pmatrix}$

(d) $\displaystyle\int_1^2 \mathbf{B}(t)\,dt = \begin{pmatrix} 3t^2 & 2t \\ \ln t & 2t^2 \end{pmatrix}\Big|_{t=1}^{t=2} = \begin{pmatrix} 9 & 2 \\ \ln 2 & 6 \end{pmatrix}$

(e) $\mathbf{A}(t)\mathbf{B}(t) = \begin{pmatrix} 6t/(t^2+1)+3 & 2/(t^2+1)+12t^2 \\ 6t^3+1 & 2t^2+4t^2 \end{pmatrix}$

(f) $\dfrac{d}{dt}\mathbf{A}(t)\mathbf{B}(t) = \begin{pmatrix} (6-6t^2)/(t^2+1)^2 & -4t/(t^2+1)^2 + 24t \\ 18t^2 & 12t \end{pmatrix}$

(g) $\displaystyle\int_1^t \mathbf{A}(s)\mathbf{B}(s)\,ds = \int_1^t \begin{pmatrix} 6s/(s^2+1)+3 & 2/(s^2+1)+12s^2 \\ 6s^3+1 & 6s^2 \end{pmatrix} ds$

$$= \begin{pmatrix} 3s + 3\ln(s^2+1) & 4s^3 + 2\tan^{-1}s \\ \frac{3}{2}s^4 + s & 2s^3 \end{pmatrix}\Bigg|_1^t$$

$$= \begin{pmatrix} 3t + 3\ln(t^2+1) - 3 - 3\ln 2 & 4t^3 + 2\tan^{-1}t - 4 - \pi/2 \\ \frac{3}{2}t^4 + t - \frac{5}{2} & 2t^3 - 2 \end{pmatrix}$$

33. $\left(\begin{array}{ccc|c} 1 & -1 & -5 & 7 \\ 5 & 4 & -16 & -10 \\ 0 & 1 & 1 & -5 \end{array}\right) \Longrightarrow \left(\begin{array}{ccc|c} 1 & -1 & -5 & 7 \\ 0 & 1 & 1 & -5 \\ 0 & 9 & 9 & -45 \end{array}\right) \Longrightarrow \left(\begin{array}{ccc|c} 1 & 0 & -4 & 2 \\ 0 & 1 & 1 & -5 \\ 0 & 0 & 0 & 0 \end{array}\right)$

Letting $z = t$ we find $y = -5 - t$, and $x = 2 + 4t$.

36. $\left(\begin{array}{ccc|c} 1 & 0 & 2 & 8 \\ 1 & 2 & -2 & 4 \\ 2 & 5 & -6 & 6 \end{array}\right) \Longrightarrow \left(\begin{array}{ccc|c} 1 & 0 & 2 & 8 \\ 0 & 2 & -4 & -4 \\ 0 & 5 & -10 & -10 \end{array}\right) \Longrightarrow \left(\begin{array}{ccc|c} 1 & 0 & 2 & 8 \\ 0 & 1 & -2 & -2 \\ 0 & 0 & 0 & 0 \end{array}\right)$

Letting $z = t$ we find $y = -2 + 2t$, and $x = 8 - 2t$.

39. $\left(\begin{array}{ccc|c} 1 & 2 & 4 & 2 \\ 2 & 4 & 3 & 1 \\ 1 & 2 & -1 & 7 \end{array}\right) \Longrightarrow \left(\begin{array}{ccc|c} 1 & 2 & 4 & 2 \\ 0 & 0 & -5 & -3 \\ 0 & 0 & -5 & 5 \end{array}\right) \Longrightarrow \left(\begin{array}{ccc|c} 1 & 2 & 0 & -2/5 \\ 0 & 0 & 1 & 3/5 \\ 0 & 0 & 0 & 8 \end{array}\right)$

There is no solution.

42. $\left(\begin{array}{ccc|ccc} 2 & 4 & -2 & 1 & 0 & 0 \\ 4 & 2 & -2 & 0 & 1 & 0 \\ 8 & 10 & -6 & 0 & 0 & 1 \end{array}\right) \xrightarrow[\text{operations}]{\text{row}} \left(\begin{array}{ccc|ccc} 1 & 2 & -1 & \frac{1}{2} & 0 & 0 \\ 0 & 1 & -\frac{1}{3} & \frac{1}{3} & -\frac{1}{6} & 0 \\ 0 & 0 & 0 & -2 & -1 & 1 \end{array}\right);$ $\mathbf{A}$ is singular.

45. $\left(\begin{array}{cccc|cccc} 1 & 2 & 3 & 1 & 1 & 0 & 0 & 0 \\ -1 & 0 & 2 & 1 & 0 & 1 & 0 & 0 \\ 2 & 1 & -3 & 0 & 0 & 0 & 1 & 0 \\ 1 & 1 & 2 & 1 & 0 & 0 & 0 & 1 \end{array}\right) \xrightarrow[\text{operations}]{\text{row}} \left(\begin{array}{cccc|cccc} 1 & 2 & 3 & 1 & 1 & 0 & 0 & 0 \\ 0 & 1 & \frac{5}{2} & 1 & \frac{1}{2} & \frac{1}{2} & 0 & 0 \\ 0 & 0 & 1 & -\frac{2}{3} & \frac{1}{3} & -1 & -\frac{2}{3} & 0 \\ 0 & 0 & 0 & 1 & -\frac{1}{2} & 1 & \frac{1}{2} & \frac{1}{2} \end{array}\right)$

$$\xrightarrow[\text{operations}]{\text{row}}
\left(\begin{array}{cccc|cccc}
1 & 0 & 0 & 0 & -\frac{1}{2} & -\frac{2}{3} & -\frac{1}{6} & \frac{7}{6} \\
0 & 1 & 0 & 0 & 1 & \frac{1}{3} & \frac{1}{3} & -\frac{4}{3} \\
0 & 0 & 1 & 0 & 0 & -\frac{1}{3} & -\frac{1}{3} & \frac{1}{3} \\
0 & 0 & 0 & 1 & -\frac{1}{2} & 1 & \frac{1}{2} & \frac{1}{2}
\end{array}\right);
\quad
\mathbf{A}^{-1} =
\left(\begin{array}{cccc}
-\frac{1}{2} & -\frac{2}{3} & -\frac{1}{6} & \frac{7}{6} \\
1 & \frac{1}{3} & \frac{1}{3} & -\frac{4}{3} \\
0 & -\frac{1}{3} & -\frac{1}{3} & \frac{1}{3} \\
-\frac{1}{2} & 1 & \frac{1}{2} & \frac{1}{2}
\end{array}\right)$$

48. We solve

$$\det(\mathbf{A} - \lambda\mathbf{I}) = \begin{vmatrix} 2-\lambda & 1 \\ 2 & 1-\lambda \end{vmatrix} = \lambda(\lambda - 3) = 0.$$

For $\lambda_1 = 0$ we have

$$\begin{pmatrix} 2 & 1 & 0 \\ 2 & 1 & 0 \end{pmatrix} \implies \begin{pmatrix} 1 & 1/2 & 0 \\ 0 & 0 & 0 \end{pmatrix}$$

so that $k_1 = -\frac{1}{2}k_2$. If $k_2 = 2$ then

$$\mathbf{K}_1 = \begin{pmatrix} -1 \\ 2 \end{pmatrix}.$$

For $\lambda_2 = 3$ we have

$$\begin{pmatrix} -1 & 1 & 0 \\ 2 & -2 & 0 \end{pmatrix} \implies \begin{pmatrix} 1 & -1 & 0 \\ 0 & 0 & 0 \end{pmatrix}$$

so that $k_1 = k_2$. If $k_2 = 1$ then

$$\mathbf{K}_2 = \begin{pmatrix} 1 \\ 1 \end{pmatrix}.$$

51. We solve

$$\det(\mathbf{A} - \lambda\mathbf{I}) = \begin{vmatrix} 5-\lambda & -1 & 0 \\ 0 & -5-\lambda & 9 \\ 5 & -1 & -\lambda \end{vmatrix} = \lambda(4-\lambda)(\lambda + 4) = 0.$$

If $\lambda_1 = 0$ then

$$\begin{pmatrix} 5 & -1 & 0 & 0 \\ 0 & -5 & 9 & 0 \\ 5 & -1 & 0 & 0 \end{pmatrix} \implies \begin{pmatrix} 1 & 0 & -9/25 & 0 \\ 0 & 1 & -9/5 & 0 \\ 0 & 0 & 0 & 0 \end{pmatrix}$$

so that $k_1 = \frac{9}{25}k_3$ and $k_2 = \frac{9}{5}k_3$. If $k_3 = 25$ then

$$\mathbf{K}_1 = \begin{pmatrix} 9 \\ 45 \\ 25 \end{pmatrix}.$$

If $\lambda_2 = 4$ then

$$\begin{pmatrix} 1 & -1 & 0 & | & 0 \\ 0 & -9 & 9 & | & 0 \\ 5 & -1 & -4 & | & 0 \end{pmatrix} \implies \begin{pmatrix} 1 & 0 & -1 & | & 0 \\ 0 & 1 & -1 & | & 0 \\ 0 & 0 & 0 & | & 0 \end{pmatrix}$$

so that $k_1 = k_3$ and $k_2 = k_3$. If $k_3 = 1$ then

$$\mathbf{K}_2 = \begin{pmatrix} 1 \\ 1 \\ 1 \end{pmatrix}.$$

If $\lambda_3 = -4$ then

$$\begin{pmatrix} 9 & -1 & 0 & | & 0 \\ 0 & -1 & 9 & | & 0 \\ 5 & -1 & 4 & | & 0 \end{pmatrix} \implies \begin{pmatrix} 1 & 0 & -1 & | & 0 \\ 0 & 1 & -9 & | & 0 \\ 0 & 0 & 0 & | & 0 \end{pmatrix}$$

so that $k_1 = k_3$ and $k_2 = 9k_3$. If $k_3 = 1$ then

$$\mathbf{K}_3 = \begin{pmatrix} 1 \\ 9 \\ 1 \end{pmatrix}.$$

54. We solve

$$\det(\mathbf{A} - \lambda\mathbf{I}) = \begin{vmatrix} 1-\lambda & 6 & 0 \\ 0 & 2-\lambda & 1 \\ 0 & 1 & 2-\lambda \end{vmatrix} = (3-\lambda)(1-\lambda)^2 = 0.$$

For $\lambda = 3$ we have

$$\begin{pmatrix} -2 & 6 & 0 & | & 0 \\ 0 & 0 & 0 & | & 0 \\ 0 & 1 & -1 & | & 0 \end{pmatrix} \implies \begin{pmatrix} 1 & 0 & -3 & | & 0 \\ 0 & 1 & -1 & | & 0 \\ 0 & 0 & 0 & | & 0 \end{pmatrix}$$

so that $k_1 = 3k_3$ and $k_2 = k_3$. If $k_3 = 1$ then

$$\mathbf{K}_1 = \begin{pmatrix} 3 \\ 1 \\ 1 \end{pmatrix}.$$

For $\lambda_2 = \lambda_3 = 1$ we have

$$\begin{pmatrix} 0 & 6 & 0 & | & 0 \\ 0 & 1 & 1 & | & 0 \\ 0 & 1 & 1 & | & 0 \end{pmatrix} \implies \begin{pmatrix} 0 & 1 & 0 & | & 0 \\ 0 & 0 & 1 & | & 0 \\ 0 & 0 & 0 & | & 0 \end{pmatrix}$$

so that $k_2 = 0$ and $k_3 = 0$. If $k_1 = 1$ then

$$\mathbf{K}_2 = \begin{pmatrix} 1 \\ 0 \\ 0 \end{pmatrix}.$$

57. Let

$$\mathbf{A} = \begin{pmatrix} a_{11} & a_{12} \\ a_{21} & a_{22} \end{pmatrix}.$$

Then

$$\frac{d}{dt}[\mathbf{A}(t)\mathbf{X}(t)] = \frac{d}{dt}\begin{pmatrix} a_1 & a_2 \\ a_3 & a_4 \end{pmatrix}\begin{pmatrix} x_1 \\ x_2 \end{pmatrix} = \frac{d}{dt}\begin{pmatrix} a_1 x_1 + a_2 x_2 \\ a_3 x_1 + a_4 x_2 \end{pmatrix} = \begin{pmatrix} a_1 x_1' + a_1' x_1 + a_2 x_2' + a_2' x_2 \\ a_3 x_1' + a_3' x_1 + a_4 x_2' + a_4' x_2 \end{pmatrix}$$

$$= \begin{pmatrix} a_1 & a_2 \\ a_3 & a_4 \end{pmatrix}\begin{pmatrix} x_1' \\ x_2' \end{pmatrix} + \begin{pmatrix} a_1' & a_2' \\ a_3' & a_4' \end{pmatrix}\begin{pmatrix} x_1 \\ x_2 \end{pmatrix} = \mathbf{A}(t)\mathbf{X}'(t) + \mathbf{A}'(t)\mathbf{X}(t).$$

60. Since

$$(\mathbf{AB})(\mathbf{B}^{-1}\mathbf{A}^{-1}) = \mathbf{A}(\mathbf{BB}^{-1})\mathbf{A}^{-1} = \mathbf{AIA}^{-1} = \mathbf{AA}^{-1} = \mathbf{I}$$

and

$$(\mathbf{B}^{-1}\mathbf{A}^{-1})(\mathbf{AB}) = \mathbf{B}^{-1}(\mathbf{A}^{-1}\mathbf{A})\mathbf{B} = \mathbf{B}^{-1}\mathbf{IB} = \mathbf{B}^{-1}\mathbf{B} = \mathbf{I}$$

we have

$$(\mathbf{AB})^{-1} = \mathbf{B}^{-1}\mathbf{A}^{-1}.$$

Table of Integrals

1. $\displaystyle\int u\,dv = uv - \int v\,du$

2. $\displaystyle\int u^n\,du = \frac{1}{n+1}u^{n+1} + C, \quad n \neq -1$

3. $\displaystyle\int \frac{du}{u} = \ln|u| + C$

4. $\displaystyle\int e^u\,du = e^u + C$

5. $\displaystyle\int a^u\,du = \frac{1}{\ln a}a^u + C$

6. $\displaystyle\int \sin u\,du = -\cos u + C$

7. $\displaystyle\int \cos u\,du = \sin u + C$

8. $\displaystyle\int \sec^2 u\,du = \tan u + C$

9. $\displaystyle\int \csc^2 u\,du = -\cot u + C$

10. $\displaystyle\int \sec u\tan u\,du = \sec u + C$

11. $\displaystyle\int \csc u\cot u\,du = -\csc u + C$

12. $\displaystyle\int \tan u\,du = -\ln|\cos u| + C$

13. $\displaystyle\int \cot u\,du = \ln|\sin u| + C$

14. $\displaystyle\int \sec u\,du = \ln|\sec u + \tan u| + C$

15. $\displaystyle\int \csc u\,du = \ln|\csc u - \cot u| + C$

16. $\displaystyle\int \frac{du}{\sqrt{a^2 - u^2}} = \sin^{-1}\frac{u}{a} + C$

17. $\displaystyle\int \frac{du}{a^2 + u^2} = \frac{1}{a}\tan^{-1}\frac{u}{a} + C$

18. $\displaystyle\int \frac{du}{u\sqrt{u^2 - a^2}} = \frac{1}{a}\sec^{-1}\frac{u}{a} + C$

19. $\displaystyle\int \frac{du}{a^2 - u^2} = \frac{1}{2a}\ln\left|\frac{u+a}{u-a}\right| + C$

20. $\displaystyle\int \frac{du}{u^2 - a^2} = \frac{1}{2a}\ln\left|\frac{u-a}{u+a}\right| + C$

21. $\displaystyle\int \sin^2 u\,du = \frac{1}{2}u - \frac{1}{4}\sin 2u + C$

22. $\displaystyle\int \cos^2 u\,du = \frac{1}{2}u + \frac{1}{4}\sin 2u + C$

23. $\displaystyle\int \tan^2 u\,du = \tan u - u + C$

24. $\displaystyle\int \cot^2 u\,du = -\cot u - u + C$

25. $\displaystyle\int \sin^3 u\,du = -\frac{1}{3}(2 + \sin^2 u)\cos u + C$

26. $\displaystyle\int \cos^3 u\,du = \frac{1}{3}(2 + \cos^2 u)\sin u + C$

27. $\displaystyle\int \tan^3 u\,du = \frac{1}{2}\tan^2 u + \ln|\cos u| + C$

28. $\displaystyle\int \cot^3 u\,du = -\frac{1}{2}\cot^2 u - \ln|\sin u| + C$

29. $\displaystyle\int \sec^3 u\,du = \frac{1}{2}\sec u\tan u + \frac{1}{2}\ln|\sec u + \tan u| + C$

30. $\displaystyle\int \csc^3 u\,du = -\frac{1}{2}\csc u\cot u + \frac{1}{2}\ln|\csc u - \cot u| + C$

31. $\displaystyle\int \sin^n u\,du = -\frac{1}{n}\sin^{n-1}u\cos u + \frac{n-1}{n}\int \sin^{n-2}u\,du$

32. $\displaystyle\int \cos^n u\,du = \frac{1}{n}\cos^{n-1}u\sin u + \frac{n-1}{n}\int \cos^{n-2}u\,du$

33. $\int \tan^n u \, du = \dfrac{1}{n-1} \tan^{n-1} u - \int \tan^{n-2} u \, du$

34. $\int \cot^n u \, du = \dfrac{-1}{n-1} \cot^{n-1} u - \int \cot^{n-2} u \, du$

35. $\int \sec^n u \, du = \dfrac{1}{n-1} \tan u \sec^{n-2} u + \dfrac{n-2}{n-1} \int \sec^{n-2} u \, du$

36. $\int \csc^n u \, du = \dfrac{-1}{n-1} \cot u \csc^{n-2} u + \dfrac{n-2}{n-1} \int \csc^{n-2} u \, du$

37. $\int \sin au \sin bu \, du = \dfrac{\sin(a-b)u}{2(a-b)} - \dfrac{\sin(a+b)u}{2(a+b)} + C, \quad a \neq b$

38. $\int \cos au \cos bu \, du = \dfrac{\sin(a-b)u}{2(a-b)} + \dfrac{\sin(a+b)u}{2(a+b)} + C, \quad a \neq b$

39. $\int \sin au \cos bu \, du = -\dfrac{\cos(a-b)u}{2(a-b)} - \dfrac{\cos(a+b)u}{2(a+b)} + C, \quad a \neq b$

40. $\int u \sin u \, du = \sin u - u \cos u + C$ **41.** $\int u \cos u \, du = \cos u + u \sin u + C$

42. $\int u^n \sin u \, du = -u^n \cos u + n \int u^{n-1} \cos u \, du$

43. $\int u^n \cos u \, du = u^n \sin u - n \int u^{n-1} \sin u \, du$

44. $\int \sin^n u \cos^m u \, du = -\dfrac{\sin^{n-1} u \cos^{m+1} u}{n+m} + \dfrac{n-1}{n+m} \int \sin^{n-2} u \cos^m u \, du$

$\qquad\qquad = \dfrac{\sin^{n+1} u \cos^{m-1} u}{n+m} + \dfrac{m-1}{n+m} \int \sin^n u \cos^{m-2} u \, du$

45. $\int \sin^{-1} u \, du = u \sin^{-1} u + \sqrt{1-u^2} + C$ **46.** $\int \cos^{-1} u \, du = u \cos^{-1} u + \sqrt{1-u^2} + C$

47. $\int \tan^{-1} u \, du = u \tan^{-1} u - \dfrac{1}{2} \ln(1+u^2) + C$ **48.** $\int u \sin^{-1} u \, du = \dfrac{2u^2-1}{4} \sin^{-1} u + \dfrac{u\sqrt{1-u^2}}{4} + C$

49. $\int u \cos^{-1} u \, du = \dfrac{2u^2-1}{4} \cos^{-1} u - \dfrac{u\sqrt{1-u^2}}{4} + C$ **50.** $\int u \tan^{-1} u \, du = \dfrac{u^2+1}{2} \tan^{-1} u - \dfrac{u}{2} + C$

51. $\int u e^{au} \, du = \dfrac{1}{a^2}(au-1)e^{au} + C$ **52.** $\int u^n e^{au} \, du = \dfrac{1}{a} u^n e^{au} - \dfrac{n}{a} \int u^{n-1} e^{au} \, du$

53. $\int e^{au} \sin bu \, du = \dfrac{e^{au}}{a^2+b^2}(a \sin bu - b \cos bu) + C$ **54.** $\int e^{au} \cos bu \, du = \dfrac{e^{au}}{a^2+b^2}(a \cos bu + b \sin bu) + C$

55. $\int \ln u \, du = u \ln u - u + C$ **56.** $\int \dfrac{1}{u \ln u} \, du = \ln|\ln u| + C$

57. $\int u^n \ln u \, du = \dfrac{u^{n+1}}{(n+1)^2}[(n+1)\ln u - 1] + C$

58. $\displaystyle\int u^m \ln^n u \, du = \frac{u^{m+1} \ln^n u}{m+1} - \frac{n}{m+1} \int u^m \ln^{n-1} u \, du, \quad m \neq -1$

59. $\displaystyle\int \ln(u^2 + a^2) \, du = u \ln(u^2 + a^2) - 2u + 2a \tan^{-1} \frac{u}{a} + C$

60. $\displaystyle\int \ln|u^2 - a^2| \, du = u \ln|u^2 - a^2| - 2u + a \ln \left| \frac{u+a}{u-a} \right| + C$

61. $\displaystyle\int \sinh u \, du = \cosh u + C$ **62.** $\displaystyle\int \cosh u \, du = \sinh u + C$

63. $\displaystyle\int \tanh u \, du = \ln \cosh u + C$ **64.** $\displaystyle\int \coth u \, du = \ln|\sinh u| + C$

65. $\displaystyle\int \operatorname{sech}^2 u \, du = \tanh u + C$ **66.** $\displaystyle\int \operatorname{csch}^2 u \, du = -\coth u + C$

67. $\displaystyle\int \operatorname{sech} u \tanh u \, du = -\operatorname{sech} u + C$ **68.** $\displaystyle\int \operatorname{csch} u \coth u \, du = -\operatorname{csch} u + C$

69. $\displaystyle\int \sqrt{a^2 + u^2} \, du = \frac{u}{2} \sqrt{a^2 + u^2} + \frac{a^2}{2} \ln \left| u + \sqrt{a^2 + u^2} \right| + C$

70. $\displaystyle\int u^2 \sqrt{a^2 + u^2} \, du = \frac{u}{8}(a^2 + 2u^2)\sqrt{a^2 + u^2} - \frac{a^4}{8} \ln \left| u + \sqrt{a^2 + u^2} \right| + C$

71. $\displaystyle\int \frac{\sqrt{a^2 + u^2}}{u} \, du = \sqrt{a^2 + u^2} - a \ln \left| \frac{a + \sqrt{a^2 + u^2}}{u} \right| + C$

72. $\displaystyle\int \frac{\sqrt{a^2 + u^2}}{u^2} \, du = -\frac{\sqrt{a^2 + u^2}}{u} + \ln \left| u + \sqrt{a^2 + u^2} \right| + C$

73. $\displaystyle\int \frac{du}{\sqrt{a^2 + u^2}} = \ln \left| u + \sqrt{a^2 + u^2} \right| + C$

74. $\displaystyle\int \frac{u^2 \, du}{\sqrt{a^2 + u^2}} = \frac{u}{2} \sqrt{a^2 + u^2} - \frac{a^2}{2} \ln \left| u + \sqrt{a^2 + u^2} \right| + C$

75. $\displaystyle\int \frac{du}{u\sqrt{a^2 + u^2}} = -\frac{1}{a} \ln \left| \frac{\sqrt{a^2 + u^2} + a}{u} \right| + C$

76. $\displaystyle\int \frac{du}{u^2 \sqrt{a^2 + u^2}} = -\frac{\sqrt{a^2 + u^2}}{a^2 u} + C$

77. $\displaystyle\int \sqrt{u^2 - a^2} \, du = \frac{u}{2} \sqrt{u^2 - a^2} - \frac{a^2}{2} \ln \left| u + \sqrt{u^2 - a^2} \right| + C$

78. $\displaystyle\int u^2 \sqrt{u^2 - a^2} \, du = \frac{u}{8}(2u^2 - a^2)\sqrt{u^2 - a^2} - \frac{a^4}{8} \ln \left| u + \sqrt{u^2 - a^2} \right| + C$

79. $\displaystyle\int \frac{\sqrt{u^2 - a^2}}{u} \, du = \sqrt{u^2 - a^2} - a \cos^{-1} \frac{a}{u} + C$

80. $\displaystyle\int \frac{\sqrt{u^2 - a^2}}{u^2} \, du = -\frac{\sqrt{u^2 - a^2}}{u} + \ln \left| u + \sqrt{u^2 - a^2} \right| + C$

81. $\displaystyle \int \frac{du}{\sqrt{u^2-a^2}} = \ln\left|u+\sqrt{u^2-a^2}\right| + C$

82. $\displaystyle \int \frac{u^2\,du}{\sqrt{u^2-a^2}} = \frac{u}{2}\sqrt{u^2-a^2} + \frac{a^2}{2}\ln\left|u+\sqrt{u^2-a^2}\right| + C$

83. $\displaystyle \int \frac{du}{u^2\sqrt{u^2-a^2}} = \frac{\sqrt{u^2-a^2}}{a^2 u} + C$

84. $\displaystyle \int \frac{du}{(u^2-a^2)^{3/2}} = -\frac{u}{a^2\sqrt{u^2-a^2}} + C$

85. $\displaystyle \int \sqrt{a^2-u^2}\,du = \frac{u}{2}\sqrt{a^2-u^2} + \frac{a^2}{2}\sin^{-1}\frac{u}{a} + C$

86. $\displaystyle \int u^2\sqrt{a^2-u^2}\,du = \frac{u}{8}(2u^2-a^2)\sqrt{a^2-u^2} + \frac{a^4}{8}\sin^{-1}\frac{u}{a} + C$

Some Basic *Mathematica* Syntax and Commands

■ *Mathematica* conventions

■ Most implementations of *Mathematica* consist of two parts, the *front end* and the *kernel*. The lines you type into the Front End of *Mathematica* are called inputs and are shown in **boldface type** (unless you specify another convention). The output from *Mathematica* is generated by the kernel and is shown in plainface type. Generally, the output is displayed unless you specifically suppress it by using a semicolon. The construct (* **comment** *) is ignored by the kernel and can be used to annotate inputs in the front end.

```
In[1] := 2
        3; (* The corresponding output of 3 is suppressed below (Out[2] is not shown). *)
        4; 5 (* The 4 is suppressed, but the 5 is shown. *)
Out[1] = 2
Out[3] = 5
```

■ An equal sign (=) is used for assignment.

```
In[4] := a = 5
Out[4] = 5
In[5] := b = 3
Out[5] = 3
In[6] := a
Out[6] = 5
In[7] := a+b
Out[7] = 8
In[8] := ab  (* This is the variable ab, not the product of a and b. *)
Out[8] = ab
In[9] := a b
Out[9] = 15
```

■ Grouping is done with parentheses only, not with square backets [ ] or braces { }.

```
In[10] := a/(a+b)
```
$$Out[10] = \frac{5}{8}$$

■ Arguments of functions are enclosed in square brackets.

In[11]:= **Sqrt[9]**

Out[11]= 3

■ Braces are used for ordered sets - called *lists* in *Mathematica*. The individual elements of a list are obtained using double square brackets [[ ]].

In[12]:= **m = {2, 7, -8}**

Out[12]= {2, 7, -8}

In[13]:= **m[[2]]**

Out[13]= 7

■ Built-in constants

■ **Pi, E, I, Infinity**

In[14]:= **Pi**

Out[14]= π

In[15]:= **N[Pi]** (* **N is a function than converts a number to decimal form.** *)

Out[15]= 3.14159

In[16]:= **E//N** (* **Using //, a function can follow its argument.** *)

Out[16]= 2.71828

In[17]:= **I^2** (* **I is the square root of** -1. *)

Out[17]= -1

In[18]:= **Infinity**

Out[18]= ∞

■ Built-in mathematical functions

■ The following functions are built into *Mathematica* with the syntax shown.
   absolute value: **Abs[x]**
   square root: **Sqrt[x]**
   trigonometric: **Sin[x], Cos[x], Tan[x], Cot[x], Sec[x], Csc[x]**
   inverse trigonometric: **ArcSin[x], ArcCos[x], ArcTan[x], ArcCot[x], ArcSec[x], ArcCsc[x]**
   exponential: **Exp[x]** or **E^x**
   natural logarithm: **Log[x]**
   hyperbolic: **Sinh[x], Cosh[x], Tanh[x], Coth[x], Sech[x], Csch[x]**

In[19]:= **Sqrt[−1]**

Out[19]= i

In[20]:= **Sqrt[90]**

Out[20]= $3\sqrt{10}$

In[21]:= **Sqrt[90]//N**

Out[21]= 9.48638

In[22]:= **N[Sin[Pi/3]]**

Out[22]= 0.866025

In[23]:= **Log[5]**

Out[23]= Log[5]

In[24]:= **Log[5]//N**

Out[24]= 1.60944

In[25]:= **Log[Exp[6]]**

Out[25]= 6

In[26]:= **Sinh[Log[3]]**

Out[26]= $\dfrac{4}{3}$

■ Derivatives

- The syntax for finding the derivative of a function is discussed in Section 1.1 of this manual. Below we show how *Mathematica* can be used to verify that $2x/(3 - x^2)$ is a solution of the differential equation $xy' - y = xy^2$.

In[27]:= **Clear[y]**
        lhs = x y'[x] - y[x]
        rhs = x y[x]^2

Out[28]= $-y[x] + x\, y'[x]$
Out[29]= $x\, y[x]^2$

In[30]:= **y[x_]:=2x/(3 − x^2)**
        y[x]

Out[31]= $\dfrac{2x}{3 - x^2}$

In[32]:= **lhs − rhs//Simplify**

Out[32]= 0

■ Rules

- We saw above under ***Mathematica* conventions** how to assign a specific value to a variable. For example, if we want the variable x to have the value 2 we input **x=2**. Once this is done, until x is either reassigned or cleared, it will have the value 2. Suppose, on the other hand, we have an expression containing x, such as $x^2 + 2x - 5$, and we want the value of this expression when $x = 2$, but we do not want to permanently assign the value 2 to x. One way to do this is to define the expression as a function **f** and evaluate **f[2]**. However, if we are only going to do this for a single value of x, a faster way is to use the following syntax.

In[33]:= **x^2 + 3x - 1 /. x ->2**

Out[33]= 9

The arrow from x to 2 is obtained by typing a *minus sign* followed by the *greater than* symbol. The expression **x ->2** is called a rule and **/.** is a replacement operator. The complete input expression **x^2 + 3x - 1 /. x ->2** says to replace x with 2 in $x^2 + 3x - 1$, but do not permanently assign the value 2 to x.

In[34]:= **x+3**

Out[34]= 3+x

■ Graphing with color and dashing

- In Section 2.2 of this manual we saw that the **Plot** command can be supplemented with options like **PlotRange** to modify the appearance of a graph. *Mathematica* also has options for colorizing graphs and drawing graphs with dashed lines or thicker lines. The **RGBColor** option is used to obtain graphs of various colors. Red is **[1,0,0]**, green is **[0,1,0]**, and blue is **[0,0,1]**. Other colors are obtained by using numbers between 0 and 1 as the arguments to **RGBColor**. On a black and white printer the colors will appear as shades of gray as seen below.

In[35]:= **Plot[{x^2, x, Sin[x], 2.5 - x^2}, {x,-3,3}, PlotStyle − > {RGBColor[1,0,0], RGBColor[0,1,0], RGBColor[0,0,1], RGBColor[1,0.5,0.8]}]**

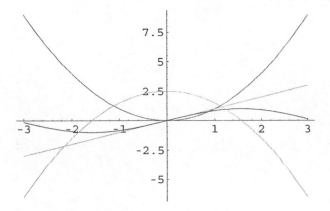

Out[35]= -Graphics-

326

■ The example below shows how to obtain a dashed graph, and thicker graphs in various shades of gray. The arguments for **Dashing** specify, alternately, the length of the dash and the length of the open space. The argument for **Thickness** is a number from 0 to 1 where 1 represents the width of the entire plot. The argument for **GrayLevel** is also a number from 0 to 1 where 0 is black and 1 is white.

In [36] := **Plot[{x^2, x, Sin[x], 2.5 - x^2},{x,-3,3}, PlotStyle − > {{Dashing[{0.05,0.05]},**
        {Dashing[{0.01,0.06]}, {Thickness[0.015], GrayLevel[0.4]},
        {Thickness[0.015], GrayLevel[0.4]}]

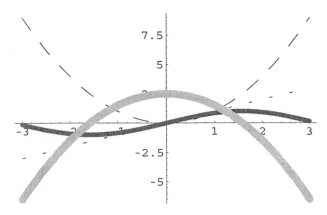

Out [36] = -Graphics-